COMPUTERS IN
MATHEMATICS

LECTURE NOTES

IN PURE AND APPLIED MATHEMATICS

1. *N. Jacobson*, Exceptional Lie Algebras
2. *L. -Å. Lindahl and F. Poulsen*, Thin Sets in Harmonic Analysis
3. *I. Satake*, Classification Theory of Semi-Simple Algebraic Groups
4. *F. Hirzebruch, W. D. Newmann, and S. S. Koh*, Differentiable Manifolds and Quadratic Forms (out of print)
5. *I. Chavel*, Riemannian Symmetric Spaces of Rank One (out of print)
6. *R. B. Burckel*, Characterization of C(X) Among Its Subalgebras
7. *B. R. McDonald, A. R. Magid, and K. C. Smith*, Ring Theory: Proceedings of the Oklahoma Conference
8. *Y.-T. Siu*, Techniques of Extension on Analytic Objects
9. *S. R. Caradus, W. E. Pfaffenberger, and B. Yood*, Calkin Algebras and Algebras of Operators on Banach Spaces
10. *E. O. Roxin, P.-T. Liu, and R. L. Sternberg*, Differential Games and Control Theory
11. *M. Orzech and C. Small*, The Brauer Group of Commutative Rings
12. *S. Thomeier*, Topology and Its Applications
13. *J. M. Lopez and K. A. Ross*, Sidon Sets
14. *W. W. Comfort and S. Negrepontis*, Continuous Pseudometrics
15. *K. McKennon and J. M. Robertson*, Locally Convex Spaces
16. *M. Carmeli and S. Malin*, Representations of the Rotation and Lorentz Groups: An Introduction
17. *G. B. Seligman*, Rational Methods in Lie Algebras
18. *D. G. de Figueiredo*, Functional Analysis: Proceedings of the Brazilian Mathematical Society Symposium
19. *L. Cesari, R. Kannan, and J. D. Schuur*, Nonlinear Functional Analysis and Differential Equations: Proceedings of the Michigan State University Conference
20. *J. J. Schäffer*, Geometry of Spheres in Normed Spaces
21. *K. Yano and M. Kon*, Anti-Invariant Submanifolds
22. *W. V. Vasconcelos*, The Rings of Dimension Two
23. *R. E. Chandler*, Hausdorff Compactifications
24. *S. P. Franklin and B. V. S. Thomas*, Topology: Proceedings of the Memphis State University Conference
25. *S. K. Jain*, Ring Theory: Proceedings of the Ohio University Conference
26. *B. R. McDonald and R. A. Morris*, Ring Theory II: Proceedings of the Second Oklahoma Conference
27. *R. B. Mura and A. Rhemtulla*, Orderable Groups
28. *J. R. Graef*, Stability of Dynamical Systems: Theory and Applications
29. *H.-C. Wang*, Homogeneous Branch Algebras
30. *E. O. Roxin, P.-T. Liu, and R. L. Sternberg*, Differential Games and Control Theory II
31. *R. D. Porter*, Introduction to Fibre Bundles
32. *M. Altman*, Contractors and Contractor Directions Theory and Applications
33. *J. S. Golan*, Decomposition and Dimension in Module Categories
34. *G. Fairweather*, Finite Element Galerkin Methods for Differential Equations
35. *J. D. Sally*, Numbers of Generators of Ideals in Local Rings
36. *S. S. Miller*, Complex Analysis: Proceedings of the S.U.N.Y. Brockport Conference
37. *R. Gordon*, Representation Theory of Algebras: Proceedings of the Philadelphia Conference
38. *M. Goto and F. D. Grosshans*, Semisimple Lie Algebras
39. *A. I. Arruda, N. C. A. da Costa, and R. Chuaqui*, Mathematical Logic: Proceedings of the First Brazilian Conference

40. *F. Van Oystaeyen*, Ring Theory: Proceedings of the 1977 Antwerp Conference
41. *F. Van Oystaeyen and A. Verschoren*, Reflectors and Localization: Application to Sheaf Theory
42. *M. Satyanarayana*, Positively Ordered Semigroups
43. *D. L. Russell*, Mathematics of Finite-Dimensional Control Systems
44. *P.-T. Liu and E. Roxin*, Differential Games and Control Theory III: Proceedings of the Third Kingston Conference, Part A
45. *A. Geramita and J. Seberry*, Orthogonal Designs: Quadratic Forms and Hadamard Matrices
46. *J. Cigler, V. Losert, and P. Michor*, Banach Modules and Functors on Categories of Banach Spaces
47. *P.-T. Liu and J. G. Sutinen*, Control Theory in Mathematical Economics: Proceedings of the Third Kingston Conference, Part B
48. *C. Byrnes*, Partial Differential Equations and Geometry
49. *G. Klambauer*, Problems and Propositions in Analysis
50. *J. Knopfmacher*, Analytic Arithmetic of Algebraic Function Fields
51. *F. Van Oystaeyen*, Ring Theory: Proceedings of the 1978 Antwerp Conference
52. *B. Kedem*, Binary Time Series
53. *J. Barros-Neto and R. A. Artino*, Hypoelliptic Boundary-Value Problems
54. *R. L. Sternberg, A. J. Kalinowski, and J. S. Papadakis*, Nonlinear Partial Differential Equations in Engineering and Applied Science
55. *B. R. McDonald*, Ring Theory and Algebra III: Proceedings of the Third Oklahoma Conference
56. *J. S. Golan*, Structure Sheaves over a Noncommutative Ring
57. *T. V. Narayana, J. G. Williams, and R. M. Mathsen*, Combinatorics, Representation Theory and Statistical Methods in Groups: YOUNG DAY Proceedings
58. *T. A. Burton*, Modeling and Differential Equations in Biology
59. *K. H. Kim and F. W. Roush*, Introduction to Mathematical Consensus Theory
60. *J. Banas and K. Goebel*, Measures of Noncompactness in Banach Spaces
61. *O. A. Nielson*, Direct Integral Theory
62. *J. E. Smith, G. O. Kenny, and R. N. Ball*, Ordered Groups: Proceedings of the Boise State Conference
63. *J. Cronin*, Mathematics of Cell Electrophysiology
64. *J. W. Brewer*, Power Series Over Commutative Rings
65. *P. K. Kamthan and M. Gupta*, Sequence Spaces and Series
66. *T. G. McLaughlin*, Regressive Sets and the Theory of Isols
67. *T. L. Herdman, S. M. Rankin, III, and H. W. Stech*, Integral and Functional Differential Equations
68. *R. Draper*, Commutative Algebra: Analytic Methods
69. *W. G. McKay and J. Patera*, Tables of Dimensions, Indices, and Branching Rules for Representations of Simple Lie Algebras
70. *R. L. Devaney and Z. H. Nitecki*, Classical Mechanics and Dynamical Systems
71. *J. Van Geel*, Places and Valuations in Noncommutative Ring Theory
72. *C. Faith*, Injective Modules and Injective Quotient Rings
73. *A. Fiacco*, Mathematical Programming with Data Perturbations I
74. *P. Schultz, C. Praeger, and R. Sullivan*, Algebraic Structures and Applications Proceedings of the First Western Australian Conference on Algebra
75. *L. Bican, T. Kepka, and P. Nemec*, Rings, Modules, and Preradicals
76. *D. C. Kay and M. Breen*, Convexity and Related Combinatorial Geometry: Proceedings of the Second University of Oklahoma Conference
77. *P. Fletcher and W. F. Lindgren*, Quasi-Uniform Spaces
78. *C.-C. Yang*, Factorization Theory of Meromorphic Functions
79. *O. Taussky*, Ternary Quadratic Forms and Norms
80. *S. P. Singh and J. H. Burry*, Nonlinear Analysis and Applications
81. *K. B. Hannsgen, T. L. Herdman, H. W. Stech, and R. L. Wheeler*, Volterra and Functional Differential Equations

82. *N. L. Johnson, M. J. Kallaher, and C. T. Long,* Finite Geometries: Proceedings of a Conference in Honor of T. G. Ostrom
83. *G. I. Zapata,* Functional Analysis, Holomorphy, and Approximation Theory
84. *S. Greco and G. Valla,* Commutative Algebra: Proceedings of the Trento Conference
85. *A. V. Fiacco,* Mathematical Programming with Data Perturbations II
86. *J.-B. Hiriart-Urruty, W. Oettli, and J. Stoer,* Optimization: Theory and Algorithms
87. *A. Figa Talamanca and M. A. Picardello,* Harmonic Analysis on Free Groups
88. *M. Harada,* Factor Categories with Applications to Direct Decomposition of Modules
89. *V. I. Istrăţescu,* Strict Convexity and Complex Strict Convexity: Theory and Applications
90. *V. Lakshmikantham,* Trends in Theory and Practice of Nonlinear Differential Equations
91. *H. L. Manocha and J. B. Srivastava,* Algebra and Its Applications
92. *D. V. Chudnovsky and G. V. Chudnovsky,* Classical and Quantum Models and Arithmetic Problems
93. *J. W. Longley,* Least Squares Computations Using Orthogonalization Methods
94. *L. P. de Alcantara,* Mathematical Logic and Formal Systems
95. *C. E. Aull,* Rings of Continuous Functions
96. *R. Chuaqui,* Analysis, Geometry, and Probability
97. *L. Fuchs and L. Salce,* Modules Over Valuation Domains
98. *P. Fischer and W. R. Smith,* Chaos, Fractals, and Dynamics
99. *W. B. Powell and C. Tsinakis,* Ordered Algebraic Structures
100. *G. M. Rassias and T. M. Rassias,* Differential Geometry, Calculus of Variations, and Their Applications
101. *R.-E. Hoffmann and K. H. Hofmann,* Continuous Lattices and Their Applications
102. *J. H. Lightbourne, III, and S. M. Rankin, III,* Physical Mathematics and Nonlinear Partial Differential Equations
103. *C. A. Baker and L. M. Batten,* Finite Geometries
104. *J. W. Brewer, J. W. Bunce, and F. S. Van Vleck,* Linear Systems Over Commutative Rings
105. *C. McCrory and T. Shifrin,* Geometry and Topology: Manifolds, Varieties, and Knots
106. *D. W. Kueker, E. G. K. Lopez-Escobar, and C. H. Smith,* Mathematical Logic and Theoretical Computer Science
107. *B.-L. Lin and S. Simons,* Nonlinear and Convex Analysis: Proceedings in Honor of Ky Fan
108. *S. J. Lee,* Operator Methods for Optimal Control Problems
109. *V. Lakshmikantham,* Nonlinear Analysis and Applications
110. *S. F. McCormick,* Multigrid Methods: Theory, Applications, and Supercomputing
111. *M. C. Tangora,* Computers in Algebra
112. *D. V. Chudnovsky and G. V. Chudnovsky,* Search Theory: Some Recent Developments
113. *D. V. Chudnovsky and R. D. Jenks,* Computer Algebra
114. *M. C. Tangora,* Computers in Geometry and Topology
115. *P. Nelson, V. Faber, T. A. Manteuffel, D. L. Seth, and A. B. White, Jr.* Transport Theory, Invariant Imbedding, and Integral Equations: Proceedings in Honor of G. M. Wing's 65th Birthday
116. *P. Clément, S. Invernizzi, E. Mitidieri, and I. I. Vrabie,* Semigroup Theory and Applications
117. *J. Vinuesa,* Orthogonal Polynomials and Their Applications: Proceedings of the International Congress
118. *C. M. Dafermos, G. Ladas, and G. Papanicolaou,* Differential Equations: Proceedings of the EQUADIFF Conference
119. *E. O. Roxin,* Modern Optimal Control: A Conference in Honor of Solomon Lefschetz and Joseph P. Lasalle
120. *J. C. Díaz,* Mathematics for Large Scale Computing
121. *P. S. Milojević,* Nonlinear Functional Analysis

122. *C. Sadosky*, Analysis and Partial Differential Equations: A Collection of Papers Dedicated to Mischa Cotlar
123. *R. M. Shortt*, General Topology and Applications: Proceedings of the 1988 Northeast Conference
124. *R. Wong*, Asymptotic and Computational Analysis: Conference in Honor of Frank W. J. Olver's 65th Birthday
125. *D. V. Chudnovsky and R. D. Jenks*, Computers in Mathematics

Other Volumes in Preparation

COMPUTERS IN MATHEMATICS

Edited by

DAVID V. CHUDNOVSKY
Columbia University
New York, New York

RICHARD D. JENKS
IBM Thomas J. Watson Research Center
Yorktown Heights, New York

CRC Press
Taylor & Francis Group
Boca Raton London New York

CRC Press is an imprint of the
Taylor & Francis Group, an **informa** business

CRC Press
Taylor & Francis Group
6000 Broken Sound Parkway NW, Suite 300
Boca Raton, FL 33487-2742

First issued in hardback 2017

© 1990 by Taylor & Francis Group, LLC
CRC Press is an imprint of Taylor & Francis Group, an Informa business

No claim to original U.S. Government works

ISBN 13: 978-1-138-41342-9 (hbk)
ISBN 13: 978-0-8247-8341-9 (pbk)

**Visit the Taylor & Francis Web site at
http://www.taylorandfrancis.com**

**and the CRC Press Web site at
http://www.crcpress.com**

LIBRARY OF CONGRESS CATALOGING-IN-PUBLICATION DATA

Computers in mathematics / edited by David V. Chudnovsky, Richard D. Jenks.
    p. cm. --(Lecture notes in pure and applied mathematics ; 125)
    Talks from the International Conference on Computers and Mathematics which took place July 29-Aug. 1, 1986 on the campus of Stanford University, sponsored by the American Association for Artificial Intelligence and the ACM Special Interest Group on Symbolic and Algebraic Manipulation.
    Includes bibliographical references.
    ISBN 0-8247-8341-7 (alk. paper)
    1. Mathematics--Data processing--Congresses. I. Chudnovsky, David V. II. Jenks, Richard D. III. International Conference on Computers and Mathematics (1986 : Stanford University) IV. American Association for Artificial Intelligence.
V. Association for Computing Machinery. Special Interest Group on Symbolic & Algebraic Manipulation. VI. Series: Lecture notes in pure and applied mathematics ; v. 125.
QA76.95.C645 1990
510' .285--dc20
                                      90-34508
                                      CIP

Preface

The International Conference on Computers and Mathematics took place July 29-August 1, 1986, on the campus of Stanford University. The conference was devoted to the exploration of the past, present, and future relationship of computers to mathematics. It served as a forum for the interaction between leading experts in the development of computer systems and methods and those in mathematics and related areas interested in their use. Some of the invited talks focused on the past and future roles of computers as a research tool in such areas as number theory, analysis, special functions, combinatorics, algebraic geometry, topology, physics, mathematical physics, and other fields. Other talks described the role of mathematics in fields that evolved from the presence of computers, such as numerical analysis, artificial intelligence, computer algebra, and theoretical computer science.

The organizing committee of the conference consisted of Woodrow Bledsoe, C. K. Chu, Gregory V. Chudnovsky, Michael E. Fisher, J. L. Lions, John McCarthy, David B. Mumford, and David Y. Y. Yun. The co-chairmen were David V. Chudnovsky and Richard D. Jenks. The poster session chairman was James H. Davenport. The local arrangement chairman was William C. Swope. The conference was sponsored by the American Association of Artificial Intelligence and the ACM Special Interest Group on Symbolic and Algebraic Manipulation (SIGSAM). Supporters of the conference included the American Association of Artificial Intelligence, the Defense Advanced Research Project Agency, Ed Fredkin, IBM Research, the National Science Foundation, Symbolics Inc., and the System Development Foundation. The program of the conference included exhaustive system tutorials, chaired by B. F. Caviness and P. J. Deuflhard. Among the systems exhibited were CAYLEY, MAPLE, REDUCE, MACSYMA, muMATH, VIEWS, SMP, MATLAB, ARCRITH, and SCRATCHPAD II. System tutorials were presented by J. Cannon for CAYLEY, K. O. Geddes and B. W. Char for MAPLE, A. C. Hearn for REDUCE, R. Pavelle for MACSYMA, D. R. Stoutemyer for muMATH, G. Cherry for VIEWS, J. Grief for SMP, C. Moler for MATLAB, S. Rump for ARCRITH, and R. S. Sutor for SCRATCHPAD II.

The conference opened with an inspiring address by E. E. David, Jr. The invited speakers included, in the order of their lectures, B. J. Birch, C. Sims, G. E. Andrews, R. S. Askey, J. R. Rice, H. Wisniewsky, D. R. Stoutemyer, R. Loos, B. M. Trager, M. J. Feigenbaum, K. G. Wilson, T. E. Regge, W. Haken, D. V. and G. V. Chudnovsky, R. W. Gosper, R. M. Karp, W. Bledsoe, G. B. Dantzig, H. W. Lenstra, R. L. Graham, P. Erdos, and D. H. Lehmer. Sessions were chaired by C. K. Chu, P. J. Deuflhard, G. Collins, B. F. Caviness, R. Hofstadter, G. B. Dantzig, R. M. Karp, J. McCarthy, J. Brillhart, and D. Y. Y. Yun.

The conference program included tutorial courses on computational algebraic geometry by D. Bayer and M. Stillman, and on symbolic and algebraic computations by J. H. Davenport, K. O. Geddes, E. Kaltofen, D. B. Saunders, M. Singer, and F. Winkler.

The number of registered participants exceeded 650. The conference attracted a broad cross-section of scientists from all over the world.

The work organized and inspired by Richard D. Jenks made the conference the success it was. All efforts were excellently coordinated by the conference secretary Tiyo Asai. Program manager R. Gebauer, with total command of the program, locations, and utilities, made it all seem smooth and easy. Equipment manager M. Mobarak performed many miracles. W. S. Swope helped greatly with local arrangements. Z. Watt served as treasurer and J. Hunter was in charge of the conference desk. Conference organizers are very grateful to B. Gatje for planning and execution of travel arrangements.

David V. Chudnovsky
Richard D. Jenks

Contents

Preface   iii

Contributors   vii

Keynote Remarks at the Conference on Computers
and Mathematics   1
Edward E. David, Jr.

SCRATCHPAD Explorations for Elliptic Theta Functions   17
George E. Andrews and R. J. Baxter

Integration and Computers   35
Richard Askey

Some Thoughts on Proof Discovery   83
W. W. Bledsoe

Computer Algebra in the Service of Mathematical
Physics and Number Theory   109
David V. Chudnovsky and Gregory V. Chudnovsky

Impact of Linear Programming on Computer Development   233
George B. Dantzig

Uses of and Limitations of Computers in Number Theory   241
P. Erdos

Strip Mining in the Abandoned Orefields of Nineteenth
Century Mathematics   261
William Gosper

Polynomial Factorization 1982-1986   285
Erich Kaltofen

Factorization Then and Now   311
D. H. Lehmer

Computer Animation in Mathematics, Science, and Art   321
Nelson L. Max

Physicists and Computer Algebra   347
Tullio Regge

Symbolic Computation: The Early Days (1950-1971)   351
Jean E. Sammet

What Computer Algebra Systems Can and Cannot Do   367
David R. Stoutemyer

Solution of Equations I: Polynomial Ideals and
 Gröbner Bases   383
Franz Winkler

Index   407

Contributors

GEORGE E. ANDREWS The Pennsylvania State University, University Park, Pennsylvania

RICHARD ASKEY University of Wisconsin-Madison, Madison, Wisconsin

R. J. BAXTER The Australian National University, Canberra, Australia

W. W. BLEDSOE* Microelectronics and Computer Technology Corporation, Austin, Texas

DAVID V. CHUDNOVSKY Columbia University, New York, New York

GREGORY V. CHUDNOVSKY Columbia University, New York, New York

GEORGE B. DANTZIG Stanford University, Stanford, California

EDWARD E. DAVID, JR. EED, Inc., Bedminster, New Jersey

P. ERDOS Mathematical Institute, Hungarian Academy of Sciences, Budapest, Hungary

WILLIAM GOSPER Symbolics, Inc., Mountain View, California

ERICH KALTOFEN Rensselaer Polytechnic Institute, Troy, New York

D. H. LEHMER† University of California at Berkeley, Berkeley, California

NELSON L. MAX Lawrence Livermore National Laboratory, Livermore, California

TULLIO E. REGGE Institute of Theoretical Physics, Torrino, Italy

JEAN E. SAMMET Federal System Division, IBM, Bethesda, Maryland

DAVID R. STOUTEMYER University of Hawaii, Honolulu, Hawaii

FRANZ WINKLER University of Delaware, Newark, Delaware

---

*Current affiliation: University of Texas at Austin, Austin, Texas

†Retired

Keynote Remarks at the Conference on Computers and Mathematics

EDWARD E. DAVID, JR. EED, Inc., Bedminster, New Jersey

Computing and mathematics must count as one of the world's most successful marriages between a tender young bride and a sour old groom. Computing is among the newest arrivals to science and technology; in fact, in some respects, such as software, it has barely begun the transition from art to science. Mathematics is the Methusalah of sciences; in fact, it antedates science, and it retains an aloof pre-scientific aura that must explain in part why it is beloved of only a select few.

No need to tell this conference that the union of computing and mathematics has created the revolution we call the information age-- and the promise of stupendous boosts in our ability to produce goods and services. But despite the country's infatuation with "high technology," neither the promise nor the unfolding challenges of the information age are well understood by the public, the media, or the movers and shakers in our political establishment.

There is little awareness of the vital link between computing and mathematics, nor of the close relationship among these and modern communications and control systems. There is little awareness that all these technologies have been essentially mathematical creations by an honor role of such mathematical greats as von Neumann, Wiener, and Shannon. There is little awareness of exactly how computing has begun to repay its debt to mathematics, not by replacing mathematicians but by giving them a major new tool.

1

Finally, there is little awareness that the benefits of the information age cannot materialize without continued, symbiotic progress in both mathematics and computer science, at both the fundamental and applied levels. As a result, both mathematics and computing are becoming more and more short-handed for the job expected of the them; and, curiously, both fields are too tight-fisted to allocate the time and effort needed to call forth the resources that would restore the situation.

In this time of Gramm-Rudman and slow economic growth, keen competition prevails among all parts of the R&D budget, both in government and industry. The academic mathematics community is concerned that it is not getting the support needed to replace itself during the inevitable wave of retirements in the 90s. The computer science community is concerned that it is not getting the support needed for it to meet the wave of growth in the field. These are real short-falls despite this nation's awesome commitment to research and development, one that will bring its total R&D spending in 1986 to almost $120 billion.

ORGANIZATION AND THEME

What we have, then, is a serious misallocation of resources. It is a misallocation that we must look to our political system to help correct. Obviously, that is not easy when, in setting the Federal budget, Congress seems to have given up politics, priorities, and prudence alike. In their place stands the demon of Gramm-Rudman, with

its automatic, across-the-board cuts. As a well-known NASA advocate
said the other day, it's the intellectual equivalent of an unmanned
budget process.

Fortunately, all is not dark. In fact, my theme today is that
you, the members of this conference, can and must make a difference.
Congress and the Administration will respond if they recognize what
are clearly national, nonpartisan needs. With your effort, the
science community can help the nation to set some different
priorities--to remove critical imbalances, even within a shrinking
budget. Today, I will focus on just three of these imbalances, each a
major threat to achieving our national aspirations in research and
development. They involve, first, the inadequate funding of certain
key fields, such as mathematics and computer science; second,
inadequate funding for fundamental research in general; and, third,
inadequate funding of the industry-university connection, that is,
research carried out by industry and the universities together. Note
that I am mostly talking about Federal funding, but not entirely. I
am certainly not talking about Federal responsibility, because many
sectors share the responsibility for getting our R&D priorities
straight--including the universities, industry, and state and local
governments.

FUNDING FOR MATHEMATICS AND COMPUTER SCIENCE

To begin with, as I said, all is not dark. Today, I am delighted
to report to you a modest, but significant breakthrough in Federal
funding of mathematics research. As you may know, I chaired an ad hoc
committee organized by the National Research Council to look into the
state of core mathematics research. In 1984 the Committee issued a
report that documented an astonishing neglect of that research for
most of the previous fifteen years. During the 70s, the real level of
Federal support did not grow at all, while support for the other
sciences grew by more than fifty percent. To remedy the situation,
the NRC report called for more than a doubling of mathematics support-
-to about $175 million in 1984 dollars by 1990.

The mathematics community had a powerful case to make. In the
last three years, the result has been a 54 percent increase in NSF
funding, helped by strong support from the NSF's director, Erich
Bloch, and an unprecedented resolution of support from the National
Science Board, NSF's governing body. Meanwhile, the Department of
Defense has increased its funding of mathematics research by more than
70 percent. The increases at DOD came about through the enlightened
efforts of the people who run the math sciences programs at DOD and
through the new $10 million program in applied mathematics set up by
the Defense Advanced Research Projects Agency (DARPA).

Even this small success hinges upon two peculiar features of
mathematics research, not shared by other fields working to increase

their take from the Federal till. These other fields include
chemistry, which has recently issued its report on opportunities in
that field, and chemical engineering and physics, which will shortly
issue reports. And of course the Computer Science Board has just
issued its own preliminary report.

The first peculiar feature of mathematics, one that sets it apart
of most other fields of research, is that financially it is a very
small field. Mathematics funding for 1986 stands at only about $105
million--small compared to other fields and still not enough to ensure
that the field can replace retiring faculty, let alone grow. The
second peculiar feature of core mathematics is that it is easily
confused with computer science. In the mathematics report, we drew a
defensible, but necessarily artificial distinction between the two.
We know the two fields are strongly tied together, as the convening of
this conference and the history of the past 50 years testify. But the
distinction is necessary to ensure that core mathematics can continue
to follow its own dynamic as well as respond to the needs of computing
and other fields.

To carry the point further, we might look beyond computer science
to an even newer member of the large "customer community" for
mathematics--namely, the life sciences. Consider the implications of
recent advances in knot theory. Now that these implications have
become clear, the obvious temptation for the National Institutes of
Health will be to tie up new money in the application of knot theory
to genetic problems. Yet in the long run NIH will better serve

mathematics and its own mission with a balance between both
fundamental and applied programs. This was a fundamental and
persuasive message of the NRC mathematics report. If we are careful
to balance our support among fields, as well as on what may seem the
knife-edge distinction between fundamental and applied work, we stand
a better chance of achieving applications that really cut Gordian
knots.

BASIC RESEARCH: THE NEED FOR BALANCE

This provides an easy transition to a closer look at the issues
posed by the Gramm-Rudman meat ax. Faced with the arbitrary cutbacks
specified by that law, balance is of course the catchword these days
in Washington. But it is also the right policy, and it always has
been.

Most fields, especially mathematics, have their own internal
driving forces. So while the proper level of support for each field
cannot be derived from first principles, it can be balanced against
opportunities for productive research in the field, which are in turn
a function of the state of knowledge and the excellence of people
available for research. And, as I have stressed, most fields have an
impact on other disciplines and on applications. Support levels must
be balanced against these external factors as well, so that the total
R&D enterprise is not held up for want of a scientific or mathematical

nail. Those are the considerations that the NRC mathematics committee
stressed.

At one time we might have hoped that with peer review and proper
advocacy, the most fruitful fields would win their share of the pot of
dollars available for fundamental research, each in proportion to
their need. But pious hopes are not enough in the current situation.
In one sense of course, it would be wrong to hope that any part of the
R&D enterprise can ever be anything but tight-fisted and short-handed.
No matter what the level of funding, the financial means and human
resources will always fall far short of the opportunities that
dedicated scientists and engineers can perceive. So there's a
continuing responsibility to prune out marginal efforts, leaving room
for new initiatives and for what might be called growth in place.
That effort is especially important in fields clamoring for bigger
pieces of the pie, like computer science.

But today the question of balance is more profound than this. In
fact, as Erich Bloch and the chairman of the National Science Board,
Roland Schmitt, have recently urged, the entire Federal, really the
entire national R&D budget, has gone badly askew. The question is not
whether we have the proper balance among the different disciplines of
fundamental research and engineering. If that were so, then funding
for a field like core mathematics could grow only at the expense of
another field of academic research. Rather the real question is
whether we have a proper balance between basic research funding and

the country's other, massive R&D undertakings. The answer is that we don't.

The effects of the existing imbalance have been apparent in the faculty and university equipment shortages that have been so much in the news. They extend to shortages of graduates in certain key fields, such as computer science. The imbalance is even more apparent in the formidable new R&D efforts on which the Federal Government has embarked. The current Administration focused initially on much more modest goals--for example, boosting support for basic research at the universities, and closer cooperation between industry and the universities. Then came a reversion to type: a returning emphasis on large, politically exciting research and development programs.

These are the megaprojects, the massive programs so beloved of central governments since the building of the pyramids and the Great Wall of China, right down to the days of our own great water projects in the West, Project Apollo, and the space shuttle. The Government has lately undertaken several new megaprojects. They notably include the Strategic Defense Initiative, but also the Super Conducting Super Collider and the Space Station, not to speak of the newest entry from last year's summit conference: the Experimental Test Reactor for Fusion Power. Partly spurred by the loss the space shuttle <u>Challenger</u>, we may also see an effort to develop a a next-generation substitute for the shuttle technology. As proposed recently by DOD, this would be a space plane that could soar anywhere on the globe within two hours, as well as provide a promised economic means for launching satellites.

Without debating the intrinsic merits of such projects, we might concede that they do give us some growth, making science and engineering a little less pinched. They do provide a politically feasible way to boost overall scientific and engineering activity. Unfortunately, the megaprojects tend to create unbalanced growth. Development crowds out fundamental research. Federal priorities crowd out disciplinary priorities. Not least, big science and engineering crowd out little science and engineering. That is, the megaprojects draw resources, especially excellent people, away from smaller-scale, commercially relevant R&D and the associated basic sciences. They crowd out efforts to ensure that U.S. industry will create the technologies needed to prevail in the teeth of fierce international competition.

One way to counteract the impact of megaprojects on basic research is by exacting a tithe on them. They money can then be used to put back what they are taking out of the R&D system. I understand that the plan is to tithe the SDI budget by 5 percent on behalf of basic research. That would amount to over $1 billion during the next five years. Better the tithe were the ten percent traditionally associated with a tithe, but 5 percent is a step in the right direction.

The next step should be SDI scholarships, fellowships, and research grants targeted especially on the small fields that will contribute most to the advance of both military technology and commercial industry--for example, computer science and core and applied mathematics. Above all, this spending must take into account that the crux of our efforts must be people. We must have a steady flow, probably an increasing flow of excellent new people into all fields of science and technology that are strategic to the nation's military, social, and commercial needs.

IMPORTANCE OF INDUSTRY-UNIVERSITY CONNECTIONS

For much the same reason, we must ensure that Federal megaprojects do not crowd out the burgeoning research relationships between universities and commercial industry. Industry must draw a balanced portion of the best and the brightest to work on its problems and opportunities. This is the third imbalance that should concern us. Again, the news here is by no means all bad.

True, some sectors of industry have been forced to cut R&D spending and are less ready to put money into these relationships. For example, I estimate that the petroleum industry has cut back its R&D spending by at least 40 percent in the past three years and correspondingly more in terms of people. In addition, attitudes are degenerating. Within industry as whole, we can observe a worsening preoccupation with short-term results. A worsening focus on the "paper entrepreneurism" of mergers and acquisitions rather than innovation, higher productivity, and international competitiveness. A

worsening reluctance to commit resources patiently to the kind of
fundamental work that gave us the transistor and the laser.

Not helping a bit in this regard is the so-called tax reform
proposed in the House and Senate bills that have just gone into
conference with great political fanfare. The President has indicated
that he will sign whatever compromise finally emerges from the
House-Senate conference, so the reform will likely become reality. By
erasing the distinction between immediate income and investment, this
reform will create a strong disincentive to investment and an even
stronger incentive to take the money and run. In no small part as a
result of this prospect, U.S. industry has gone on a capital
investment strike--witness the effect in LTV's bankruptcy and IBM's
latest financial results.

The good news here is that, despite the storm clouds gathering
over the economy, industry spending for R&D has continued to climb--to
a record $58 billion in 1986. That's up more than 9 percent over last
year. And, even though industry still casts a small shadow on campus
compared to the Federal Government, industry continues to be the
fastest growing source of support for academic research. The demand
for innovation is so intense worldwide that industry will be hesitant
to cut back its level of R&D spending, even with economic slowdown.
And that includes its spending on research with the universities.

Your concern here is mathematics and the computer, but a recent
report in <u>Science</u> magazine about research biotechnology provides some
useful insights on how the relationship between the universities and

industry is developing. The report is based on a survey of more than 1200 faculty members at 40 major universities. The findings?; University biotechnologists with industrial support are more productive than their colleagues without such support. They are stimulated by their interaction with their counterparts in industry and by the practical problems of industry. They publish more, get more patents, and earn more. Their results are more quickly translated into useful goods and services. There are some effects to watch warily; more proprietary results to be held closely; and more influence by immediate market goals on the directions of academic research.

In short, despite noting these risks, the report in <u>Science</u> confirms my own belief that university-industry relationships are not only here to stay, but here to grow, and here to grow more rapidly than anyone even now appreciates. They are becoming a critical means for more swiftly translating this nation's excellence in basic research into greater competitiveness in the international marketplace. And they are providing added resources for helping academic research become less tight-fisted and short-handed.

This is not the place to survey the many things that are happening on this scene, except to note that we have barely started. Again, in seeking to develop these relationships, the important consideration is balance. We will benefit from a diversity of participants pursuing a diversity of approaches in a diversity of fields.

For example, diversity will involve not just hot fields like
biotechnology, microelectronics, or computer science, but others like
metallurgy, combustion, water resources, core mathematics and, yes,
soft sciences like economics and sociology: involving not just large
companies, but the small and mid-sized companies that are the biggest
source of jobs and economic growth in this country. And involving not
just initiatives from the Federal Government, but initiatives from
state and local governments, which have in recent years taken the lead
in fostering relationships between industry and the universities.
Thus, in my own state of New Jersey, I served on a Governor's
Commission that proposed and saw adopted a $90 million bond issue with
this goal. Among other things, it has been used to support "centers
of excellence" in such fields as ceramics, food processing,
biotechnology, and the treatment of toxic waste.

PERCEPTIONS, PROSPECTS, AND CONCLUSIONS

So, despite the budgetary limits to growth that we hear so much
about, there is room for you to play your part--to help restore the
balance, to help get our R&D priorities straight, even with the
Gramm-Rudman meat ax overhead.

The effort I have in mind is of course an educational effort, one
directed not at grass roots "science literacy," but at the leadership
in government, industry and the universities. My experience with the
NRC math report tells me that the job is tough but doable. It also

tells me that the bigger the amounts needed, the more (exponentially?) difficult the job. The mathematicians have only achieved about half their very modest objectives. Given the present situation, a tight-fisted approach to lobbying won't work.

The educational message must be carefully tailored to both perceptions and prospects. As I said earlier, in the case of your own fields, much remains to be done. Very few people in this country perceive the full prospects of the unfolding information age. The number of components on an electronic chip is currently doubling about every 18 months. By the turn of the century, the industry will squeeze at least 100 million components on a chip, compared to a million today. The price of a megabyte of dynamic memory is falling by more than 20 percent a year, the price of all hardware is falling by 30 percent a year, and a foot of fiber optic cable now costs less than lamp cord.

One obvious result of all this is that in the next few years, we will have desktop, or "under-the-desk," microcomputers with the processing power of a mainframe today. At the same time, we are beginning to see the shape of computing to come--parallel and other new computer architectures, new semi-conductor materials, fuzzy logic, and perhaps neural networks and optical computers.

In communications, the capacity of lightwave systems--measured as a function of pulse rate times unrepeated distance--is doubling each year. Researchers at AT&T Bell Labs recently transmitted the equivalent of 20 gigabits over 42 miles of optical fiber. But even

that achievement represents only about one percent of what is
theoretically possible. In setting up data networks and processing
systems, distance is fast losing all relevance.

Not even the experts can fully appreciate what this furious pace
of progress means, because few can know just what we can build with
these components--whether robotic vision systems, voice recognition
systems, expert systems, symbolic machines, computer-integrated
manufacturing and distribution systems, and so on. By the same token,
few people appreciate that to create these systems we must achieve
vastly greater understanding, not only of computer science and key
branches of mathematics, but of products and processes, and of the
sciences that underlie them. We have just heard that a manufacturer
near here will shortly introduce a hand-held symbol-manipulating
calculator. If that is indeed true, the implications for education
and research are mind-boggling. However, the understanding of such
matters is advancing more slowly than the hardware and the promises.

The challenge is to make clear to the powers that be that
advances in microelectronics and photonics will go for nought without
equivalent advances in mathematics, computer science, and other fields
of basic research. Our society routinely celebrates the rate of
progress in microelectronics and computer hardware. You should strive
to instill greater appreciation of the little-known fact that our
computational ability over the last decade has in fact benefited much
more from improvement in software algorithms than from improvements in
hardware speed.

A recent case in point is of course the new linear programming algorithm developed by Narendra Karmarkar at Bell Labs. AT&T applied the algorithm to the ten-year Pacific Facilities Network plan, a problem which has 42,500 variables and 15,000 constraints. It ran the problem in less than four minutes, compared to 76 minutes using prior simplex methods. Now that <u>is</u> a productivity improvement! It's one to bear in mind when addressing complaints about the mundane, if critical, problem of increasing software productivity.

The Karmarkar algorithm scored another kind of victory, too. News of its discovery made the front page of the New York Times. So enlightened appreciation for the role of mathematics and computer science is growing. We can see more awareness that advances like parallel processing, neural networks--even a truly user-friendly microcomputer using voice recognition--will depend heavily upon advances in mathematics and computer science. Again, the challenge is to make that awareness grow. Because with it, we may hope that core mathematics and computing will become less short-handed . . . and willing even to be less tight-fisted with the effort to communicate about new opportunities and the need for a balanced R&D enterprise. For that will be the key to progress. Let me leave you with that challenge and wish you a most productive conference.

SCRATCHPAD Explorations for Elliptic Theta Functions

GEORGE E. ANDREWS [1] The Pennsylvania State University, University Park, Pennsylvania

R. J. BAXTER The Australian National University, Canberra, Australia

1. Introduction. In a recent series of papers [5], [6], [9] we have examined a two-particle generalization of hard hexagon model in statistical mechanics. Our object in this paper is to examine the known algorithms that failed us in our research, the lucky accidents that provided our insight for obtaining our solutions and the developments in exploratory algorithms we think might systematize problems like this.

The explicit problem from [5], [6] and [9] that we wish to consider is easily stated. Let $X_m(a,b,c;q)$ be a three parameter sequence of polynomials in q where $1 \leq a,b,c \leq 3$, $b + c \geq 4$ (a,b,c integers) with

---

[1]Partially supported by National Science Foundation Grant MCS8201733.

(1.1) $X_0(a,b,c;q) = \begin{cases} 1 & \text{for } a = b \\ 0 & \text{otherwise} \end{cases}.$

(1.2) $X_m(a,b,c;q) = \sum_{h=4-b}^{3} q^{\frac{1}{2}|\nu(h)-\nu(c)|m} X_{m-1}(a,h,b;q) ,$

where $\nu(1) = 0$, $\nu(2) = 2$ and $\nu(3) = 1$. What we want is $X_\infty(a,b,c;q)$ as
a modular form of some sort. However initially we have little to go on
except our faith arising from the history of exactly solved models that there
is a positive answer to our quest.

In Section 2 we briefly consider classical algorithms that were useful
in the past. Section 3 is devoted to the lucky accidents that led to the
right answers. Section 4 provides a brief discussion of open problems
concerning exploratory algorithms in computer algebra.

The symbolic algebra package used throughout this work was IBM's
SCRATCHPAD.

2.   Classical Algorithms for Solving the Hard Hexagon Model.

We refer the reader to [8; Ch. 14] for a complete treatment of the Hard
Hexagon Model which was originally solved by one of us (RJB) in 1979 [7]. In
this section we shall concentrate on that aspect of it concerning a limit of
polynomial sequences. Instead of the sequence described by (1.1) and (1.2),
we consider

(2.1) $D_0(q) = D_1(q) = 1 .$

(2.2) $D_n(q) = D_{n-1}(q) + q^{n-1}D_{n-2} , \quad n \geq 2.$

Again we would like to find $D_\infty(q)$ as essentially a modular form.

We note that the coefficients b_n in

$$(2.3) \qquad\qquad D_\infty(q) = \sum_{n=0}^\infty b_n q^n \ ,$$

are easily computed using (2.1) and (2.2) (indeed $D_\infty(q)$ and $D_n(q)$ are

identical in coefficients of q^m for $m < n - 1$).

Once a large number of b_n's have been computed we may insert them into

Euler's infinite product representation algorithm (EIPRA) [2; p. 98], [3; p. 104]:

E.I.P.R.A. If

$$(2.4) \qquad\qquad f(q) \equiv 1 + \sum_{n=1}^\infty b_n q^n = \prod_{n=1}^\infty (1-q^n)^{-a_n} \ ,$$

where the a_n and b_n are integers, then

$$(2.5) \qquad\qquad nb_n = \sum_{j=1}^n D_j b_{n-j} \ ,$$

where

$$(2.6) \qquad\qquad D_j = \sum_{d \mid j} d a_d \ .$$

Furthermore if either sequence a_n or b_n is given, the other is uniquely

determined by (2.5) and (2.6).

Using EIPRA, we find that for as many a_n as we care to compute (RJB

computed a few hundred in 1979) the sequence is periodic modulo 5 and the

first five terms are $a_1 = 1$, $a_2 = a_3 = 0$, $a_4 = 1$, $a_5 = 0$. Thus EIPRA yields

the following conjecture (which is in fact equivalent to the first
Rogers-Ramanujan identity):

$$(2.7) \qquad D_\infty(q) = \prod_{n=1}^{\infty} \frac{1}{(1-q^{5n-4})(1-q^{5n-1})} \; .$$

Next we ask: how can we prove (2.7)? Instead of following the original
method, we try to approximate $D_n(q)$ by Gaussian polynomials. The Gaussian

polynomials $\begin{bmatrix} n \\ m \end{bmatrix}$ are defined by

$$(2.8) \qquad \begin{bmatrix} n \\ m \end{bmatrix} = \begin{bmatrix} n \\ m \end{bmatrix}_q = \begin{cases} \dfrac{(1-q^n)(1-q^{n-1})\dots(1-q^{n-m+1})}{(1-q^m)(1-q^{m-1})\dots(1-q)} \; , & n \geq 0, \; m \geq 0 \\[2mm] 0, \; n \geq 0, \; m < 0. \end{cases}$$

They are simply encoded in SCRATCHPAD directly from the definition
(2.8).

So we encode the $D_n(q)$ from (2.2) and the $\begin{bmatrix} n \\ m \end{bmatrix}$ from (2.8). We first
ask for several $D_n(q)$:

$(2.9) \qquad D_0(Q) : 1$

$(2.10) \qquad D_1(Q) : Q + 1$

$(2.11) \qquad D_2(Q) : Q^2 + Q + 1$

$(2.12) \qquad D_3(Q) : Q^4 + Q^3 + Q^2 + Q + 1$

$(2.13) \qquad D_4(Q) : Q^6 + Q^5 + 2Q^4 + Q^3 + Q^2 + Q + 1$

(2.14) $D_5(Q) : Q^9 + Q^8 + Q^7 + 2Q^6 + 2Q^5 + 2Q^4 + Q^3 + Q^2 + Q + 1$

(2.15) $D_6(Q)$

$Q^{12} + Q^{11} + 2Q^{10} + 2Q^9 + 2Q^8 + 2Q^7 \, 3Q^6 + 2Q^5 + 2Q^4 + Q^3$

$+$

$Q^2 + Q + 1$

For purposes of illustration we restrict our attention to the $D_n(q)$ with odd subscripts; those with even subscripts are treated similarly.

We note that $D_{2n-1}(q)$ appears to be of degree n^2. Since $\begin{bmatrix} 2n \\ n \end{bmatrix}$ is also, we consider $D_{2n-1}(q) - \begin{bmatrix} 2n \\ n \end{bmatrix}$ and ask SCRATCHPAD for these polynomials with $n = 1,2,3,4$ and 5: The results are

(2.16)   0

(2.17) $(-Q^2)$

(2.18) $- Q^7 - Q^6 - Q^5 - Q^4 - 2Q^3 - Q^2$

(2.19) $- Q^{14} - Q^{13} - 2Q^{12} - 2Q^{11} - 4Q^{10} - 4Q^9 - 5Q^8 - 4Q^7 - 4Q^6 - 3Q^5$

$+$

$- 3Q^4 - 2Q^3 - Q^2$

(2.20) $- Q^{23} - Q^{22} - 2Q^{21} - 3Q^{20} - 5Q^{19} - 6Q^{18} - 9Q^{17} - 10Q^{18} - 12Q^{15}$

$+$

$- 13Q^{14} - 14Q^{13} - 14Q^{12} - 14Q^{11} - 13Q^{10} - 11Q^9 - 10Q^8 - 8Q^7$

$+$

$- 6Q^6 - 5Q^5 - 3Q^4 - 2Q^3 - Q^2$

For each $n \geq 2$ it appears that the resulting polynomial is q^2 times polynomials of degree $n^2 - 4 = (n-2)(n+2)$. This suggests that for our next approximation to $D_{2n-1}(q)$ we use $\begin{bmatrix} 2n \\ n \end{bmatrix} - q^2 \begin{bmatrix} 2n \\ n+2 \end{bmatrix}$.

We now ask SCRATCHPAD to give us $D_{2n-1}(q) - \begin{bmatrix} 2n \\ n \end{bmatrix} + q^2 \begin{bmatrix} 2n \\ n+2 \end{bmatrix}$ and the results for $n = 1, 2, 3, 4$ and 5 are:

(2.21)   0

(2.22)   0

(2.23) $(-Q^3)$

(2.24) $- Q^{10} - Q^9 - Q^8 - Q^7 - Q^6 - Q^5 - Q^4 - Q^3$

(2.25) $- Q^{19} - Q^{18} - 2Q^{17} - 2Q^{16} - 3Q^{15} - 3Q^{14} - 4Q^{13} - 4Q^{12} - 4Q^{11}$
          $+$
          $- 4Q^{10} - 3Q^9 - 3Q^8 - 3Q^7 - 2Q^6 - 2Q^5 - Q^4 - Q^3$

In exactly the same manner as above, we see that

$$\begin{bmatrix} 2n \\ n \end{bmatrix} - q^2 \begin{bmatrix} 2n \\ n+2 \end{bmatrix} - q^3 \begin{bmatrix} 2n \\ n+3 \end{bmatrix}$$

is a still better approximation to $D_{2n-1}(q)$.

It should be clear by now that with a symbolic algebra package like SCRATCHPAD this process can be carried on indefinitely, and the inevitable conclusion is the following conjecture (found and proved by Schur [11]):

(2.26) $D_{2n-1}(q) = \sum_{\lambda=-\infty}^{\infty} q^{\lambda(10\lambda+1)} \begin{bmatrix} 2n \\ n-5\lambda \end{bmatrix} - \sum_{\lambda=-\infty}^{\infty} q^{(2\lambda-1)(5\lambda-2)} \begin{bmatrix} 2n \\ n-5\lambda+2 \end{bmatrix}$.

Once you have conjectured (2.26) and the counterpart for $D_{2n}(q)$ the proof is quite straightforward (see e.g. [1]).

Indeed the general result proved by Schur [11] for all n is

(2.27) $$D_n(q) = \sum_{\lambda=-\infty}^{\infty} (-1)^\lambda q^{\lambda(5\lambda+1)/2} \begin{bmatrix} n+1 \\ \left[\frac{n+1-5\lambda}{2} \right] \end{bmatrix} ,$$

where $\begin{bmatrix} a \\ b \end{bmatrix}$ is the Gaussian polynomial and $[a]$ is the greatest integer in a.

The above procedure of approximating polynomials by Gaussian polynomials of the same degree and then iterating we shall call Schur's Gaussian Polynomial Approximation Algorithm (SGPAA).

3. The limiting behavior of $X_m(3,3,3;q)$.

We return now to the problem posed in the introduction. We want $X_\infty(a,b,c;q)$ expressed as a modular form, presumably some nice linear combination of infinite products and theta functions. In the following we restrict our consideration to $X_\infty(3,3,3;q)$.

We begin following the lead of Section 2. Consider

(3.1) $$X_\infty(3,3,3;q) = \prod_{n=1}^{\infty} (1-q^n)^{-a_n} .$$

The first thirty-six a_n are as follows:

Table 1.

n	a_n	n	a_n	n	a_n
1	2	13	2	25	2
2	1	14	1	26	4
3	1	15	0	27	-3
4	1	16	1	28	1
5	2	17	3	29	7
6	0	18	0	30	-5
7	2	19	2	31	2
8	1	20	2	32	9
9	1	21	-1	33	-8
10	1	22	1	34	0
11	2	23	4	35	13
12	0	24	-2	36	-12

Clearly Table 1 suggests that EIPRA unaided will not tell us the nice answer we want. Let us try an alternative; this new approach was successful in treating the more complicated cases of Regime II in the original hard hexagon model [8; p. 437, eq. (14.5.22a), (14.5.22b)]. Namely if we multiply both sides of (2.7) by $\prod_{n \geq 1}(1-q^n)$ and apply Jacobi's Triple Product [2; p. 22, Cor. 2.9], we obtain

(3.2)
$$\prod_{n \geq 1}(1-q^n) \cdot D_\infty(q) = \sum_{\lambda=-\infty}^{\infty} (-1)^\lambda q^{\lambda(5\lambda+1)/2}$$

$$= 1 - q^2 - q^3 + q^9 + q^{11} - q^{21} - q^{24} + \ldots$$

Once one contemplates (3.2) it is easy to deduce the rule of formation of the exponents in the series. So we consider

(3.3)
$$\prod_{n \geq 1}(1-q^n)X_\infty(3,3,3;q) = \sum_{n \geq 0} c_n q^n .$$

The first twenty c_n are as follows:

n	c_n	n	c_n
0	1	10	3
1	1	11	3
2	1	12	3
3	1	13	3
4	1	14	4
5	2	15	4
6	1	16	4
7	2	17	5
8	2	18	6
9	2	19	6

Again we face unrevealing data. Undaunted we return to Table 1. While we see eventual trouble, we nonetheless note that the a_n looked initially like they were going to be periodic with period 6. Indeed for fourteen terms of the power series we have the following approximation:

(3.4) $X_\infty(3,3,3;q)$

$$\approx \prod_{n=1}^{\infty} \frac{(1-q^{6n})}{(1-q^n)(1-q^{6n-1})(1-q^{6n-5})}$$

$$= \prod_{n=1}^{\infty} \frac{(1-q^{6n-3})(1-q^{6n})}{(1-q^n)(1-q^{2n-1})}$$

$$= \prod_{n=1}^{\infty} \frac{(1+q^n)(1-q^{3n})}{(1-q^n)}$$

$$= \frac{\prod_{n=1}^{\infty}(1-q^{3n})}{1-2q+2q^4-2q^9+2q^{16}-2q^{25}+\ldots} \quad .$$

$$= \frac{1-q^3-q^6+q^{15}+q^{21}-q^{36}-q^{45}+\ldots}{1-2q+2q^4-2q^9+2q^{16}-2q^{25}+\ldots} \quad .$$

Thus we see that this last expression would become a power series with very few nonzero terms if we were to multiply by $(1 + 2\sum_{n\geq1}(-1)^n q^{n^2})$. Hence, in desperation we consider

(3.5) $(1 + 2\sum_{n\geq1}(-1)^n q^{n^2}) X_\infty(3,3,3;q)$.

$$= 1 - q^3 - q^6 + q^{17} + q^{18}\ldots$$

$$= \sum_{\lambda=-\infty}^{\infty} q^{\lambda(35\lambda+1)/2} - q^3\sum_{\lambda=-\infty}^{\infty} q^{\lambda(35\lambda-29\lambda)/2} \quad ,$$

valid at least for the first 200 terms!!!

The device of multiplication by $(1 + 2 \sum_{n \geq 1} (-1)^n q^{n^2})$ works well for

$X_m(a,b,c;q)$ provided $b = c$; however when $b \neq c$ things remain mysterious.

Hence we hope that Schur's Gaussian Polynomial Approximation Algorithm (SGPAA) may be appliable. The following table gives the first few $X_m(3,3,3;q)$:

Table 3

n	$X_n(3,3,3;q)$
0	1
1	1
2	$1 + 2q$
3	$1 + 2q + 2q^2 + q^3$
4	$1 + 2q + 4q^2 + 3q^3 + 3q^4 + q^5$
5	$1 + 2q + 4q^2 + 5q^3 + 6q^4 + 6q^5 + 4q^6 + 2q^7 + q^8$

In analogy with the work in Section 2, we want families of polynomials $p_i(m,\lambda,q)$ $(i=1$ or $2)$ such that

$$(3.6) \qquad X_m(3,3,3;q) = \sum_{\lambda=-\infty}^{\infty} q^{\lambda(35\lambda+1)/2} \, p_1(m,\lambda,q)$$

$$- \sum_{\lambda=-\infty}^{\infty} q^{3+\lambda(35\lambda-29)/2} \, p_2(m,\lambda,q) \; ,$$

where

(3.7) $$\lim_{m \to \infty} p_i(m,\lambda,q) = \frac{1}{1-2q+2q^4-2q^9+2q^{16}\cdots} \ .$$

When we try the SGPAA directly (i.e. taking the $p_i(m,\lambda,q)$ to be Gaussian polynomials) nothing sensible emerges (as we well expect).

Again we have to resort to ad hoc methods. Examining the very useful representation of $D_n(q)$ given by (2.27), we might hope that

(3.8) $$g_n(q) = \sum_{\lambda=-\infty}^{\infty} (-1)^\lambda q^{\lambda(7\lambda+1)/2} \begin{bmatrix} N \\ \left[\frac{N-7\lambda}{2}\right] \end{bmatrix}$$

would somehow apply to our work here. A close study of $g_N(q)$ yields the following:

(3.9) $$q^{N^2} g_{2N}(q^{-1}) = \sum_{\lambda=-\infty}^{\infty} q^{35\lambda^2+\lambda} \begin{bmatrix} 2N \\ N-7\lambda \end{bmatrix}$$

$$- \sum_{\lambda=-\infty}^{\infty} q^{35\lambda^2-29\lambda+6} \begin{bmatrix} 2N \\ N-7\lambda+3 \end{bmatrix} \ .$$

Let us, therefore, define

(3.10) $$G(q) \equiv \lim_{N \to \infty} q^{N^2} g_{2N}(q^{-1})$$

$$= \frac{1}{\prod_{n=1}^{\infty}(1-q^n)} \left\{ \sum_{\lambda=-\infty}^{\infty} q^{35\lambda^2+\lambda} - \sum_{\lambda=-\infty}^{\infty} q^{35\lambda^2-29\lambda+6} \right\} \ .$$

Thus our conjectured formula (3.5) may be rewritten as

(3.11)
$$X_\infty(3,3,3;q^2) = \left[\prod_{n=1}^{\infty} \frac{1}{1+q^{2n-1}} \right] G(q)$$

Now both $G(q)$ and $X_\infty(3,3,3;q^2)$ are limits of polynomial sequences. Hence we wish to formulate a polynomial version of (3.11) that converges to (3.11) in the limit. A few hours of dodging and weaving with SCRATCHPAD led to the following conjecture which was easily checked for $m \leq 7$:

(3.12)
$$X_m(3,3,3;,q^2) = \sum_{j=0}^{m} (-q)^j {\begin{bmatrix} m \\ j \end{bmatrix}}_{q^2} q^{(m-j)^2} g_{2(m-j)}(q^{-1}).$$

Now (3.12) is certainly a major step forward. If we can prove it we have (3.11) immediately by [2; p. 153, limit equation]. However (3.12) is much different from the representation (3.6) we were hoping for. Thus we would like to rewrite (3.12) so that it resembles (3.6). This is easily accomplished by using (3.8) for the $q^{(m-j)^2} g_{2(m-j)}(q^{-1})$. Consequently (3.12) may be rewritten as

(3.13) $X_m(3,3,3,q^2)$

$$= \sum_{\lambda=-\infty}^{\infty} q^{35\lambda^2+\lambda} \left[\sum_{i=0}^{m} (-q)^i {\begin{bmatrix} m \\ i \end{bmatrix}}_{q^2} {\begin{bmatrix} 2m-2i \\ m-i-7\lambda \end{bmatrix}} \right]$$

$$- q^6 \sum_{\lambda=-\infty}^{\infty} q^{35\lambda^2-29\lambda} \left[\sum_{i=0}^{m} (-q)^i {\begin{bmatrix} m \\ i \end{bmatrix}}_{q^2} {\begin{bmatrix} 2m-2i \\ m-i-7\lambda+3 \end{bmatrix}} \right].$$

Let us consider the first intereior sum in the case $\lambda = 0$, $q = 1$; it is thus just a sequence of integers say T_m, whose values are:

m	T_m	m	T_m
0	1	4	19
1	1	5	51
2	3	6	141
3	7	7	393

A check of Neal Sloane's book on integer sequences [12] reveals that T_m is the sequence 1070, namely the central term in the expansion of $(1 + x + x^2)^m$.

This last observation is the crucial one in unravelling the behavior of $X_\infty(a,b,c;q)$ in general. The polynomials we have been searching for are q-analogs of trinomial coefficients (i.e. the coefficients in $(1 + x + x^2)^n$). The literature on the trinomial coefficients themselves is sparse to say the least [10; p. 78, p. 163]. · q-Trinomial coefficients have never arisen before to our knowledge. Indeed all the polynomials $X_m(a,b,c;q^2)$ have expansions like (3.6) wherein the $p_i(m,\lambda,q)$ are replaced by various q-analogs of trinomial coefficients.

In [6] we have developed an extensive account of the q-trinomial coefficients. It turns out that they have numerous surprising and subtle properties. For example, let us examine the polynomial from the first sum in (3.13) with $\lambda = 0$:

(3.14)
$$T_m(q) = \sum_{i=0}^{m} (-q)^i \begin{bmatrix} m \\ i \end{bmatrix}_{q^2} \begin{bmatrix} 2m-2i \\ m-i \end{bmatrix} .$$

we obtain easily using SCRATCHPAD that

m	T_m
0	1
1	1
2	$1 + 2q^2$
3	$1 + 2q^2 + 2q^4 + 2q^6$
4	$1 + 2q^2 + 4q^4 + 4q^6 + 4q^8 + 2q^{10} + 2q^{12}$
5	$1 + 2q^2 + 4q^4 + 6q^6 + 8q^8 + 8q^{10} + 8q^{12} + 6q^{14}$
	$+ 4q^{16} + 2q^{18} + 2q^{20}$

Clearly $T_m(q)$ is an even function of q (i.e. a polynomial in q^2);
however, this is not at all obvious in the definition (3.14). The evenness
of $T_m(q)$ turns out to be important, and we had a hard time proving it. We
refer the reader to [6; Section 2.4, eq. (2.4.2)] for details.

4. Open Problems. In this paper we have examined the explorations via
SCRATCHPAD that led to our complete solution of a generalization of the hard
hexagon model. In Section 2 we examined two very useful algorithms: EIPRA
and SGPAA. If instead of EIPRA, we had had an algorithm to search for

$$(4.1) \qquad f(q) \equiv 1 + \sum_{n=1}^{\infty} b_n q^n$$

$$= \prod_{n=1}^{\infty} (1-q^n)^{-a_n} \pm q^d \prod_{n=1}^{\infty} (1-q^n)^{-c_n} \, ,$$

then we could have guessed (3.5) without difficulty. To determine <u>useful</u>
a_n, d, and c_n in (4.1) given b_n is a tall order since there is no longer
uniqueness of the representation. If some sensible algorithm can be
developed for (4.1), we can obviously ask whether such a thing can be done
for larger linear combinations of infinite products. The number-theoretic
implications of this problem are discussed in [4].

In Section 3, our discovery of q-trinomial coefficients suggests clearly
that SGPAA can be extended to approximations using q-trinomial coefficients.
Indeed we are now reasonably certain of what the useful q-analogs of the
coefficients in $(1 + x + x^2 + \ldots + x^k)^n$ are. Whether these will prove
useful in approximation algorithms like SGPAA remains to be seen.

References

1. G. E. Andrews, A polynomial identity which implies the Rogers-Ramanujan identities, Scripta Math., 28 (1970), 297-305.

2. G. E. Andrews, The Theory of Partitions, Encyclopedia of Mathematics and Its Applications, Vol. 2., Addison-Wesley, Reading, 1976. (Reissued: Cambridge University Press, London and New York, 1984).

3. G. E. Andrews, q-Series: Their Development and Application in Analysis, Number Theory, Combinatorics, Physics, and Computer Algebra, CBMS Regional Conference Series, No. 66, American Math. Soc., Providence, 1986.

4. G. E. Andrews, Further problems on partitions, Amer. Math. Monthly, (to appear).

5. G. E. Andrews and R. J. Baxter, Lattice gas generalization of the hard hexagon model: II. The local densities as elliptic functions, J. Stat. Phys., 44 (1986), 713-728.

6. G. E. Andrews and R. J. Baxter, Lattice gas generalization of the hard hexagon model: III. q-trinomial coefficients (to appear).

7. R. J. Baxter, Hard hexagons: exact solution, J. Phys., A13 (1980), L61-L70.

8. R. J. Baxter, Exactly Solved Models in Statistical Mechanics, Academic Press, London and New York, 1982.

9. R. J. Baxter and G. E. Andrews, Lattice gas generalization of the hard hexagon model: I. star-triangle relation and local densities, J. Stat. Phys., 44 (1986), 249-271.

10. L. Comtet, Advaned Combinatorics, Reidel, Dordrecht and Boston, 1974.

11. I. Schur, Ein Beitrag zur additiven Zahlentheorie und zur Theorie der Kettenbruche, S.-B. Preuss. Akad. Wiss. Phys.-Math. Kl., 1917, pp. 302-321. (Reprinted in I. Schur: Gesammelte Abhandlungen, Vol. 2., pp. 117-136, Springer, Berlin, 1973).

12. N. J. A. Sloane, A Handbook of Integer Sequences, Academic Press, New York, 1973.

Integration and Computers

RICHARD ASKEY [1] University of Wisconsin-Madison, Madison, Wisconsin

The role of computer algebra in the discovery of a geometric setting of certain multiple integrals, and the extension to some as yet unproven conjectures is outlined. So far the main role was played by pure thought, but the data provided by computer algebra systems was probably essential. Now that a natural setting has been found, it is easy to find other problems where computer algebra will provide additional useful data.

1. <u>Introduction</u>. At one time there would have been only one topic that would be considered under this title. That is numerical integration. The methods used there vary from very old ideas like Newton's idea to approximate the function being integrated, possibly with Gauss's refinement to do polynomial interpolation at appropriate points to integrate exactly polynomials of twice the degree that one can integrate by arbitrary interpolation, or the beautiful and very important idea of using indefinite integrals of step functions that goes under the name of spline functions, to Monte Carlo methods that made it possible to approximate some multiple integrals of high dimension. I will say nothing about this.

The next topic that many people will think of is exact indefinite integration. The Risch algorithm has been implemented in a number of

(1) Supported in part by NSF grant MCS 840071

computer algebra systems, and various refinements are being developed
where other indefinite integrals that can not be done in terms of
elementary functions are added to the class of primitive functions.
Again I will say nothing about this.

The only thing that seems to be left is the class of functions
whose indefinite integrals can not be evaluated in terms of elementary
functions, yet certain definite integrals can be evaluated in terms of
known quantities. A classical example is

$$(1.1) \qquad\qquad \int_{-\infty}^{\infty} e^{-x^2}\, dx = \sqrt{\pi} \, ,$$

with

$$(1.2) \qquad\qquad \int_{0}^{\infty} \frac{t^{x-1}}{1+t}\, dt = \frac{\pi}{\sin \pi x} \, , \quad 0 < \operatorname{Re} x < 1$$

as another example which many mathematics students see on an M.A.
exam or a Ph.D preliminary exam.

The first question that needs to be considered is whether such
integrals are important or not. They can be. Here is an example which
arose in some multivariate statistical work of Mehta and Dyson. Over
twenty years ago they considered a problem in nuclear physics, and
needed the value of

$$(1.3) \qquad \frac{1}{(2\pi)^{n/2}} \int_{-\infty}^{\infty} \prod_{1 \leq i < j \leq n} |t_i - t_j|^{2z} \prod_{i=1}^{n} e^{-t_i^2/2} \, dt_i = F_n(z).$$

They could evaluate this when $z = 1/2$ (it was known, as they pointed out), when $z = 1$ and when $z = 2$. They could also evaluate the integral for all z with $\mathrm{Re}\, z > -1/2$ when $n = 2$. However they needed $F_n'(1/2)$ for large n. There is no way to do this numerically, so the only hope was to evaluate $F_n(z)$ exactly. The only reasonable conjecture that fit all their data was

$$(1.4) \qquad F_n(z) = \prod_{j=1}^{n} \frac{\Gamma(\tfrac{1}{2}z+1)}{\Gamma(z+1)} .$$

See Mehta's book [34] for more on this integral. If this conjecture was true, then it was easy to find $F_n'(1/2)$. For

$$\frac{d}{dz} \Gamma(z) = \Gamma(z)\psi(z)$$

with

$$(1.5) \qquad \psi(z) = -\gamma - 1/2 + \sum_{n=1}^{\infty} \frac{z}{n(n+z)} .$$

and

$$(1.6) \qquad \gamma = -\Gamma'(1).$$

Gauss showed that

$$\psi(p/q) = -\gamma - \log q - \pi/2 \cot \pi p/q$$

$$+ \sum_{n=1}^{q-1} \cos(2\pi pn/q)\log(2 \sin n\pi/q),$$

when p and q are positive integers, p < q, so

$$\psi(1/2) = -1/2 - 2 \log 2.$$

Also $\psi(z+1) = \psi(z) + 1/z$ and $\psi(1) = -\gamma$. For all these facts on
$\psi(z)$ see [14, §1.7–§1.7.2].

The identity (1.3) remained as a conjecture for many years until
E. Bombieri came across an integral like this, and also could not
evaluate it. He asked a colleague, A. Selberg, about the integral, and
Selberg brought out a reprint of an old paper of his [40]. This paper
contained a proof of

$$(1.6) \qquad \int_0^1 \cdots \int_0^1 \prod_{1 \leq i < j \leq n} |t_i - t_j|^{2z} \prod_{i=1}^n t_i^{x-1}(1-t_i)^{y-1}dt_i$$

$$= \prod_{i=1}^n \frac{\Gamma(x+(n-i)z)\Gamma(y+(n-i)z)\Gamma(iz+1)}{\Gamma(x+y+(2n-i-1)z)\Gamma(z+1)} ,$$

It is easy to see that (1.6) implies (1.3). There is still no direct proof of (1.3), but there is a second proof of (1.6). This is a recent result of Aomoto [4] which will be mentioned later.

In some other work in nuclear physics Dyson came across a second integral. It is

$$(1.7) \qquad \frac{1}{(2\pi)^n} \int_{-\pi}^{\pi} \cdots \int_{-\pi}^{\pi} \prod_{1 \leq j < k \leq n} [2 - 2\cos(\theta_k - \theta_j)]^{\ell} \prod_{j=1}^{n} d\theta_j$$

$$= \frac{1}{(2\pi)^n} \int_{-\pi}^{\pi} \cdots \int_{-\pi}^{\pi} \prod_{j \neq k} [1 - e^{i(\theta_k - \theta_j)}]^{\ell} \prod_{j=1}^{n} d\theta_j$$

Dyson [13] observed that this integral is the constant term in the following Laurent polynomial

$$\prod_{1 \leq j \neq k \leq n} (1 - t_j t_k^{-1})^{\ell} \ .$$

When $n = 3$ he was able to find this constant term. It is $(3\ell)!/(\ell!)^3$. He also had read Bailey's book [10] on hypergeometric series when he was a student and knew that the identity he used to find this constant term was extended by Dixon. Using Dixon's more general sum he showed that

$$\text{C.T.} \quad \prod_{1 \leq j \neq k \leq 3} (1 - t_j t_k^{-1})^{a_j} = \frac{(a_1 + a_2 + a_3)!}{a_1! \, a_2! \, a_3!} \ .$$

From this it is easy to conjecture that

$$(1.8) \qquad \text{C.T.} \prod_{1 \le j \ne k \le n} (1 - t_j t_k^{-1})^{a_j} = \frac{(a_1 + \cdots + a_n)!}{a_1! \cdots a_n!}$$

when $a_1, \cdots, a_n$ are nonnegative integers. Here C.T. $f(t_1, \cdots, t_n)$ is the constant term in the Laurent expansion of $f(t_1, \cdots, t_n)$ when f is a polynomial in $t_1, t_1^{-1}, \cdots, t_n, t_n^{-1}$. Proofs of (1.8) were found by Gunson [19] and Wilson [44], and later Good [18] found the proof of (1.8) from The Book. Since his proof illustrates a point I want to make, it will be given here. The right hand side of (1.8) satisfies the difference equation

$$(1.9) \qquad R_n(a_1, \cdots, a_n) = \sum_{i=1}^{n} R_n(a_1, \cdots, a_{i-1}, a_i - 1, a_{i+1}, \cdots, a_n).$$

It also satisfies

$$(1.10) \qquad R_n(a_1, \cdots, a_{i-1}, 0, a_{i+1}, \cdots, a_n) = R_{n-1}(a_1, \cdots, a_{i-1}, a_{i+1}, \cdots, a_n)$$

and

$$(1.11) \qquad R_n(0, \cdots, 0) = 1.$$

Good was greedy, and showed that not only the constant term on the left satisfied (1.9), but that the full function on the left satisfied (1.9). This follows from the Lagrange interpolation expansion

(1.12)
$$1 - \sum_{k=1}^{n} \prod_{\substack{j=1 \\ j \neq k}}^{n} \frac{x-t_k}{t_j-t_k}$$

with $x = 0$. For this becomes

$$1 - \sum_{k=1}^{n} \prod_{\substack{j=1 \\ j \neq k}}^{n} \frac{1}{1-t_j t_k^{-1}}$$

and then

$$G_n(t_1, \cdots, t_n) = \prod_{1 \leq j \neq k \leq n} (1-t_j t_k^{-1})^{a_j}$$

satisfies (1.9). Equations (1.10) and (1.11) are also satisfied by the constant term of $G_n(t_1, \cdots, t_n)$, so by induction (1.8) is satisfied.

The point about this proof is the degree to which the extra freedom Dyson introduced with the a_j's was used. When $a_j = \ell$, $j = 1, \cdots, n$, then the right hand side is $(n\ell)!/(\ell!)^n$. A proof of this case of (1.8) by induction will be harder, and I once thought it would probably be impossible. Stembridge has shown that to be wrong [41]. However, the extra freedom that comes with more free parameters is often very useful.

2. <u>Hypergeometric and basic hypergeometric series and q-extensions</u>.

A hypergeometric series is a series $\sum c_k$ with c_{k+1}/c_k a
rational function of k. A basic hypergeometric series has c_{k+1}/c_k a
rational function of q^k. The binomial theorem can be written as

$$(2.1) \qquad (1-x)^{-a} = \sum_{n=0}^{\infty} \frac{(a)_n}{n!} x^n$$

with the shifted factorial $(a)_n$ defined by

$$(2.2) \qquad (a)_n = \Gamma(n+a)/\Gamma(a).$$

The term ratio is $(n+a)/(n+1)$. One q-extension has

$$\frac{c_{n+1}}{c_n} = \frac{(1-q^{n+a})}{(1-q^{n+1})}$$

and if $c_0 = 1$ the sum corresponding to (2.1) is

$$(2.3) \qquad \frac{(q^a x;q)_{\infty}}{(x;q)_{\infty}} = \sum_{n=0}^{\infty} \frac{(q^a;q)_n}{(q;q)_n} x^n$$

when $|q| < 1$, $|x| < 1$. The undefined expressions in (2.3) are
defined by

$$(2.4) \qquad (a;q)_n = (1-a) \cdots (1-aq^{n-1}), \quad n = 1,2,\ldots,$$
$$= 1 \qquad\qquad\qquad, \quad n = 0,$$

and when $|q| < 1$,

$$(2.5) \qquad (a;q)_\infty = \prod_{n=0}^{\infty} (1-aq^n).$$

Observe that

$$\lim_{q \to 1} \frac{(1-q^a)}{(1-q)} = a$$

so the sum in (2.3) reduces to the sum in (2.1) when $q \to 1$. We will usually write (2.3) with "a" instead of q^a for a number of reasons. These include the possibility of having "a" independent of q, which is useful at times, and to make printing easier.

The first q-extension of the results in section 1 was found by Andrews [1] as a theorem for $n = 2$ and 3 and as a conjecture for $n > 3$. The case $n = 2$ of the Dyson-Gunson-Wilson identity (1.8) is the constant term of

$$(2.6) \qquad (1 - \frac{x_1}{x_2})^{a_1} (1 - \frac{x_2}{x_1})^{a_2} \ .$$

Since $(1 - \frac{x_2}{x_1})^{a_2} = (-\frac{x_2}{x_1})^{a_2}(1 - \frac{x_1}{x_2})^{a_2}$, all the terms of the Laurent expansion of (2.6) can be found from the binomial theorem. There are two natural ways to extend this to q-series and products. One is to consider

$$(2.7) \qquad (\frac{x_1}{x_2} ; q)_{a_1} (\frac{x_2}{x_1} ; q)_{a_2} .$$

The second is to consider

$$(2.8) \qquad (\frac{x_1}{x_2} ; q)_{a_1} (\frac{qx_2}{x_1} ; q)_{a_2} .$$

The extra factor q in (2.8) can be thought of as the fudge factor that allows (2.8) to be a direct part of the q-binomial theorem. For

$$(\frac{qx_2}{x_1} ; q)_{a_2} = (-\frac{qx_2}{x_1})^{a_2} q^{a_2(a_2-1)/2} (\frac{q^{-a_2}x_1}{x_2} ; q)_{a_2}$$

so

$$(\frac{x_1}{x_2} ; q)_{a_1} (\frac{qx_2}{x_1} ; q)_{a_2} = (-\frac{x_2}{x_1})^{a_2} q^{a_2(a_2+1)/2} (\frac{q^{-a_2}x_1}{x_2} ; q)_{a_1+a_2} .$$

If (2.7) is used there is a duplication of the factor $(1 - \frac{x_1}{x_2})$. Andrews used the second extension and made the following conjecture

(2.9)   C.T. $\displaystyle\prod_{1 \le j < k \le n} (\frac{x_j}{x_k} ; q)_{a_j} (\frac{qx_k}{x_j} ; q)_{a_k} = \frac{(q;q)_{a_1 + \cdots + a_n}}{(q;q)_{a_1} \cdots (q;q)_{a_n}}$.

This conjecture has now been proven by Zeilberger and Bressoud [46]
Slightly before Andrews formulated the conjecture (2.9) Macdonald [13]
made a major discovery which is related to all the integrals in section
1, although this was not known then. A limiting case of the
q-binomial theorem is Jacobi's triple product identity

(2.10) $\displaystyle (x;q)_\infty \ (\frac{q}{x} ; q)_\infty (q;q)_\infty = \sum_{-\infty}^{\infty} (-1)^n \ q^{\binom{n}{2}} \ x^n$.

Macdonald found infinitely many extensions of this identity, one
attached to each affine root system. A bit will be said about this in
the next section.

   The next q extensions were formulated in the late 1970's. The
first was a refinement of the Macdonald identities. These are some
conjectures of Macdonald [14]. They will be mentioned in the next
section. The other one is a set of q-extensions of Selberg's
integral. To formulate these we define a q-integral. Following
Fermat and Wallis, who used this idea to integrate x^k on [0,1],
define

(2.11) $\displaystyle \int_0^a f(x)d_q x := a(1-q) \sum_{n=0}^{\infty} f(aq^n)q^n$.

This was introduced independently by Thomae [42] and Jackson [25]. When trying to extend Selberg's integral, the first thing one needs is an extension of the beta integral. The q-binomial theorem (2.3) can be rewritten as

$$(2.12) \qquad \int_0^1 t^{x-1} \frac{(tq;q)_\infty}{(tq^{x+y};q)_\infty} \, d_q t = \frac{\Gamma_q(x)\Gamma_q(y)}{\Gamma_q(x+y)}$$

where

$$(2.13) \qquad \Gamma_q(x) = \frac{(q;q)_\infty}{(q^x;q)_\infty} (1-q)^{1-x} \ .$$

For simplicity take $0 < q < 1$. The q-gamma function satisfies

$$(2.14) \qquad \Gamma_q(x+1) = \frac{(1-q^x)}{(1-q)} \Gamma_q(x)$$

and

$$(2.15) \qquad \lim_{q \to 1^-} \Gamma_q(x) = \Gamma(x).$$

See [5] and [2, Appendix A]. The second has a very nice proof of (2.15) due to Gosper.

The other problem in extending Selberg's integral is to decide what takes the place of the power of the discriminant

$$D_n(t_1, \cdots, t_n) = \prod_{1 \leq j < k \leq n} (t_j - t_k)^2 .$$

Motivated by Andrews's choice of function in (2.9) I used

$$\prod_{1 \leq j < k \leq n} t_k^{2\ell} \, (q^{1-\ell} t_j t_k^{-1} ; q)_{2\ell}$$

to replace $[D_n(t_1, \cdots, t_n)]^\ell$, and conjectured that

$$(2.16) \qquad \int_0^1 \cdots \int_0^1 \prod_{1 \leq j < k \leq n} t_k^{2\ell}(q^{1-\ell} t_j t_k^{-1}; q)_{2\ell} \prod_{j=1}^n t_j^{x-1} \frac{(t_j q; q)_\infty}{(t_j q^y; q)_\infty} \, d_q t_j$$

$$= q^{\ell\binom{n}{2} + 2\ell^2 \binom{n}{3}} \prod_{j=1}^n \frac{\Gamma_q(x + (n-j)\ell)\Gamma_q(y + (n-j)\ell)\Gamma_q(j\ell + 1)}{\Gamma_q(x + y + (2n - j - 1)\ell)\Gamma_q(\ell + 1)} .$$

Both this conjecture and conjecture (2.8) of Andrews were done for the first nonelementary case, $n = 2$ for (2.16) and $n = 3$ for (2.8), by showing they are equivalent to the following identity of F.H. Jackson [26].

$$(2.17) \qquad \sum_{j=0}^{2k} \frac{(q^{-2k}; q)_j (a; q)_j (b; q)_j}{(q; q)_j (\frac{q^{1-2k}}{a}; q)_j (\frac{q^{1-2k}}{b}; q)_j} \left(\frac{q^{1-k}}{ab} \right)^j$$

$$= \frac{(q^{k+1}; q)_k (abq^k; q)_k}{(aq^k; q)_k (bq^k; q)_k} \, q^{-k} .$$

This is a q-extension of the result of Dixon that Dyson used to prove the case n - 3 of (1.8). I was unable to prove (2.16) for larger n, so a natural question is how did I guess the exponent $2\ell^2(\frac{n}{3})$ which vanishes when n - 1 and n - 2. The answer is that there are other extensions of the beta integral, and many of them seem to generate Selberg like multivariate integrals that have simple values. Two were found by Ramanujan. The first he wrote as an integral

$$(2.18) \qquad \int_0^\infty t^{x-1} \frac{(-t;q)_\infty}{(-tq^{x+y};q)_\infty} \, dt = \frac{\Gamma(x)\Gamma(1-x)\Gamma_q(y)}{\Gamma_q(1-x)\Gamma_q(x+y)} \,.$$

The second is usually written as a sum

$$(2.19) \qquad \sum_{-\infty}^\infty \frac{(a;q)_n}{(b;q)_n} \, x^n = \frac{(ax;q)_\infty(\frac{q}{ax};q)_\infty (q;q)_\infty(\frac{b}{a};q)_\infty}{(x;q)_\infty(\frac{b}{ax};q)_\infty (b;q)_\infty(\frac{q}{a};q)_\infty}$$

when $|b/a| < |x| < 1$, where

$$(2.20) \qquad (a;q)_n = (a;q)_\infty/(aq^n;q)_\infty \,.$$

This can be rewritten as

$$(2.21) \qquad \int_0^\infty t^{x-1} \frac{(-tc;q)_\infty}{(-tcq^{x+y};q)_\infty} \, d_q t = \frac{\Gamma_q(x)\Gamma_q(y)}{\Gamma_q(x+y)} \frac{(-cq^x;q)_\infty(-q^{1-x}/c;q)_\infty}{(-c;q)_\infty(-q/c;q)_\infty}$$

where

$$(2.22) \qquad \int_0^\infty f(t)d_q t := (1-q) \sum_{-\infty}^\infty f(q^n)q^n.$$

It is easy to see that

$$(2.23) \qquad \lim_{n\to\infty} \int_0^{a^{-N}} f(t)d_q t = \int_0^\infty f(t)d_q t$$

and Ismail's proof of (2.19) from the q-binomial theorem [24] really uses (2.23). See [6] for simple proofs of (2.18) and (2.19).

There are many other q-extensions of beta integrals. The classical beta integral can be changed into other integrals by changing variables. That is not possible for a q-integral (which is just an infinite series), but there seem to be q-extensions for almost all the useful changes of variable of the beta integral. Two are the following.

A q-integral that is supported on both sides of zero can be defined by

$$(2.24) \qquad \int_c^d f(t)d_q t = \int_0^d f(t)d_q t - \int_0^c f(t)d_q t.$$

Then

$$(2.25) \qquad \int_{-d}^{c} \frac{(-qx/d;q)_{\infty}(qx/c;q)_{\infty}}{(-q^{\beta}x/d;q)_{\infty}(q^{\alpha}x/c;q)_{\infty}} \, d_q t$$

$$= \frac{\Gamma_q(\alpha)\Gamma_q(\beta)cd(-d/c;q)_{\infty}(-c/d;q)_{\infty}}{\Gamma_q(\alpha+\beta)(c+d)(-dq^{\alpha}/c;q)_{\infty}(-cq^{\beta}/d;q)_{\infty}} \, .$$

See Sears [38] for this when written as the sum of two series and
Andrews and Askey [3] in this form. This also exists when the limits
c and d are infinite. See [8].

The second is more general than given below, since the
restrictions on a,b,c, and d can be relaxed at the cost of some
discrete masses or a contour integral. See Askey and Wilson [9].

$$(2.26) \qquad \frac{1}{2\pi} \int_{-1}^{1} \frac{h(x;1)h(x;q^{1/2})h(-x;1)h(-x;q^{1/2})}{h(x;a)h(x;b)h(-x;c)h(-x;d)} \frac{dx}{\sqrt{1-x^2}}$$

$$= \frac{(abcd;q)_{\infty}}{(q;q)_{\infty}(ab;q)_{\infty}(cd;q)_{\infty}(-ac;q)_{\infty}(-ad;q)_{\infty}(-bc;q)_{\infty}(-bd;q)_{\infty}}$$

when $\max(|q|,|a|,|b|,|c|,|d|) < 1$ and

$$(2.27) \qquad h(x;a) = \prod_{n=0}^{\infty} (1-2axq^n + a^2q^{2n}).$$

There are a few other q-beta integrals, but these are a
representative sample, and probably contain the most important
examples.

See Rahman [36] for a two dimensional q-Selberg integral using a special case of (2.26) as the beta part. He has two degrees of freedom from (2.26). He also formulated an n dimensional conjecture.

3. Root systems and further extensions.

While completing the classification of semisimple Lie algebras, a set of vectors in Euclidean n space was isolated. These are called root systems. To define them take a real Euclidean space E with an inner product (α,β). To a nonzero vector $\alpha \in E$ define the orthogonal hyperplane $P_\alpha = \{\beta \in E : (\beta,\alpha) = 0\}$. A reflection σ_α can be defined by

$$\sigma_\alpha(\beta) = \beta - \frac{2(\beta,\alpha)}{(\alpha,\alpha)}\,\alpha.$$

This sends α to $-\alpha$ and leaves fixed all vectors in P_α. A root system R is a set of vectors in E satisfying the following conditions.

3.1a R is finite, spans E and 0 is not in R.

3.1b   If $\alpha \in R$, the reflection σ_α leaves R invariant.

3.1c   If $\alpha,\beta \in R$ then $2(\beta,\alpha)/(\alpha,\alpha)$ is an integer.

Sometimes another condition is added that says that the only multiples of $\alpha \in R$ that are in R are $\pm\alpha$. We will not use this, since it removes the root system corresponding to Selberg's integral.

The three innocent looking conditions above are actually very strong, and there are only a few root systems. The irreducible ones come in five infinite families and five exceptional ones. The two that are most important for us are called A_n and BC_n. These can be defined as follows.

(3.2) A_n is the set of vectors $\pm(t_j - t_k)$, $j, k = 0, 1, \cdots, n$.

(3.3) BC_n is the set of vectors $\pm t_j$, $\pm 2t_j$, $\pm(t_j \pm t_k)$,

$j, k = 1, 2, \cdots, n$, where the choice of plus or minus signs is independent.

There are three root systems which are subsets of BC_n . B_n is missing the vectors $\pm 2t_j$, C_n is missing the vectors $\pm t_j$, and D_n is missing both of these sets. In addition there are five isolated sets, G_2, F_4, E_6, E_7, E_8 . Pictures of the root systems in two dimensions follow.

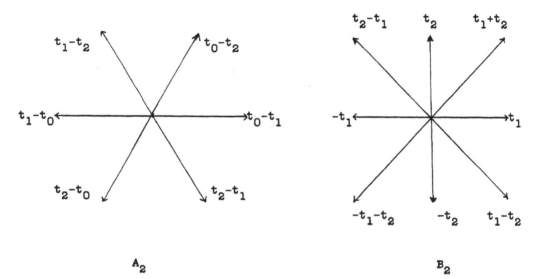

A_2 B_2

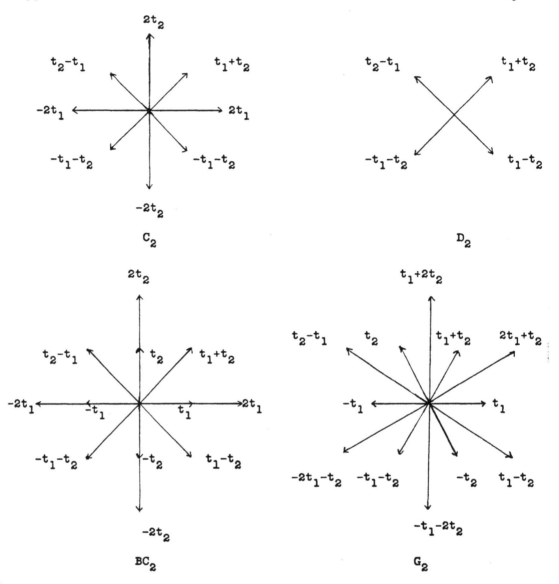

C_2

D_2

BC_2

G_2

In two dimensions there are two reductions. Clearly B_2 and C_2 are isomorphic. D_2 is reducible, since it is isomorphic to $A_1 \times A_1$.

D_3 is isomorphic to A_3. These are the only reductions in the basic families $A_n, B_n, C_n, D_n, E_6, E_7, E_8, F_4$ and G_2.

For a treatment of root systems see Bourbaki [11] or Humphreys [23]. Notice the lengths of the roots are very restricted, with BC_n, $n \geq 2$, the only ones having roots of three different lengths. Macdonald was the first to realize that the integrals of Selberg and Dyson can be given an interpretation related to root systems. He formed a formal exponential e^r where r is a root, set $s_i = e^{r_i}$, and formed the product

$$(3.4) \qquad \prod_{r \in R} (1 - e^r)^{\ell}$$

where R is one of the root systems. The product is taken over all roots in R and ℓ is a nonnegative integer. Macdonald's first conjecture, before Selberg's integral came to light, was that the constant term in (3.4) is the product of certain binomial coefficients. The constant term is the term in the formal expansion of (3.4) that is independent of the roots $r \in R$. In particular he conjectured that

$$\text{C.T.} \quad \prod_{r \in R} (1 - e^r)^{\ell} = \prod_i \binom{d_i \ell}{\ell}$$

where the integers d_i are associated with each root system. They can be defined in many ways. One of the easiest to explain is that they are the degrees of the basic polynomial invariants for the reflection group defined by the root system. These numbers are the following.

$$A_n: \qquad\qquad i+1, \quad i = 1,2,\cdots,n$$

$$B_n: \qquad\qquad 2i, \quad i = 1,2,\cdots,n$$

$$C_n: \qquad\qquad 2i, \quad i = 1,2,\cdots,n$$

$$D_n: \quad n \text{ and } 2i, \quad i = 1,2,\cdots,n-1,$$

(3.5)
$$BC_n: \qquad\qquad 2i+2, \quad i = 1,2,\cdots,n$$

$$G_2: \qquad 2,6$$

$$F_4: \qquad 2,6,8,12$$

$$E_6: \qquad 2,5,6,8,9,12$$

$$E_7: \qquad 2,6,8,10,12,14,18$$

$$E_8: \qquad 2,8,12,14,18,20,24,30$$

The Dyson-Gunson-Wilson identity with all $a_i = \ell$ reduces to this conjecture for A_{n-1} , and this was one of the facts that led Macdonald to this conjecture. After Selberg's integral became known it was possible to obtain some of these conjectures. Macdonald showed that Selberg's integral proves his conjecture for B_n, C_n, D_n and BC_n . He and a few others used the same argument to add some freedom to his conjecture. What was changed was the exponent ℓ. For A_n it was possible to have n degrees of freedom. For the other root systems this seems to be too much to hope for except for low dimensional cases. Selberg's integral has three degrees of freedom, and the roots in BC_n have three different lengths. This is not an accident. For BC_n, Selberg's integral implies.

(3.6) $\quad$ C.T. $\displaystyle\prod_{j=1}^{n} (1-s_j)^a(1-s_j^{-1})^a(1-s_j^2)^b(1-s_j^{-2})^b$

$\displaystyle\cdot \prod_{1\le j<k\le n} (1-s_j/s_k)^c(1-s_k/s_j)^c(1-s_js_k)^c(1-s_j^{-1}s_k^{-1})^c$

$\displaystyle= \prod_{j=0}^{n-1} \frac{(2a+2b+2cj)! \; (2b+2cj)! \; (c(j+1))!}{(a+2b+(n+j-1)c)! \; (a+b+cj)! \; (b+cj)! c!}$

The factors $(1-s_j^{\pm 1})$ correspond to the short roots t_j, the factors $(1-s_j^{\pm 2})$ correspond to the double roots $2t_j$, and the remaining factors correspond to the cross roots. To see how Selberg's integral implies (3.6), set $I_n(n,b,c)$ equal to this constant term. Then

$\displaystyle I_n(a,b,c) = \frac{1}{(2\pi)^n} \int_{-\pi}^{\pi} \cdots \int_{-\pi}^{\pi} \prod_{j=1}^{n} (1-e^{i\theta_j})^a(1-e^{i\theta_j})^a$

$\displaystyle \cdot (1-e^{2i\theta_j})^b(1-e^{-2i\theta_j})^b \prod_{1\le j<k\le n} (1-e^{i(\theta_j+\theta_k)})^c(1-e^{-i(\theta_j+\theta_k)})^c$

$\displaystyle \cdot (1-e^{i(\theta_j-\theta_k)})^c(1-e^{-i(\theta_j-\theta_k)})^c \prod_{j=1}^{n} d\theta_j$

$\displaystyle = \frac{1}{(2\pi)^n} \int_{-\pi}^{\pi} \cdots \int_{-\pi}^{\pi} \prod_{j=1}^{n} (2-2\cos\theta_j)^a(2-2\cos 2\theta_j)^b$

$\displaystyle \cdot \prod_{1\le j<k\le n} (2-2\cos(\theta_j+\theta_k))^c(2-2\cos(\theta_j-\theta_k))^c \prod_{j=1}^{n} d\theta_j$

$$= \frac{1}{\pi^n} \int_0^\pi \cdots \int_0^\pi \prod_{1 \le j < k \le n} \left[2\sin\frac{(\theta_j+\theta_k)}{2} \cdot 2\sin\frac{(\theta_j-\theta_k)}{2} \right]^{2c}$$

$$\cdot \prod_{j=1}^n \left[2\sin\frac{\theta_j}{2} \right]^{2a} [2\sin\theta_j]^{2b} d\theta_j$$

$$= \frac{1}{\pi^n} \int_0^\pi \cdots \int_0^\pi \prod_{1 \le j < k \le n} [2\cos\theta_k - 2\cos\theta_j]^{2c} \prod_{j=1}^n \left[2\sin\frac{\theta_j}{2} \right]^{2a} [2\sin\theta_j]^{2b} d\theta_j$$

$$= \frac{2^n}{\pi^n} \int_0^{\pi/2} \cdots \int_0^{\pi/2} \prod_{1 \le j < k \le n} [2\cos2\varphi_k - 2\cos2\varphi_j]^{2c}$$

$$\cdot \prod_{j=1}^n [2\sin\varphi_j]^{2a} [4\sin\varphi_j\cos\varphi_j]^{2b} d\varphi_j$$

$$= \frac{2^n}{\pi^n} \int_0^{\pi/2} \cdots \int_0^{\pi/2} \prod_{1 \le j < k \le n} [4\sin^2\varphi_j - 4\sin^2\varphi_k]^{2c}$$

$$\cdot \prod_{j=1}^n [2\sin\varphi_j]^{2a+2b} [2\cos\varphi_j]^{2b} d\varphi_j \ .$$

Now let $t_j = \sin^2\varphi_j$ to obtain

$$I_n(a,b,c) = \frac{2^{2n[a+2b+(n-1)c]}}{\pi^n} \int_0^1 \cdots \int_0^1 \prod_{1 \leq j < k \leq n} |t_j - t_k|^{2c} \prod_{j=1}^n t_j^{a+b-\frac{1}{2}}$$

$$\cdot (1-t_j)^{b-\frac{1}{2}} dt_j$$

$$= \frac{2^{2n[a+2b+(n-1)c]}}{\pi^n} \prod_{j=1}^n \frac{\Gamma(a+b-\frac{1}{2}+(n-j)c)\Gamma(b-\frac{1}{2}+(n-j)c)\Gamma(jc+1)}{\Gamma(a+2b-1+(2n-j-1)c)\Gamma(c+1)} \ .$$

Now use the duplication formula for $\Gamma(x)$, i.e.

$$2^{2x-1}\Gamma(x)\Gamma(x+\tfrac{1}{2}) = \sqrt{\pi}\ \Gamma(2x)\ ,$$

to obtain (3.6).

It is then an easy task to obtain Macdonald's conjecture for B_n, C_n, D_n and BC_n by specialization of a, b and c. With this as a model it was then natural to ask if it was possible to add more degrees of freedom in the exceptional cases G_2, F_4, E_6, E_7, E_8. W. Morris considered this question for G_2 and obtained the following numbers for the constant term problem:

Set

(3.7) $G(a,b) = \text{C.T.}(1-t)^a(1-t^{-1})^a(1-s)^a(1-s^{-1})^a(1-ts)^a$

$$\cdot (1-(ts)^{-1})^a(1-t/s)^b(1-s/t)^b(1-t^2s)^b(1-t^{-2}s^{-1})^b(1-ts^2)^b(1-t^{-1}s^{-2})^b.$$

Then G(a,b) is given as follows:

a \ b	0	1	2	3	4
0	1	6	90	1680	34650
1	6	12	108	1440	23100
2	90	72	396	3780	47124
3	1680	660	2340	16320	159600
4	34650	7644	18360	95760	743820

These were obtained using an Apple II. It is far from clear that there is a formula giving these numbers, but after they are factored it starts to look like they can be obtained as quotients of factorials, as in the case of the infinite families.

An evening of work led to the following conjecture.

$$(3.8) \qquad G(a,b) = \frac{(3a+3b)!\,(3b)!\,(2a)!\,(2b)!}{(2a+3b)!\,(a+2b)!\,(a+b)!\,a!\,b!\,b!} \,.$$

The way to find this is to look for prime factors, such as 11 which occurs when a = b = 2, and so try to figure out what the largest factorial in the numerator is. Put in the factors it implies, and see what is the largest factorial in the bottom that is needed to cancel other relatively large primes. This is tedious, but Morris showed me his new data one afternoon, and told me his conjecture the next morning. A few days later he told me a more general conjecture that corresponds to Andrews's conjectured extension of the Dyson-Gunson-Wilson identity. This time he used MACSYMA to do the

symbolic work. One part of the problem of getting a conjecture was
much easier. In this case the factors that appear in the factorials
are replaced by polynomials in q, with n being replaced by
$1 + q + \ldots + q^{n-1}$. There is less cancellation in this case, and it is
easier to see that factors $1 + q$, $1 + q + q^2$ and $1 + q^3 + q^6$ should
be put together as $1 + q + \ldots + q^8$ and $1 + q$, while when $q = 1$
it is not clear if 2,3,3 should be 2 and 9 or 3 and 6.

Morris then looked at a q-version of BC_2 and BC_3 and found
enough data, again using MACSYMA, to make the following conjecture
[35].

$$(3.9) \quad \text{C.T} \prod_{j=1}^{n} (s_j;q)_a (qs_j^{-1};q)_a (qs_j^2;q^2)_b (qs_j^{-2};q^2)_b$$

$$\cdot \prod_{1 \leq j < k \leq n} (s_j s_j;q)_c (qs_j^{-1}s_k^{-1};q)_c (s_j s_k^{-1};q)_c (qs_k s_j^{-1};q)_c$$

$$= \prod_{j=0}^{n-1} \frac{(q;q)_{2a+2b+2cj} (q;q)_{2b+2cj} (q;q)_{c(j+1)}}{(q;q)_{a+2b+(n+j-1)c} (q;q)_{2b+cj} (q;q)_{a+c} (q;q)_c}$$

$$\cdot \frac{(q^2;q^2)_{2b+cj} (q^2;q^2)_{a+cj}}{(q^2;q^2)_{a+b+cj} (q^2;q^2)_{b+cj}}$$

Then he looked at F_4. Everything had been going nicely, so it
was a bit of a surprise to find out that the mainframe he was using was
too small to obtain any nontrivial data. He was sending his
conjectures and data to Ian Macdonald, and Macdonald found a way around

the impase brought on by the size of the problem for F_4. He
reformulated the BC_n conjecture intrinsically using the weights of
the roots. This made sense for all the root systems, and so Morris was
able to write down the following conjecture for F_4.

(3.10)   C.T. $\displaystyle\prod_{i=1}^{4}(1-x_i^2)^a(1-x_i^{-2})^a \prod_{1\leq i<j\leq4}(1-x_i^2x_j^2)^b(1-x_i^{-2}x_j^{-2})^b$

$$\cdot (1-x_i^2x_j^{-2})^b(1-x_i^{-2}x_j^2)^b \prod_{r_1,r_2,r_3,r_4=\pm1}(1-x_1^{r_1}x_2^{r_2}x_3^{r_3}x_4^{r_4})^a$$

$$= \frac{(6a+6b)!(4a+4b)!(2a+6b)!(4a+2b)!(2a+4b)!(4b)!(3a)!}{(5a+6b)!(3a+5b)!(3a+4b)!(3a+3b)!(2a+3b)!(a+3b)!(2a+b)!}$$

$$\cdot \frac{(3b)!(2a)!(2b)!}{(a+2b)!(a+b)!a!a!b!b!b!}$$

A careful reader might be curious how Macdonald could use roots in
the q case. They exist, and the root systems are called affine root
systems. These root systems are associated with the Euclidean
Kac-Moody algebras. A nice introduction to these algebras is [33], and
a more detailed treatment is in [27]. However Kac has changed the
notation that Macdonald and others used. I have followed Macdonald's
notation. Euclidean Kac-Moody algebras, or as they were once known,
affine Lie algebras, are the setting where Macdonald found his
identities [31]. These identities are Laurent expansions of infinite
products that arise from the finite products being considered in the
Macdonald conjectures by letting the parameters be either infinite or

zero. This can not be done in the case when $q = 1$, but it makes
sense in the case when $|q| < 1$. There are two parts to the proof of
the Macdonald identities, and they can be illustrated nicely with the
Jacobi triple product, which is the Macdonald identity for A_1.
Consider

$$(3.11) \qquad (x;q)_\infty (qx^{-1};q)_\infty = \sum_{-\infty}^{\infty} c_n x^n .$$

Replacing x by qx gives

$$(qx;q)_\infty (x^{-1};q)_\infty = \sum_{-\infty}^{\infty} c_n q^n x^n .$$

But the left hand side is

$$(1-x^{-1})(qx;q)_\infty (qx^{-1};q)_\infty = -x^{-1}(x;q)_\infty (qx^{-1};q)_\infty$$

so

$$\sum_{-\infty}^{\infty} c_n q^n x^n = -\sum_{-\infty}^{\infty} c_n x^{n-1} = -\sum_{-\infty}^{\infty} c_{n+1} x^n .$$

This gives

$$c_{n+1} = -q^n c_n$$

or

$$c_n = (-1)^n q^{n(n-1)/2} c_0 .$$

Thus to find c_n in (3.11) it is sufficient to find the constant term c_0. There are a number of ways of doing this. For A_1 one of my favorite ways is to find the constant term in

(3.12) $$(x;q)_a (q/x;q)_a = \sum_{j=-a}^{a} c_j(a) x^j .$$

As was remarked before, all the terms in (3.12) can be found by the q-binomial theorem. In the case of the other affine root systems it does not seem possible to find all the coefficients in the Laurent expansion, but the constant term seems to have a nice expression.

Morris [35] made a few more observations. A_n is really the exceptional root system, since it is much nicer than the others. In addition to the labeling Dyson found, there are two other ways to label roots of A_2 so the constant term has a nice form. These are

(3.13) $\quad$ C.T.$(1-\frac{t_0}{t_1})^a(1-\frac{t_0}{t_2})^a(1-\frac{t_1}{t_0})^b(1-\frac{t_2}{t_0})^b(1-\frac{t_1}{t_2})^c(1-\frac{t_2}{t_1})^c$

$$= \frac{(a+b+c)!(a+b)!(2c)!}{(a+c)!(b+c)!a!c!} \ .$$

(3.14) $\quad$ C.T.$(1-\frac{t_0}{t_1})^a(1-\frac{t_1}{t_0})^a(1-\frac{t_0}{t_2})^b(1-\frac{t_2}{t_0})^b(1-\frac{t_1}{t_2})^c(1-\frac{t_2}{t_1})^c$

$$= \frac{(a+b+c)!(2a)!(2b)!(2c)!}{(a+b)!(a+c)!(b+c)!a!b!c!} \ .$$

The labels for (3.13) can be extended to A_n as follows.

(3.15) $\quad$ C.T. $\displaystyle\prod_{j=1}^{n}(1-\frac{t_0}{t_j})^a(1-\frac{t_j}{t_0})^b \prod_{1\leq j<k\leq n}(1-\frac{t_j}{t_k})^c(1-\frac{t_k}{t_j})^c$

$$= \prod_{j=0}^{n-1} \frac{(a+b+jc)!((j+1)c)!}{(a+jc)!(b+jc)!c!} \ .$$

Morris proved (3.15) by showing it is equivalent to a Selberg type integral, but this time using Cauchy's beta integral

(3.16) $\quad \displaystyle\frac{1}{2\pi}\int_{-\infty}^{\infty} \frac{dt}{(1+it)^a(1-it)^b} = \frac{\Gamma(a+b-1)}{\Gamma(a)\Gamma(b)} 2^{-a-b-1} \ .$

He also formulated a q-extension

$$(3.17) \quad \text{C.T.} \prod_{j=1}^{n} (t_j/t_0;q)_a (qt_0/t_j;a)_b \prod_{1 \leq j < k \leq n} (t_j/t_k;q)_c (qt_k/t_j;q)_c$$

$$= \prod_{j=0}^{n-1} \frac{(q;q)_{a+b+jc}(q;q)_{(j+1)c}}{(q;q)_{a+jc}(q;q)_{b+jc}(q;q)_c} \; .$$

Recently Kadell [28] and Habsieger [21] have proven conjecture (2.16), and each one has also observed that this conjecture implies conjecture (3.17). Their arguments to prove (2.16) are different, and so are their arguments to give (3.17). Habsieger's proof of (2.16) follows Selberg's proof, with a major change in the one part of Selberg's argument that looked like it would not extend. Kadell's proof is different, and he proves more than (2.16). See the next section.

4. <u>**Aomoto's extension of Selberg's integral**</u> In Feynman's delightful
autobiographical work [17, pages 86-87], he tells his secret of doing
integrals that others could not do. Mathematics students knew Cauchy's
theorem, and a couple of other tricks, and when they could not evaluate
an integral that looked like it could be evaluated and asked Feynman,
he could often do it. His secret was that he knew one trick they did
not know, to differentiate with respect to a free parameter. We teach
this for power series, and regularly tell our students that integrals
and series are essentially the same thing, but Feynman is right that
this method is not one that is in every mathematician's bag of tricks.
There is a related method that is also useful in evaluating some
important integrals that should be better known than it is. This is to
integrate by parts, or use the fundamental theorem of calculus, to set
up a difference equation in a free parameter. Euler [15] was one of
the first to do this. Here is one version of this argument.

$$0 = \int_0^1 \frac{d}{dx} t^x(1-t)^y dt = x \int_0^1 t^{x-1}(1-t)^y dt - y \int_0^1 t^x(1-t)^{y-1} dt$$

$$= -y \int_0^1 t^{x-1}(1-t)^{y-1} dt + (x+y) \int_0^1 t^{x-1}(1-t)^y dt \ .$$

If

$$(4.1) \qquad B(x,y) := \int_0^1 t^{x-1}(1-t)^{y-1} dt$$

then

$$B(x,y) = \frac{x+y}{y} B(x,y+1) = \frac{(x+y)_n}{(y)_n} B(x,y+n)$$

$$= \frac{(x+y)_n}{n!} n^{1-x-y} \frac{n!}{(y)_n} n^{y-1} \int_0^n t^{x-1}(1-\frac{t}{n})^{n+y-1}dt.$$

When $n \to \infty$ this gives

(4.2) $$B(x,y) = \frac{\Gamma(y)}{\Gamma(x+y)} \int_0^\infty t^{x-1}e^{-t}dt \; ,$$

since

(4.3) $$\frac{1}{\Gamma(x)} = \lim_{n\to\infty} \frac{(x)_n}{n!} n^{1-x} \; .$$

Then $y = 1$ gives Euler's integral representation of $\Gamma(x)$ for Re $x > 0$.

   It does not seem to be possible to use this argument on Selberg's integral, but Aomoto [4] found an extension of Selberg's integral where it works. In particular he considered

(4.4) $$A_n(k;x,y,z) = \int_{C_n} \prod_{i=1}^{k} t_i w(t;x,y,z)dt$$

where $k = 1,2,\ldots,n-1$,

$$w(t;x,y,z) = \prod_{1 \le i < j \le n} |t_i - t_j|^{2z} \prod_{i=1}^{n} t_i^{x-1}(1-t_i)^{y-1},$$

$$t = (t_1,\ldots,t_n), \quad dt = dt_1 \ldots dt_n,$$

and $C^n = [0,1]^n$ is the unit cube $0 \le t_i \le 1$, $i = 1,2,\ldots,n$.

As above

$$0 = \int_{C_n} \frac{\partial}{\partial t_1} \prod_{i=1}^{k} t_i \, w(t;x,y,z)dt$$

$$= x \int_{C_n} \prod_{i=2}^{k} t_i \, w(t;x,y,z)dt + (y-1) \int_{C_n} \prod_{i=1}^{k} t_i \, \frac{w(t;x,y,z)dt}{t_1-1}$$

$$+ 2z \sum_{j=2}^{n} \int_{C_n} \prod_{i=1}^{k} t_i \, \frac{w(t;x,y,z)}{t_1-t_j}dt \ .$$

Since $w(t;x,y,z)$ is symmetric in the t_i's, the first integral on the right is $A(k-1;x,y,z)$. In the second integral on the right use $t_1 = t_1 - 1 + 1$ to obtain an integral that adds to the first and a new one. In the third integral, use

Lemma 3.1. $\displaystyle\int_{C_n}\prod_{i=1}^{k}t_i\,\frac{w(t;x,y,z)}{x_1-x_j}dt = 0$ if $2 \le j \le k$

$$= -\frac{1}{2}A_n(k-1) \quad \text{if} \quad k < j \le n .$$

Proof. When $2 \le j \le k$ the transposition $t_1 \longleftrightarrow t_k$ changes the sign of the integrand, so the integral vanishes. When $k < j \le n$, $t_1 \longleftrightarrow t_j$ gives

$$\frac{t_1}{t_1-t_j} \to \frac{t_j}{t_j-t_1} = \frac{t_j-t_1+t_1}{t_j-t_1} = 1 - \frac{t_1}{t_1-t_j}$$

so

$$2\int_{C_n}\prod_{i=1}^{k}t_i\,\frac{w(t;x,y,z)}{t_1-t_j}dt = \int_{C_n}\prod_{i=2}^{k}x_i\,w(t;x,y,z)dt = A_n(k-1).$$

These combine to give

Lemma 3.2. If $y > 1$ then

$$(y-1)\int_{C_n}\prod_{i=2}^{k}t_i\,\frac{w(t;x,y,z)}{1-t_1}dt = [x+y-1+(n-k)z]A_n(k-1) .$$

As above

(4.5) $\quad 0 = \int\limits_{C_n} \frac{\partial}{\partial t_1} t_1^2 \prod_{i=2}^{k} t_i w(t;x,y,z)dt$

$= (x+1)A_n(k) + (y-1) \int\limits_{C_n} t_1 \prod_{i=1}^{k} t_i \frac{w(t;x,y,z)}{t_1-1}dt$

$+ 2z \int\limits_{C_n} \sum_{j=2}^{n} t_1 \prod_{i=1}^{k} t_i \frac{w(t;x,y,z)}{t_1-t_j}dt$

$= (x+1)A_n(k) + (y-1)A_n(k) + (y-1) \int\limits_{C_n} \prod_{i=1}^{k} t_i \frac{w(t;x,y,z)}{t_1-1}dt$

$+ 2z \sum_{j=2}^{n} \int\limits_{C_n} \frac{t_1 \prod_{i=1}^{k} t_i}{t_1-t_j} w(t;x,y,z)dt$

$= (x+y)A_n(k) + (y-1)A_n(k-1) - [x+y-1+(n-k)z]A_n(k-1) + 2z \sum_{i=2}^{n}$

from $\ t_1 = t_1 - 1 + 1\ $ and Lemma 3.2.

To handle the integral in the last sum, when $\ 2 \le j \le k\ $ the transposition $\ t_1 \leftrightarrow t_j\ $ gives

$$\frac{t_1^2 t_j}{t_1-t_j} \rightarrow \frac{t_1 t_j^2}{t_j-t_1} - \frac{t_1 t_j[t_j-t_1]}{t_j-t_1} - \frac{t_1^2 t_j}{t_1-t_j} \ .$$

When $k < j \leq n$

$$\frac{t_1^2}{t_1-t_j} - \frac{t_1(t_1-t_j)}{t_1-t_j} + \frac{t_1 t_j}{t_1-t_j} = t_1 + \frac{t_1 t_j}{t_1-t_j} \ ,$$

and the second term changes sign when $t_1 \longleftrightarrow t_j$. Together these give

Lemma 3.3 $\displaystyle\int_{C_n} t_1 \prod_{i=1}^{k} t_i \frac{w(t;x,y,z)}{t_1-t_j}dt = \frac{1}{2} A_n(k), \quad 2 \leq j \leq k$,

$$= A_n(k), \quad k < j \leq n \ .$$

Using this in (4.5) gives

(4.6) $A_n(K)[x+y+(2n-k-1)z] = [x+(n-k)z]A_n(k-1)$.

Iteration gives

(4.7) $\displaystyle A_n(k) = \prod_{j=1}^{k} \frac{[x+(n-j)z]}{[x+y+(2n-j-1)z]} S_n(x,y,z)$

where $S_n(x,y,z)$ is Selberg's integral

(4.8) $\displaystyle S_n(x,y,z) = \int_{C_n} w(t;x,y,z)dt$.

When $k = n$, formula (4.7) gives

(4.9) $\qquad S_n(x+1,y,z) - \displaystyle\prod_{j-1}^{n} \dfrac{[x+(n-j)z]}{[x+y+(2n-j-1)z]} S_n(x,y\ z)$.

Symmetry then leads to

(4.10) $\qquad S_n(x,y+1,z) - \displaystyle\prod_{j-1}^{n} \dfrac{[y+(n-j)z]}{[x+y+(2n-j-1)z]} S_n(x,y,z)$.

This can be turned around to give

$$S_n(x,y,z) - \prod_{j-1}^{n} \left[\frac{x+y+(2n-j-1)z}{y+(n-j)z} \right] S_n(x,y+1,z)$$

$$- \prod_{j-1}^{n} \frac{(x+y+(2n-j-1)z)_\ell}{(y+(n-j)z)_\ell} S_n(x,y+\ell,z)$$

$$- \prod_{j-1}^{n} \frac{(x+y+(2n-j-1)z)_\ell}{\ell!\ e^{xn+yn+(2n-j-1)z-1}} \frac{\ell!\ e^{yn+(n-j)z-1}}{(y+(n-j)z)_\ell}$$

$$\cdot \int_0^\ell \cdots \int_0^\ell \prod_{1 \leq i < j \leq n} |t_i - t_j|^{2z} \prod_{i-1}^{n} t_i^{x-1}\left(1-\frac{t_i}{\ell}\right)^{y+\ell-1} dt_i .$$

When $\ell \to \infty$ this gives

(4.11) $\qquad S_n(x,y,z) - \displaystyle\prod_{j-1}^{n} \dfrac{\Gamma(y+(n-j)z)}{\Gamma(x+y+(2n-j-1)z)} R_n(x,z)$

with

$$(4.12) \quad R_n(x,z) = \int_0^\infty \cdots \int_0^\infty \prod_{1 \le i < j \le n} |t_i - t_j|^{2z} \prod_{i=1}^n t_i^{x-1} e^{-t_i} dt_i \ .$$

By symmetry

$$(4.13) \qquad S_n(x,y,z) = \prod_{j=1}^n \frac{\Gamma(x+(n-j)z)}{\Gamma(x+y+(2n-j-1)z)} R_n(y,z) \ .$$

Together (4.11) and (4.13) give

$$(4.14) \qquad S_n(x,y,z) = \prod_{j=1}^n \frac{\Gamma(x+(n-j)z)\Gamma(y+(n-j)z)}{\Gamma(x+y+(2n-j-1)z)} T_n(z) \ .$$

Selberg's argument [40] or the one given in [2, Chapter 5] can be used to show that

$$(4.15) \qquad\qquad T_n(z) = \prod_{j=1}^n \frac{\Gamma(jz+1)}{\Gamma(z+1)} \ .$$

These combine to give

$$(4.16) \qquad A_n(k;x,y,z) = \prod_{j=1}^k \left[\frac{x+(n-j)z}{x+y+(2n-j-1)t} \right]$$

$$\cdot \prod_{j=1}^n \frac{\Gamma(x+(n-j)z)\Gamma(y+(n-j)z)\Gamma(jz+1)}{\Gamma(x+y+(2n-j-1)z)\Gamma(z+1)} \ .$$

This argument was given in detail because it shows how the extra freedom that Aomoto introduced made it possible to find a relatively simple proof of Selberg's result. Contrast his argument with Selberg's argument from [40], or from [2, Chapter 5] or [21]. In the case of the Mehta-Dyson integral (1.3), there is still no direct evaluation. It would be nice to say that computer algebra played a central role in the introduction of this extra freedom, but that is not the case. It did play an essential role in formulating the conjectures for q-extensions of Selberg's result, and for G_2. However, it only provided the data that Morris and Macdonald were able to make sense of. I gave Morris's data for G_2 to H. Simon to see if BACON could come up with a formula that would match it. He has not sent me a way of making sense of this data in the year and a half time since it was sent to him. He did not know that factorials were involved, while Morris did. This made it possible for Morris to get a conjecture in less than a day, while we are still waiting for BACON's conjecture. This is just one of what I assume are many instances where it helps very much to know a little bit about what to look for. Here is another instance. After seeing Aomoto's argument, and translating his extension back to the root system BC_n, it became clear that one should look at G_2 again and try to see if further degrees of freedom could be added to (3.7) and (3.8). Here is a conjecture that was formulated from data that was obtained using REDUCE.

Let

$$(4.17) \quad H(s,t;a,b) = (1-t)^a(1-t^{-1})^a(1-s)^a(1-s^{-1})^a(1-ts)^a(1-t^{-1}s^{-1})^a$$

$$\cdot (1-ts^{-1})^b(1-st^{-1})^b(1-t^2s)^b(1-t^{-2}s^{-1})^b(1-st^2)^b(1-s^{-1}t^{-2})^b$$

and G(a,b) the function defined in (3.8).

Conjecture: C.T. $(1-t)(1-t^{-1})H(s,t;a,b) = \frac{2(3a+3b+1)}{2a+3b+1}G(a,b)$

C.T. $(1-t)(1-t^{-1})(1-s)(1-s^{-1})H(s,t;a,b) = \frac{2(3a+3b+1)(3a+3b+2)}{(2a+3b+1)(a+2b+1)}G(a,b)$

C.T.$(1-t^2 s)(1-t^{-2}s^{-1})H(s,t;a,b) = \frac{2(3a+3b+1)(3b+1)}{(2a+3b+1)(a+2b+1)}G(a,b)$

C.T. $(1-t^2 s)(1-t^{-2}s^{-1})(1-st^2)(1-s^{-1}t^{-2})H(s,t;a,b)$

$= \frac{6(3a+3b+1)(3a+3b+2)(3b+1)(3b+2)}{(2a+3b+1)(2a+3b+2)(2a+3b+3)(a+2b+1)}G(a,b)$

I would like to thank William Long of the University of Wisconsin Physics Department for providing access to REDUCE and for writing a program that allowed me to get data that led to this conjecture.

5. <u>Further remarks.</u> Now that we have more of an idea about what is happening to those multiple integrals, it is easy to find further problems where computer algebra will be useful. The first one is to see if it is possible to add partial factors in G_2 corresponding to the short roots and long roots at the same time and still have a constant term that factors. Next, one wants to know what the factors of 2 and 3 in the above conjecture really represent. To find out one should consider the two affine versions of G_2. In one of these, Morris's conjecture has been proven independently by Habsieger [20] and Zeilberger [45]. In the other, where the short roots are on base q and the long roots on base q^3, the Macdonald Morris conjecture is still open. Here the added degrees of freedom might be useful in helping us find a proof.

Next, one needs to look at the various ways of moving through the roots of BC_n and G_2 to see if there is a conjecture there, like the full Macdonald conjecture [32], which will work for all root systems. At present we do not know how to get started proving the Macdonald conjectures for F_4, E_6, E_7 and E_8, and it is almost certain that we can not get enough data directly from E_8 to come up with a conjecture that will allow added degrees of freedom. The only way to do this is to generate data in low dimensional cases, and then try to understand it. As was remarked at this meeting, the window where computer algebra is useful can be small (below a certain level calculations can be done by hand, and above a certain level the problems are to too big for current systems), but this window can sometimes provide enough new information to make it possible to understand and/or prove the general case.

One should look at the remaining one dimensional beta integrals to
see what other multidimensional integrals can be formed from them that
can be evaluated as products. Some conjectures were given in [7] and
[36], and it is likely there are many more. There is also a finite
field version of the gamma function, called a Gauss sum, and of a beta
integral, called a Jacobi sum. There seem to be multidimensional sums
of Selberg type in this setting. See Evans [16] for a theorem in two
dimensions and a conjecture in n dimensions.

For a long time there seemed to have only been one application of
Selberg's integral beyond Selberg's original application in [39]. This
was the use of a special case in moment spaces by Karlin and Shapley
[29]. In the last few years there have been a number of applications.
See Dotšenko and Fateev [12], Hanlon [22] plus unpublished work,
Koranyi [30], Regev [37] and Tsuohiya and Kanie [43].

Finally, I would like to thank R. William Gosper for showing me
how powerful symbolic algebra systems are, and to thank G. Collins, A.
Hearn, R. Jenks, J. Moses and the many other people who have developed
these systems. G. Andrews and D. and G. Chudnovsky have used
SCRATCHPAD to solve some problems I would have liked to solve. This is
the best inducement to help one invest the time necessary to learn to
use these important tools that have been added to our tool box.
L. Solomon saved me from an embarrassing error.

REFERENCES

[1] G. E. Andrews, Problems and prospects for basic hypergeometric functions, in The Theory and Application of Special Functions, ed. R. Askey, Academic Press, New York, 1975, 191-224.

[2] G. E. Andrews, q-Series: Their Development and Application in Analysis, Number Theory, Combinatorics, Physics, and Computer Algebra, Regional Conference Series in Mathematics, 66, Amer. Math. Soc., Providence, R.I., 1986.

[3] G. E. Andrews and R. Askey, Another q-extension of the beta function, Proc. Amer. Math. Soc. 81 (1981), 97-100.

[4] K. Aomoto, Jacobi polynomials associated with Selberg integrals, SIAM J. Math. Anal. to appear.

[5] R. Askey, The q-gamma and q-beta functions, Appl. Anal. 8(1978), 125-141.

[6] R. Askey, Ramanujan's extensions of the gamma and beta functions, Amer. Math. Monthly 87(1980), 346-359.

[7] R. Askey, Some basic hypergeometric extensions of integrals of Selberg and Andrews, SIAM J. Math. Anal. 11(1980), 938-951.

[8] R. Askey, A q-extension of Cauchy's form of the beta integral, Quart. J. Math. Oxford (2) 32(1981), 255-266.

[9] R. Askey and J. Wilson, Some basic hypergeometric orthogonal polynomials that generalize Jacobi polynomials, Mem. Amer. Math. Soc. #319 (1985).

[10] W. N. Bailey, Generalized Hypergeometric Series, Cambridge Math. Tract No. 32, Cambridge Univ. Press, 1935; Reprinted: Hafner, New York, 1964.

[11] N. Bourbaki, Groupes et Algèbres de Lie, chapitres IV, V, VI, Hermann, Paris, 1968.

[12] V. S. Dotsenko and V.A. Fateev, Conformal algebra and multipoint correlation functions in 2D statistical models, Nuclear Phys. B240 (1984), 312-348.

[13] F. J. Dyson, Statistical theory of the energy levels of complex systems, I, J. Math. Phys. 3(1962), 140-156.

[14] A. Erdelyi, et al, Higher Transcendental Functions, Vol. 1, McGraw-Hill, New York, 1953; Reprinted Krieger, Malabar, Florida, 1981.

[15] L. Euler, De productis ex infinitis factoribus ortis, Commentarii academiae scientiarum Petropolitanae 11(1739), 1750, 3-31, reprinted in Opera Omnia, ser. 1, vol. 14, 260-290.

[16] R. Evans, Identities for products of Gauss sums over finite fields, L'Enseignement Math. 27(1981), 197-209.

[17] R. P. Feynman, "Surely You're Joking, Mr. Feynman!", W. W. Norton, New York and London, 1985.

[18] I. J. Good, Short proof of a conjecture of Dyson, J. Math. Phys. 11(1970), 1884.

[19] J. Gunson, Proof of a conjecture of Dyson in the statistical theory of energy levels, J. Math. Phys. 3(1962), 752-753.

[20] L. Habsieger, La q-conjecture de Macdonald-Morris pour G_2, C. R. Acad. Sci., 303(1986), 211-213.

[21] L. Habsieger, Une q-intégrale de Selberg-Askey, SIAM J. Math. Anal., to appear.

[22] P. Hanlon, Cyclic homology and the Macdonald conjectures, Invent. Math. 86(1986), 131-159.

[23] J. E. Humphreys, Introduction to Lie Algebras and Representation Theory, Graduate Texts in Mathematics, No. 9, Springer-Verlag, New York and Berlin, 1972.

[24] M. E.-H. Ismail, A simple proof of Ramanujan's $_1\psi_1$ sum, Proc. Amer. Math. Soc. 63(1977), 185-186.

[25] F. H. Jackson, On q-definite integrals, Quart. J. Pure Appl. Math. 41(1910), 193-203.

[26] F. H. Jackson, Certain q-identities, Quart. J. Math. 12(1941), 167-172.

[27] V. Kac, Infinite Dimensional Lie Algebras, second edition, Cambridge Univ. Press, Cambridge, 1985.

[28] K. Kadell, A proof of Askey's conjectured q-analog of Selberg's integral and a conjecture of Morris, SIAM J. Math. Anal., to appear.

[29] S. Karlin and L. S. Shapley, Geometry of moment spaces, Mem. Amer. Math. Soc. 12, 1953.

[30] A. Koranyi, The volume of symmetric domains, the Koecher gamma function, and an integral of Selberg, Studia. Sci. Math. Hung. 17(1982), 129-133.

[31] I. G. Macdonald, Affine root systems and Dedekind's η-function, Invent. Math. 15(1972), 91-143.

[32] I. G. Macdonald, Some conjectures for root systems, SIAM J. Math. Anal. 13(1982), 988-1007.

[33] I. G. Macdonald, Kac-Moody-Algebras, in Lie Algebras and Related Topics, ed. D. J. Britten, F. W. Lemire and R. V. Moody, Canadian Math. Soc. Conference Proc. 5, Amer. Math. Soc., Providence, R.I., 1986, 69-109.

[34] M. L. Mehta, Random Matrices and the Statistical Theory of Energy Levels, Academic Press, New York, 1967.

[35] W. G. Morris, Constant term identities for finite and affine root systems: conjectures and theorems, Ph.D. thesis, Univ. of Wisconsin-Madison, 1982.

[36] M. Rahman, Another conjectured q-Selberg integral, SIAM J. Math. Anal. 17(1986), 1267-1279.

[37] A. Regev, Asymptotic values for degrees associated with strips of Young diagrams, Advances in Math. 41(1981), 115-136.

[38] D. B. Sears, Transformation of basic hypergeometric functions of special type, Proc. London Math. Soc. 52(1951), 467-483.

[39] A. Selberg, Über einen Satz von A. Gelfond, Arch. Math. Naturvid, 44(1941), 159-171.

[40] A. Selberg, Bemerkninger om et multipelt integral, Norsk Mat. Tidsskr. 26(1944), 71-78.

[41] J. Stembridge, A short proof of Macdonald's conjecture for the root systems of type A, to appear.

[42] J. Thomae, Beiträge zur Theorie der durch die Heinesche Reihe: $1 + ((1-q^{\alpha})(1-q^{\beta})/(1-q)(1-q^{\gamma}))x + \ldots$ darstellbaren Functionen, J. Reine Angew. Math. 70(1869), 258-281.

[43] A. Tsuchiya and Y. Kanie, Fock space representations of the Virasoro algebra, Publ. Res. Inst. Math. Sci., Kyoto, 22(1986), 259-327.

[44] K. Wilson, Proof of a conjecture of Dyson, J. Math. Phys. 3(1962), 1040-1043.

[45] D. Zeilberger, A proof of the G_2 case of Macdonald's root
     system-Dyson conjecture, SIAM J. Math. Anal., to appear.

[46] D. Zeilberger and D. M. Bressoud, A proof of Andrews' q-Dyson
     conjecture, Discrete Math. 54(1985), 201-224.

Some Thoughts on Proof Discovery

W. W. BLEDSOE* Microelectronics and Computer Technology Corporation, Austin, Texas

In this talk we first mention some of the exciting new efforts to speed up the search process in Automated Theorem Proving (ATP), and express the belief that these new "speed demons" will play a crucial role in modern technology, where a great deal of reasoning power is needed. We then argue that speed alone can never cope with the really difficult theorems that arise from mathematics and challenging application areas, theorems that are nevertheless fairly easy for gifted mathematicians. We suggest that a part of the research effort in ATP be directed toward methods which have proven useful (powerful) for human mathematicians, but which have been largely neglected by the ATP community. Finally we list and briefly discuss some of these research areas.

This is an exciting time for Automated Theorem Proving (ATP). A subject that looked like a weak appendage to computer science a few years ago has emerged as a formidable subarea. A large percentage of the papers at recent meetings, such as AAAI and IJCAI, has been devoted to Automated Reasoning (AR) in one form or another. A new journal, the Journal for Automated Reasoning, has appeared. Applications for ATP are popping up all over the place and the trend is for even more to emerge.

AR is the heart of any "intelligent" system, especially as these systems are called upon to do complicated "humanlike" things such as: duplicate expert behavior, verify and synthesize computer programs, draw conclusions from large logic data bases, play board games, prove mathematical theorems, recognize scenes, learn automatically, and serve as a computer language. A "requirement" for AR is present in all of these systems and the need for faster, more powerful AR will increase as our programs continue their trend toward more *causality*, deeper reasoning, reasoning from basic

---

*Current affiliation: University of Texas at Austin, Austin, Texas

principles.

Also with these needs has come new methods to satisfy them, especially those that have brought increases in speed and functionality. From the PROLOG community has come a great increase in the speed of unification by the use of novel compilation techniques. These methods which were designed originally for horn clause theorems, have been extended to handle the full first order logic (with the occurs check). Two or three orders of magnitude in speed enhancements over earlier provers seem to be obtainable. Increases of speed have also been achieved by automatic inheritance mechanisms, and by methods for quickly retrieving from a data base those clauses whose literals are likely to unify with a current goal. And we have seen advances in the use of *rewrite rules* and prudent handling of equality through the use of *complete sets of reductions*.

So this is indeed an exciting time for ATP (or AR). This is a viable field, good workers here will be well rewarded. There is much still to be done and a waiting audience to receive the results. The last 40 years have been productive and I am proud to have been part of it.

So what is the problem? Why am I concerned? Before I get into that let me mention that I am mainly concerned with *Proof Discovery*, not the rest of AR.

Some Subfields of Automated Reasoning.

Automatic Proof Discovery

> This is my main interest; the rest of the presentation will center on it.

Proof Checking [WT, deB, BM]

> Some good work here but we still need systems that will accept proofs which are produced naturally by mathematicians.

Man-Machine Interactive Systems

- Argonne Automated Reasoning Lab [Wo2], proving open theorems in Mathematics.

- Boyer-Moore Prover [BM]. Recently used by Shankar to mechanically prover the Goëdel Incompleteness Theorem.

- The prover for these systems has to be strong in its own right.

Provers for Special Theories

- Inequality Provers [Sh, Op, BKS]

- Wu Geometry Prover [Wu1, Wu2, Ch]

- MACSYMA [Ma, Mo] for solving Mathematics problems (Integrals, DE's, Algebraic Manipulation, etc.)

Much more – This is not a review paper.

What is the Problem

So what *is* the problem? Why change anything? In spite of these successes I believe that we are headed for some *radical* changes in the way that ATP is being done. Why? The sobering truth is that these provers, for all their speed, are *not* able to compete favorably with *humans* on most truly difficult theorems in mathematics and application areas.

Let us analyze the process.

Discovering the proof of a theorem in Mathematics can be viewed as searching a tree (Figure 1). There are usually many *goal nodes* ⊛ in the tree. Finding one of them finishes the proof.

These trees tend to be very branchy and sometimes have enormous depth. The simplest automatic provers search these trees from left to right, depth first, until a goal node is found. Others do a breadth first search, exploring all branches down to a certain level. The object of most research in ATP (Automated Theorem Proving) is to speed up this search.

The principal tactic for speeding up search, is *pruning*, whereby unproductive branches – those that do not contain a goal node – are eliminated, thereby enabling the prover to reach deeper levels on selected branches. This is often called merit-first searching. Pruning sometimes has an adverse effect; it is possible that all branches containing goal nodes are eliminated, in which case the proof procedure is *incomplete*; or that all shallow goal

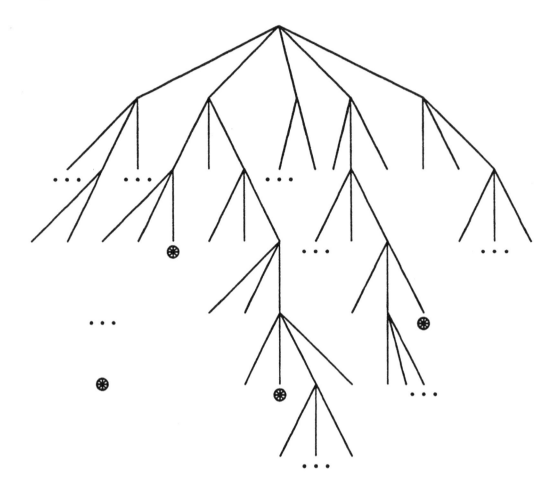

Figure 1. A Proof Search Tree

nodes are excluded and the search is driven to very deep levels. But, on the whole, pruning *is* a good idea; the problem lies in selecting the best pruning algorithms.

As was mentioned earlier, recent years has seen the emergence of a great deal of activity on *fast provers*. These techniques avoid pruning altogether but make up for that by greatly speeding up the search. It is clear to some of us that speed alone can *never* solve the proof search problem (even though as we said, there are many applications where these fast provers can be enormously valuable, and this exciting work should be pushed with full vigor.)

The truth is that our provers (automatic provers, versus human provers) are *not* very powerful, they are not able to prove difficult theorems, they tend to be *syntactic* in nature, they tend to be *unmotivated* in their search, they compare poorly with beginning mathematicians on proving all but the simplest theorems, they beat out expert mathematicians only on problems that require a great deal of syntactic combinatorial search.[1]

This is hurtful to our hopes and ambitions because we wish for our provers to be powerful reasoning devices, to compete with people on all but the hardest theorems, and to join forces with mathematicians in a man-machine symbiosis to tackle the frontiers of mathematics.

To better understand our ambition, the role we hope for our provers, let us mention the *four color problem* in topology which was finally proved in recent years [FC] with the help of a computer. The proof was done in two steps: a person analyzed and transformed the problem into a very large set of calculations; a computer carried out these calculations (several hundred thousand). This is a commendable result, a milestone in the use of computers in mathematics, but it is *not* what we have in mind. We have in mind increasing the computer's role: the computer program should do the first step as well as the second. It is such ambitions that drive us on with the work.

---

[1] For certain theories, researchers have developed specific algorithms for proving statements in that theory. Two examples of this are Wu's geometry theorem prover [Wu, Ch], and MACSYMA [Ma, Mo], which routinely solves problems beyond the reach of the best mathematicians. Also some impressive results have been obtained where human assistance has helped control the search [Wo2], and where the prover acts principally as a proof checker.

Now let's go back to our problem of finding the goal nodes (or *a* goal node) in the search tree. Why are people so much better at this search than are machines? My answer to that question is simple: because our computer provers have not used the methods that have proven so effective for mathematicians over the centuries. People have accumulated knowledge about problem solving and have used it to accomplish various tasks including theorem proving. It seems incredible that we would attempt the same tasks without using the tools that offer the only known solutions.

We will shortly try to enumerate some of these tools, these methods used by the mathematician in proof search, and indicate how they might be implemented in our provers. Indeed some of this, a small part, has already been realized in our present provers, but the bulk has not.

The Lodestone Fixation.

The stumbling block to prover designers is the desire for a "magical" solution, a "lodestone", a simple algorithm that will search out and eventually find a proof. But alas, it requires more complicated mechanisms to do complicated reasoning; the human prover uses a whole collection of methods of different sorts. Where the prover might ask itself "which clause will I resolve[2] next?", the person would ask "*what* will I *do* next?". Where the prover would use a series of homogeneous steps, the person would use a hierarchy of actions. The prover might set to work immediately "making progress" by resolving clauses, the human would think about the problem, try to understand it, reflect upon it, rearrange it, reformulate it, try to see a similarity to other problems, before proceeding. He usually determines a broad plan of action, based upon one or two major theorems or definitions in his theory. He draws diagrams and uses examples to help see how the proof might go, and often conjecture intermediate goals.

If he does decide to use one of these major theorems (lemmas) thereby generating for himself a new subgoal which now must be proved, he might pause to "try a counterexample". That is, he wants to ensure that this new

---

[2]It is not our intent here to downgrade Resolution [Ro]; indeed we believe that Resolution, and the work leading up to it, represents the most important milestone in the history of ATP research. Most of the deficiencies that we attribute here to resolution are also suffered by *All* existing automatic provers.

subgoal is indeed valid, that he has not gotten off on the wrong track. If he can show that the new subgoal is invalid (by exhibiting a counterexample on which it fails) then he can safely prune away that branch of the search tree, greatly speeding up the search.

In other words, there is a myriad of actions that he can and might take. And here is the place where the prover designer is faced with a dilemma: he had hoped for a simple algorithm, but he is faced with the need for a very complicated one, one based on all the methods and tricks of a human mathematician. How can he cope with all of this? The only methods that we know for these kind of situations are those from the field of *Artificial Intelligence*, especially the subfield of *Expert Systems*. What we need is an expert system that simulates a mathematician.

This is not a new thought but one that has not been carried out. Some inroads have been made but they are far from satisfactory.

We cannot just apply expert system technology and expect everything to fall into place. There are major problems to be overcome, like the encoding of mathematical knowledge in such a way that it is readily available to the prover when needed, in a useful form. We not only must store the thousands of known mathematical theorems, definitions, and axioms, but more importantly: the "know how", techniques and tricks; the choice examples; the methods needed for generating examples, generating conjectures, evaluating counterexamples, carrying out procedures such as limits and derivatives, evaluating arithmetic expressions. We also need to face the issue of "pattern recognition", automatic recognition of what we have in terms of what we already know.

One of the things that a person does best is to jump around in the search tree. When working on one branch he discovers something that prompts him to try something different on another. Or he might decide that a certain branch of the tree is not promising, even though he has no counterexample with which to exclude it, because, perhaps, he thinks it is getting "too messy". Needless to say, the knowledge base we spoke of earlier must contain methods for determining when a branch is not promising, *i.e.*, for *evaluating* nodes in the tree; these evaluations (or *priorities*) must be allowed to evolve with time, as the proof search proceeds. Also an *agenda* mechanism must be provided whereby the prover can be directed from one part of the tree to another, and back again when the priorities are right.

Some Examples

Let us look at a couple of example of theorems being proved, and compare the methods used by a "standard" computer prover (a *Resolution Program*) with those used by a person.

First let us consider the following problem:

The Truthtellers and the Liars Problem.

On a certain island the inhabitants are partitioned into those who always tell the truth and those who always lie. I landed on the island and met three inhabitants A, B, and C. I asked A, 'Are you a truthteller or a liar?' He mumbled something which I couldn't make out. I asked B what A had said. B replied, 'A said he was a liar'. C then volunteered, 'Don't believe B, he's lying!' What can you tell about A, B, and C.

This can be clausified as follows:

$$1 \quad P(T(x1)) \; P(L(x1)) \qquad\qquad \text{(input)}$$

$$2 \quad \neg P(T(x1)) \; \neg P(L(x1)) \qquad\qquad \text{(input)}$$

$$3 \quad \neg P(T(x1)) \; \neg P(Says(x1,x2)) \; P(x2) \qquad \text{(input)}$$

$$4 \quad \neg P(L(x1)) \; \neg P(Says(x1,x2)) \; \neg P(x2) \quad \text{(input)}$$

$$7 \quad P(Says(A, Unknown) \qquad\qquad \text{(input)}$$

$$8 \quad P(Says(B, Says(A, L(A)))) \qquad\qquad \text{(input)}$$

$$9 \quad P(Says(C, L(B))) \qquad\qquad \text{(input)}$$

$$10 \quad \neg P(T(C))$$

Clauses 1-10 represent the problem. (See Lusk and Overbeek, [Lu]). With ordinary resolution there are 29 resolvents generated in the first round and 204 in the second. The proof is completed in about 18 rounds after thousands of resolvents have been obtained.

On the other hand a person does not search by resolving clauses; instead he studies the clauses individually and notes from Clause 8 that B is a liar (since A would never say that he was a liar, because if he was a liar he would lie about it), and hence C is a truthteller.

It is completely foreign to a mathematician's nature to generate 29 clauses, let alone 204, or hundreds of thousands! He is too smart for that!

Lim+ and Lim*

We will now look at Lim+, the theorem from calculus which states that the limit of a sum of two functions is the sum of their limits. See Figure 2. Figure 3 shows the *skolemized form* of lim+, whereby the quantifiers have been removed by using *skolem* function symbols and skolem variables[3].

---

[3]The variables ε_1, ε_2, δ, and x, can be instantiated by terms during the proof. The expression δ_ε denotes the unary skolem function δ applied to the variable ε; similarly for the expressions δ_{ε_2} and x_δ.

$$\forall \varepsilon_1 > 0 \ \exists \delta_1 > 0 \ \forall x_1(|x_1 - a| < \delta_1 \rightarrow |f(x_1) - l| < \varepsilon_1)$$

$$\wedge$$

$$\forall \varepsilon_2 > 0 \ \exists \delta_2 > 0 \ \forall x_2(|x_2 - a| < \delta_2 \rightarrow |g(x_2) - k| < \varepsilon_2)$$

$$\longrightarrow$$

$$\forall \varepsilon > 0 \ \exists \delta > 0 \ \forall x(|x - a| < \delta \rightarrow |(f(x) + g(x)) - (l + k)| < \varepsilon)$$

Figure 2
Lim +

$$(0 < \varepsilon_1 \rightarrow (0 < \delta_{\varepsilon_1} \wedge (|x_1 - a| < \delta_{\varepsilon_1} \rightarrow |f(x_1) - l| < \varepsilon_1)))$$

$$\wedge$$

$$(0 < \varepsilon_2 \rightarrow (0 < \delta_{\varepsilon_2} \wedge (|x_2 - a| < \delta_{\varepsilon_2} \rightarrow |g(x_2) - k| < \varepsilon_2)))$$

$$\longrightarrow$$

$$(0 < \varepsilon_0 \rightarrow (0 < \delta \wedge (|x_\delta - a| < \delta \rightarrow |(f(x_\delta) + g(x_\delta)) - (l + k)| < \varepsilon_0)))$$

Figure 3
Skolemized form of Lim+

Of course, other facts are needed in order to prove the theorem: Axioms for the real numbers and for the < predicate; and lemmas such as the triangle inequality.

Lim+ has been a difficult theorem for automatic provers; only recently has it been proved by a *general purpose* prover[4]. Much earlier it and other limit theorems of calculus were proved using an incomplete mechanism, the *limit heuristic*. But we will ignore these two methods here (VE-STR and the limit heuristic) and concentrate on how a resolution prover would attempt to prove Lim+ and compare that with the efforts of a person.

Before we do, let me say a word about "fetching lemmas". Before an automatic prover can prove a theorem it must be provided (usually by people) the list of all axioms, definitions, and lemmas needed in the proof. The human prover, by contrast, is able to "fetch" these from his store of knowledge, or to make them up (and prove them if needed) as he proceeds. Very little has been published on automatically fetching and conjecturing lemmas [Pl1, Pl2, Le1]; this remains a neglected, but crucially important, part of automated theorem proving.

Now let us first discuss how a person would prove Lim+. There are many ways for doing this; we show parts of one such scenario, and compare that with what a resolution prover might do.

---

[4]Using the Variable-elimination, shielding-term-removal, and restricted chaining (VE-STR) methods of [Bl1].

HUMAN PROOF SUMMARY

A. Recognition Phase.

He might make the following observations:

- This is a formula with a hypothesis and conclusion, involving the arithmetic predicate $<$ and the function symbols $+$ and $-$, and $||$.

- The *Major Actors* here are: f and g. There are also minor actors: l, k, a.

These are symbols that are *unique* to this theorem.

B. To start the proof, he might

- Select the main conclusion to prove

    (3) $\quad |(f(x_6) + g(x_6)) - (l + k)| < \varepsilon_0$

- It contains all the actors f, g, l, k. So he realizes that he must select hypotheses containing these (Because these symbols are peculiar to this theorem!).

    · He selects the main conclusions from the two hypotheses

        (1) $\quad |f(x_1) - l| < \varepsilon_1$

        (2) $\quad |g(x_2) - k| < \varepsilon_2$

    · He realizes that these are the only hypotheses which contain the major actors f and g, and knows for certain that his must prove the goal

        (4) $\quad$ (1) $\wedge$ (2) $\longrightarrow$ (3)

    · He also knows immediately that the variables x_1 and x_2 must be given the value x_6, by a limited matching of $f(x_1)$ with $f(x_6)$, etc.[5]

---

[5] He has already noticed that f and g are dual symbols, what is done for one must be done analogously for the other.

· He sees that (3) does *not* match either (1) or (2), but nevertheless he does not go thrashing about trying to match (3) with a lot of other hypothesis parts (which do not even contain the main actors f and g) or with other lemmas stored in his knowledge base. Rather he tries to understand how he might transform (3) into a form that is "like" (1) and (2).

· Noting that (1) contains the pair f, l and (2) the pair g, k, he sets the subgoal of expressing the interior of (3) in terms of the interiors of (1) and (2),

SUBGOAL: Express $((f(x_\delta) + g(x_\delta)) - (l + k))$ in terms of $(f(x_\delta) - l)$ and $(g(x_\delta) - k)$

and *solves* this to get

(3') $|[f(x_\delta) - l] + [g(x_\delta)) - k]| < \varepsilon_0$

· His goal is now

(4')   (1) $\wedge$ (2) $\longrightarrow$ (3')

Which he abstracts to

(4'') $|A| < \varepsilon_1 \wedge |B| < \varepsilon_2 \longrightarrow |A + B| < \varepsilon_0$

which is free of the actors. He is now clearly ready to seek help from outside lemmas, which he does, finding and using the triangle inequality: $|A + B| < |A| + |B|$[6].

· This leads to the goal

(1) $\wedge$ (2) $\longrightarrow |f(x_\delta) - l| + |g(x_\delta)) - k| < \varepsilon_0$

Continuing with this kind of reasoning leads him to put $\varepsilon_0/2$ for ε_1 and for ε_2, and to finish the solution of (4''). In a similar way he "cleans up" the rest of the proof of Lim+, eventually using the substitution $\delta = Min(\delta_1, \delta_2)$.

---

[6] Actually, he probably anticipated the use of the triangle inequality back in the proof where he rearranged the interior of (3).

Notice that the person performs a number of *common sense* tasks. For example, the fact that his goal involved the functions f and g, motivates him to look for hypotheses with these function symbols in them. Notice also that he sometimes does *loose* matching: "Look for something containing f and g". On the other hand, as we shall see the computer is looking for *absolute* matches between the goal and parts of the hypotheses and lemmas. It is forced to thrash wildly, "looking in the small", rather than being gently guided in a correct direction by loose matches on the most important concepts. And to imitate people the program would need to have some recognition capability, to determine which predicate and function symbols are the most important and/or peculiar to the given theorem.

The resolution proof would start with a list of *clauses* representing the theorem (Clauses 1-7, Figure 4).

lim+ clauses

1. $\varepsilon \leq 0 \qquad 0 < \delta_\varepsilon$

2. $\varepsilon \leq 0 \qquad \delta_\varepsilon \leq |x - a| \qquad |f(x) - l| < \varepsilon$

3. $\varepsilon' \leq 0 \qquad 0 < \delta'_{\varepsilon'}$

4. $\varepsilon' \leq 0 \qquad \delta'_{\varepsilon'} \leq |x - a| \qquad |g(x) - k| < \varepsilon'$

5. $0 < \varepsilon_0$

6. $\delta \leq 0 \qquad |x_\delta - a| < \delta$

7. $\delta \leq 0 \qquad \varepsilon_0 \leq |f(x) + g(x) - (l + k)|$

lim* clauses

7'. $\delta \leq 0 \qquad\qquad\qquad \varepsilon_0 \leq |f(x) * g(x) - l * k|$

Extra needed Lemmas (clauses)

8. $|A + B| \leq |A| + |B|$

9. $\varepsilon/2 \leq A \qquad\qquad \varepsilon/2 \leq B \qquad A + B < \varepsilon$

10. $\varepsilon \leq 0 \qquad\qquad\quad 0 < \varepsilon/2$

11. $\varepsilon/M \leq |A| \qquad\qquad M < |B| \qquad M \leq 0 \qquad |B \cdot A| < \varepsilon$

12. $|B| - |A - B| \leq |A|$

Figure 4
Clauses for Lim+ and Lim*

Since the prover has no ability whatever to fetch needed lemmas we must give it extra clauses, (Clauses 8-12, Figure 4). On the one hand, this list must contain *all* the needed lemmas, but on the other hand, we must not include too many that are not needed for fear of "choking" the prover with too many options. (The often suggested, but ill advised, strategy of including *all* known theorems, definitions, and axioms, as hypotheses, will quickly lead to a disastrous combinatorial explosion).

We start the proof process with Clause 7.[7] The first round generates 149 clauses, even without including the clauses for the equality axioms. The second round produces thousands. Needless to say this is a far cry from what happened in the human case. Should a person be proud of producing such a proof? (Even if it were somehow possible with a fast prover.) We think not. What, for example, would be needed for a slightly harder theorem, like Lim* (the limit of a product theorem) or the Heine Borel Theorem?

This is not the beautiful process we know as *mathematics*. This is "cover your eyes with blinders and hunt through a cornfield for a diamond-shaped grain of corn". Mathematicians have given us a great deal of direction over the last two or three millennia. Let us pay attention to it. We are faced with a decision: either look at these "people methods" or never do very well at the automatic proving game.

Shall we be cheated out of the chance of seeing a computer prove a truly difficult theorem during our lifetime? We hope not. If it did not seem so possible then we would not be so concerned.

Some Important Research Areas in ATP.

We will now mention some research areas in ATP that would seem to be important to the future of this field. Much of this was alluded to in our earlier remarks.

Central in all of this is *knowledge*. We see no satisfactory solution to the ATP performance problem without a substantial *knowledge base*, which contains much of known mathematics, and a great deal of supporting information including the so called *common sense* knowledge, which is so useful

---

[7] Using the main conclusion as a *set of support*.

in all we do. It is not enough to collect together all known theorems, axioms, definitions, and proofs. We need also, *methods, procedures,* and *tricks of the trade,* which have been used so successfully by the great mathematicians over the years, and also the *diagrams, constructions, figures,* and *examples* of all sorts which play key roles in proof discovery.

These components cannot be just thrown together in some kinds of lists or trees, but must be organized into a *structured knowledge base* that can be powerfully used in actual applications. The building of this mathematical knowledge base (MKB) must go hand in hand with development and use of proof discovery methods. The knowledge representation problem is a formidable one, whose solution cannot be completely foreseen ahead of time, but hopefully can be solved by repeated use of our evolving provers on difficult mathematical theorems.

The field of AI has long recognized the importance of knowledge, and knowledge bases which contain "how-to" *methods* as well as facts. This is the key ingredient of *Expert Systems,* which attempt to duplicate expert behavior. Some researchers have advocated for a number of years, the use of "common sense" reasoning [Mc1, Mc2] and worked on the representation of common sense knowledge [Ho], while others are developing large encyclopedic knowledge bases [Bo] and large structured common-sense knowledge bases [Le3].

ATP Research Areas

Now let us make a listing of several other research areas. These are related to each other in various ways, and all draw heavily on an adequate mathematical knowledge base.

- Using *Analogy* in proof discovery.

- Generating and using *Examples* in proof discovery

- The use of *Counterexamples* to prune the search tree

- Automatic *Conjecturing* of lemmas and subgoals

- Automatic *Fetching* of useful lemmas and definitions from a large knowledge base

- *Agenda* mechanisms for controlling the proof search

- Mathematical Pattern Recognition -
  Recognizing a "scene" in terms of what is already known.

- Automatic *Learning*

- *Planning* and *Abstraction* [Ne, Pla]

- Higher-level reasoning, meta reasoning, higher order logic

One could also mention other important areas such as the use of *rewrite rules* (demodulators, reductions) and complete sets of reductions, which are already in use and are being adequately researched.

Analogy is the heart and soul of all intelligent behavior, especially mathematical behavior. Why have we made so little use of analogy in ATP? The answer, we believe, is quite simple: to use analogy we must have something with which to analogize. When your prover is *essentially devoid of knowledge*[8] how can there by any automatic analogy? So we seem to have the maxim

> There can be no effective proofs by analogy without an adequate knowledge base.

The use of analogy that we have in mind is as follows: Given a theorem *A* that we wish to prove, fetch the proof of another theorem *A'* which is similar to *A*, and use it as a guide for proving *A*. There is no real problem here when the proofs of *A'* and *A* are essentially the same; but when a sizable variation is encountered there is a sudden and dramatic need for knowledge to handle the new parts of the proof of *A* while keeping to the general plan of the proof of *A'*.

Of course, it takes more than a knowledge base, but also effective control mechanisms for carrying out the whole analogue making procedure. Some

---

[8]Actually the existing provers do have knowledge in terms of (a few) theorems and definitions, but essentially no store of "methods".

work has been done on analogy in ATP and problem solving [Kl, Gr, Wi, Ca],[9] but the surface has hardly been scratched. This work goes hand in hand with *Generalization* and *Machine Learning* [ML1, ML2].

We predict no substantial advance in ATP until our provers begin to effectively use Analogy with the help of an adequate MKB.

Examples also play a crucial role in all problem solving activity, especially in solving mathematical problems and proving mathematical theorems. We often hear: "Let's try a special case". "What happens on this example?"

These examples somehow offer insight leading to the *conjecturing* of an important subgoal, a "halfway house" to the eventual solution. They also often help us see additional results and generalizations that we had not expected. This is also closely related to the generation and use of *counterexamples* to prune unproductive branches from the search tree. Again we must confess that very little has been done in these areas [Ba, Wa] and in conjecturing [Le1, Le2, Bl2]. An important part of *conjecturing* is the evaluation of *relevance*. Again this difficult activity (automatically determining the relevance of a derived lemma) requires substantial support from an adequate knowledge base. Lenat's milestone paper [Le1] bears this out.

As we mentioned earlier, the *fetching problem*, the problem of automatically selecting from a large knowledge base, the lemmas, axioms, and definitions, needed for the current goal, is crucially important but an essentially neglected area. [Pl1, Pl2, Ov]. Here we can make use of results from the Data Base research community [DB]. One difficulty with automating this process is that often, for a human mathematician, he is not really fetching a lemma from a knowledge base but *making up* one to satisfy a current need (and possibly proving it).

Machine Learning is becoming a formidable subarea of AI [ML1, ML2]. Many of these methods can be applied to ATP, but more is needed. In the long run this too is a *must* for ATP.

The use of an *agenda* to control search is a familiar occurrence in AI, but it has as yet found little use in ATP [Ty]. The problem here is that in order to "bounce around" in the search tree, we must have effective ways of *eval-*

---

[9]Since this is not a review paper, no attempt is made to include all pertinent references. See [Da,BL, Lo, Wo1] for some recent reviews.

uating positions in the tree. This is not unlike evaluating board positions in chess but much harder. Each branch of the tree must have a priority number associated with it that is allowed to change as the proof proceeds, and activity on one branch is discontinued when its priority becomes too low, but might be resumed later.

Abstraction (Planning)

In their classic paper [Ne], Newell, Simon, and Shaw suggested a form of abstraction, whereby the given theorem is abstracted to an easier one, the proof of which then acts as a guide to the proof of the given theorem. Plaisted [Pla] has greatly enhanced this theory. For example we might abstract the formula $P(x, f(x, y))$ to $P(x)$ or $P(x, f(y))$ or even P.

Much of the activity of a mathematician is at the *meta level*: "I can finish this proof like I did the last"; "Putting these two methods together we get a solution of an earlier problem". We believe that such meta level procedures will become commonplace in ATP [Wey, Go]. Also most of it involved *Higher Order Logic*; we must somehow cope with the enormous search problem that seems to plague this area [An].

A problem with most of the advanced research in ATP, especially in the areas that we have been discussing, is the lack of experimentation. Often, when new methods are proposed they are *not* accompanied by results which show that any real power is gained by their use. Some of this is very formal, and formidable looking, much pomp and little action. This again is a symptom of the "disease" of seeking to get by without the use of an extensive knowledge base. Unfortunately, the use of such large knowledge bases requires complicated methods that don't always satisfy our aesthetic yearnings.

One wonders whether truly exceptional ATP can be accomplished by a series of Ph.D. dissertations. Each such contribution is simply too small. What is needed is a few larger efforts, that extend over many years with several participants, with some working on gathering and encoding knowledge, while other devise new proof methods and exercise them on hard theorems with each effort greatly influencing the other. Such a team approach has been very successful at some laboratories (Argonne National Lab, SRI, MCC,...) where fast proving methods are being developed, but we know of no comparable effort for the "people methods" we have been discussing here, supported by a large knowledge base.

One final word about knowledge bases. When a young person sets out to become a mathematician he/she spends many years in acquiring knowledge, knowledge of mathematical results, but more importantly knowledge about how to *do* mathematics. And we note that this training progresses through all the major areas we listed above. How can we expect our automatic prover to compete with such a young person if it too is not also given comparable knowledge.

So it is time to make a significant change in direction. But the newcomer should not think that he can ignore the foundations that have been lain by earlier work: skolemization, clausal representations, resolution (and variations of it), completeness proofs, equality techniques, inequality methods, natural-deduction provers, rewrite rules, fast unification, and much, much, more. We must not reinvent these "wheels". These basic material can be quickly assembled and should be.

Some of these ideas have been around for a number of years [Bl3]. For example, the signal papers of Newell, Simon, and Shaw [Ne] and Gelernter [Ge] suggested and used many of them. We believe that this trend should have been continued and enlarged, but it was soon overshadowed by the emergence of other techniques emphasizing speed, and thus the field took a turn away from people methods.

In Summary

We hope that these observations and suggestions will be helpful, but only time will tell. We are excited about much of what is going on in ATP right now and see an important role for it in modern technology, but we also feel strongly that a marked change of direction is needed if we are to achieve a mechanical prover which can compete favorably with mathematicians.

References.

[An] Andrews, P. B., *et al*, "Automating Higher-order Logic", in *Automated Theorem Proving: After 25 Years*, W. W. Bledsoe and Donald Loveland (eds.), Am. Math. Soc., Contemporary Mathematics Series, 19 (1984).

[Ba]   Ballantyne, M. and Bledsoe, W. W., "On Generating and Using Examples in Proof Discovery", *Machine Intelligence 10*, Harwood, Chichester, (1982), pp. 3-39.

[BKS]   Bledsoe, W. W., Kunen, K. and Shostak, R., "Completeness Results for Inequality Provers", *J. Artif. Intel. 27*, (1985), pp. 255-288.

[BL]   Bledsoe, W. W. and Loveland, D., (eds.) *Automated Theorem Proving: After 25 Years*, Am. Math. Soc., Contemporary Mathematics Series, 19 (1984).

[Bl1]   Bledsoe, W. W. and Hines, L. M., "Variable Elimination and Chaining in a Resolution-Based Prover for Inequalities", *Proc. CADE-5*, Les Arcs, France (1980), Springer-Verlag.

[Bl2]   Bledsoe, W. W., "Using Examples to Generate Instantiations for Set Variables", *Proc. IJCAI-83*, Karlsruhe, Germany, (Aug. 1983), pp. 892-901.

[Bl3]   Bledsoe, W. W., "Non-resolution Theorem Proving", *Artificial Intelligence*, 1977, pp. 1-35. Also in *Readings in Artificial Intelligence* (Webber and Nilsson, Eds.), Tioga, Palo Alto, (1981), pp. 91-108.

[BM]   Boyer, R. S. and Moore, J. S., *A Computational Logic*, Academic Press, New York, (1979).

[Bo]   Weyer, S. and Borning, A., "A Prototype Electronic Encyclopedia", *ACM Transactions on Office Information Systems*, Vol. 3, No. 1, (1985), pp. 63-88.

[Ca]   Carbonell, J. G., "Learning by Analogy: Formulating and Generalizing Plans from Past Experience", *Machine Learning*, Michalski, Carbonell, Mitchell (eds.), Tioga Pub. (1983), pp. 137-161.

[Ch]   Chou, S. C., "Proving Geometry Theorems Using Wu's Method", Tech. Report 50, Institute for Computing Science, University of Texas, (July 1986).

[Da] Davis, M., "The Prehistory and Early History of Automated Deduction", *The Automation of Reasoning I & II*, (Siekmann and Wrightman, eds.), Springer-Verlag, (1983).

[DB] Fagin, R., Nievergelt, J., Pippenger, N. and Strong, H. R., "Extendible Hashing-A Fast Access Method for Dynamic Files", *ACM Transactions on Database Systems*, Vol. 4, No. 3, (1979), pp. 315-344.

Nievergelt, J., "Binary Search Trees and File Organizations", *Computing Surveys*, Vol. 6, No. 3, (1974), pp. 195-207.

[deB] deBruijn, N. G., "A Survey of The AUTOMATH Project", Department of Mathematics, Eindhoven University of Technology, The Netherlands, (1980).

[FC] Saaty, T. L. and Kainen, P. C., *The Four Color Problem: Assault and Conquest*, McGraw-Hill (1977), Dover (1986)

[Ge] Gelernter, H., "Realization of a Geometry Theorem-Proving Machine", in *Computers and Thought*, Feigenbaum and Feldman (eds.), McGraw-Hill, (1963).

[Go] Gordon, M., Milner, A., and Wadsworth, C., "Edinburgh LCF: A Mechanized Logic of Computation", *Lecture Notes in Computer Science*, Vol. 78, Springer-Verlag, New York, (1979).

[Gr] Greiner, R., "Learning by Understanding Analogies", Ph.D. Dissertation, Stanford University, Dept. of Computer Science, (Sept. 1985).

[Ho] Hobbs, J. R., *et al*, "Commonsense Summer: Final Report", Report No. CSLI-85-35, October 1985, Center for Study of Language and Information, Stanford University.

[Kl] Kling, R. E., "A Paradigm for Reasoning by Analogy", *Artificial Intelligence*, Vol. 2 (1971), pp. 147-178.

[Le1] Lenat, D. B., "AM: An Artificial Intelligence Approach to Discovery in Mathematics as Heuristic Search", SAIL AIM-286, Artificial Intelligence Laboratory, Stanford University, (July 1976).

[Le2] Lenat, D. B., "EURISKO: A Program That Learns New Heuristics
       and Domain Concepts, the Nature of Heuristics III: Program Design
       and Results", *Artificial Intelligence*, 21, (1,2) (1983), pp. 61-98.

[Le3] Lenat, D. B., Prakash, M., and Shepherd, M., "CYC: Using Common
       Sense Knowledge to Overcome Brittleness and Knowledge Acquisition
       Bottlenecks", *AI Magazine 6*, (1986), pp. 65-85.

[Lo]   Loveland, D. W., "Automated Theorem-Proving: A Quarter-Century
       Review", in *Automated Theorem Proving: After 25 Years*, W. W.
       Bledsoe and Donald Loveland (eds.}, Am. Math. Soc., Contemporary
       Mathematics Series, 19 (1984), pp. 1-45.

[Lu]   Lusk, E. and Overbeek, R., "Non-Horn Problems", *Jour. of Auto-
       mated Reasoning*, 1, (1985), pp. 104-105.

[Ma]   Martin, W. A. and Fateman, R. J., "The MACSYMA System", *ACM
       Proceedings of the Second Symposium on Symbolic and Algebraic Ma-
       nipulation*, Los Angeles, CA, (March 1971).

[Mc1] McCarthy, J., "Some Expert Systems Need Common Sense", *Annals
       of the New York Academy of Sciences* 426 (1983), pp. 129-137.

[Mc2] McCarthy, J. and Hayes, P. J., "Some Philosphical Problems from the
       Standpoint of Artificial Intelligence", *Machine Intelligence 4*, (1969)
       (B. Meltzer and D. Michie, eds.) American Elsevier, New York, pp.
       463-502.

[ML1] Michalski, R. S., Carbonell, J. G., and Mitchell, T. M. (eds.), *Machine
       Learning: An Artificial Intelligence Approach*, Tiago, Palo Alto, CA
       (1983).

[ML2] Michalski, R. S., Carbonell, J. G., and Mitchell, T. M. (eds), *Ma-
       chine Learning: An Artificial Intelligence Approach, Vol. II*, Morgan
       Kaufman, Los Altos, CA (1986).

[Mo]   Moses, J., "Algebraic Simplification – A Guide for the Perplexed",
       *ACM Proceedings of the Second Symposium on Symbolic and Algebraic
       Manipulation*, Los Angeles, CA, (March 1971).

[Ne] Newell, A., Shaw, J. C., and Simon, H. A., "Empirical Explorations of the Logic Theory Machine: A Case Study in Heuristics", *Computers and Thought*, Feigenbaum and Feldman (eds.), McGraw-Hill, (1963).

[Op] Nelson, G. and Oppen, D., "A Simplifier Based on Efficient Decision Algorithms", *Proc. 5th ACM Symposium on Prin. Prog. Lang*, (1978).

[Ov] Overbeek, Ross, "An Implementation of Hyper-resolution", *Computer and Mathematics with Applications 1* (1975), pp. 201-214.

[Pla] Plaisted, D.A. "Abstraction Mappings in Mechanical Theorem Proving", *Proc. CADE-5*, Les Arcs, France (1980), Springer-Verlag.

[Pl1] Plummer, D. and Bundy, A., "Gazing: Identifying Potentially Useful Inferences", Working Paper 160, Department of Artificial Intelligence, University of Edinburgh, (February 1984).

[Pl2] Plummer, D., "Gazing: A Technique for Controlling the Use of Rewrite Rules in a Natural Deduction Theorem", Department of Artificial Intelligence, University of Edinburgh, Ph.D. Thesis, to be defended Summer 1986.

[Ro] Robinson, J. A., "A Machine-Oriented Logic Based on the Resolution Principle", *Jour. Assoc. for Comput. Mach.*, (1965), pp. 23-41.

[Sh] Shostak, R., "On the SUP-INF Method for Proving Presburger Formulas, *JACM 24*, (1977), pp. 529-543.

[Ty] Tyson, Mabry, "APRVR: A Priority-Ordered Agenda Theorem Prover", *Proc. AAAI National Conference*, Pittsburgh, PA, (Aug. 1982), pp. 225-228.

[Wa] Wang, T.-C., "Designing Examples for Semantically Guided Hierarchical Deduction", Tech. Report ATP-80, ATP, University of Texas at Austin. *Proc. IJCAI 9* (1985), pp. 1201-1207.

[Wey] Weyhrauch, R. W., "Prolegomena to a Theory of Mechanical Formal Reasoning", *Artificial Intelligence 13*, (1980), pp. 133-171.

[Wi]   Winston, P. H., "Learning and Reasoning by Analogy", *CACM*, Vol. 23, No. 12, (1979), pp. 689-703.

[Wo1] Wos. L., *et al*, "An Overview of Automated Reasoning and Related Fields", *Jour. of Automated Reasoning 1*, (1985), pp. 5-48.

[Wo2] Wos, L., Overbeek, R., Lusk, E., Boyle, J., *Automated Reasoning: Introduction and Applications*, Prentice-Hall, Englewood Cliffs, NJ (1984).

[WT]   Weyhrauch, R. W., "FOL: A Proof Checker for First-order-logic", Stanford Ai Lab Memo AIM-235, (September 1974).

Computer Algebra in the Service of Mathematical Physics and Number Theory

DAVID V. CHUDNOVSKY AND GREGORY V. CHUDNOVSKY   Columbia University,
New York, New York

1   Introduction.

Methods of computer algebra become more familiar to a wide audience of
theoretical mathematicians and physicists. The environment of computer
algebra system leads to a greater acceptance of computer instruments in
the mathematical research. Methods of symbolic manipulation provided by
computer algebra systems in combination with high-power number crunch-
ing abilities of traditional hardware and software open the way to truly large
scale computations often needed by mathematicians and physicists.

These possibilities of better servicing the everyday needs of the researcher
emphasize as a priority the development of new algorithms and methods of
efficient programming solution of basic computational tasks of theoretical
and applied mathematics. Considerable progress in this area is associated
with classes of problems, when explicit complexity description exists, like
the study of algebraic complexities of the basic algebraic tasks: polynomial
operations, matrix multiplication, solution of problems of linear algebra and
linear programming. In some basic number-theoretic problems important
progress in the construction of new algorithms is associated with the use
of algebraic geometry (primality testing and factorization). Many of new
algorithms and methods of number-theoretic computations and computer
algebra programming are discussed in detail in the talks presented at this
conference. We decided to focus on a few important computational problems, whose implementation often requires considerable number-crunching
effort. At the cornerstone of these problems lies the task of evaluation and
tabulation of values of transcendental and algebraic functions—a task of
classical numerical analysis. Our specific interest in these problems arose
from our work with computer algebra systems, whose environment gives
a use a wonderful capability of multiple precision computations. This capability of computing with full precision (for big integers), and with very
high precision to satisfy the curiosity of a diophantine geometer, led to the
responsibility of efficient programming. We describe new low complexity
(both operational and logical) methods and algorithms for computations
of solutions of differential equations and their efficient evaluation, and of
solution of algebraic equations. These computations are often basic in applied problems, but even more so in a variety of problems of number theory
and algebraic geometry (topology). This includes computations (in positive

its multidimensional generalization to systems of equations, is widely recognized as one of the most cumbersome issues in computational mathematics (cf. [B1]). We touch upon the complexity issues for these and other closely related problems. The results we present in this lecture together with applications of our computational methods to specific mathematical problems can be used in a variety of computer algebra systems and were themselves born within the SCRATCHPAD (IBM) programming and computational environment. We use this opportunity to express our fascination with powers of this well-developed and fully-grown beautiful system, and to express our thanks to the Computer Algebra Group at T.J. Watson Research Center of IBM for the possibility of working in the SCRATCHPAD environment. Particular thanks go to R. Jenks, B. Trager and R. Sutor.

The paper that follows our talk and expands on it, is organized as follows. In §2 we describe the methods of computation of solutions of (linear) differential equations via the method of regular power series expansions. Various methods of computation of these expansions and related difference equations with polynomial coefficients are presented. They significantly improve on our previous algorithms [C4]. In §3 the power series expansions are used for fast evaluation of solutions of linear differential equations with an arbitrary precision n of leading digits. All solutions with given initial conditions can be analytically continued along a path of the length L in at most $O(M(n) \cdot (\log^3 n + \log L))$ of boolean operations, where $M(n)$ is the boolean complexity of M-bit multiplication. If a solution (or an equation) possesses additional arithmetic properties, $\log^3 n$ can be further decreased to $\log^2 n$ or $\log n$. These low complexity computational methods are particularly efficient, in practice as well as in theory, in computation of invariants of linear differential equations, such as invariants of the monodromy group. Methods of continued fraction expansions and the parallelization methods for the presented algorithms are described in §§4-5. In §§6-7 we describe applications of fast computation of the monodromy groups to the classical uniformization problem for the Riemann surfaces á la Klein-Poincaré [K2], [P2]. Our interest here lies in the algebraicity of the invariants of this uniformization (Fricke, accessory or other parameters on the corresponding Teichmüller spaces), whenever the Riemann surface itself is defined over **Q** or $\overline{\mathbf{Q}}$. Our computations in this area highlight a special role played by the Riemann surfaces uniformized by arithmetic subgroups (and of blowups of hypergeometric equations). It seems that in all other cases there is an absence of nontrivial algebraic relations for accessory, Fricke and other parameter of uniformization of the Riemann surfaces defined over **Q**. In particular, a simple Whittaker conjecture [W2] is generically incorrect. In §8 we describe applications of the study of nearly integral power series and formal groups over **Z** and **Z**$_p$. These applications include new results in algebraic geometry, new complexity bounds and explicit expression for the universal characteristic class associated with spin manifolds with S^1 action. §§9-10 are devoted to the polynomial root finding methods. After the review in §9, we describe in §10 new algorithms of root finding for sparse and general polynomials of one variable. Our high-precision methods are particularly efficient in the vector and parallel environments. Some of these algorithms were implemented on the CRAY-II (Minnesota Center) and IBM 3090VF with degrees up to 15,000.

2 Computations of power series expansions.

Power series manipulations over various rings and fields are one of the most important features of advanced computer algebra systems. Power series expansions are crucial in the study of solutions of nonlinear (algebraic), differential or integral equations, where no close form solution can be found. Power series expansions are important in various mathematical and physical problems. To name but a few: 1) power series expansions serve as a basis for construction of rational (Padé) approximations; 2) power series expansions give an efficient way of computation of algebraic functions (numbers); 3) analytic continuation method based on power series expansions provides an efficient way of computation of Galois and monodromy groups of linear differential equations and algebraic function fields; 4) power series manipulations are crucial in the study of formal groups and various identities of hypergeometric functions and modular forms; 5) while traditional methods of computer algebra allow us to study indefinite integrals of elementary functions, the constants of classical analysis, represented by definite integrals, can be expressed by and very well approximated by power series expansions—among these constants are elements of the monodromy groups, whose arithmetic nature one wants to study.

The crucial problem in the implementation of power series packages is the development of fast algorithms that can handle thousands of terms in power series expansions. In applications the most important classes of functions that are to be expanded are classes of algebraic functions and solutions of linear differential equations with polynomial (rational function or algebraic) coefficients. As we see in §3 the same algorithms provide with

efficient methods of evaluation and of analytic continuation of solutions of
these equations. Our algorithms are all based on the analysis of (linear) recurrences on coefficients of power series expansions. But before we present
these algorithms we have to define the measures of complexity. Since we are
dealing with numbers of variable sizes and precision we have to differentiate
between the operational and the total (boolean) complexity. By operational
complexity of an algorithm one understands the number of primitive operations (most notably additions and multiplications), independent of the sizes
of numbers involved, needed to complete this algorithm. By the total (or
boolean) complexity we understand the *total* number of primitive operations
(on short or single-bit data) needed to complete a given program. The main
distinction between the conversion from the operational to the total complexity, depending on the size of numbers involved, is described by the total
complexity of multiplication of big numbers.

Let us denote hereinafter by $M(n)$ the total complexity of multiplication
of two n-bit integers. Then the best known upper bound on $M(n)$ belongs
to Schonhage-Strassen [SS2]:

$$M(n) = O(n \log n \log \log n).$$

In comparison, a total complexity of addition is relatively simple: it is only
$0(n)$ (in the scalar case on numbers of n bits). The multiplication of bigfloats
is, obviously, as bit complex as that of bignums.

All algebraic operations on bigfloats have bit complexity of the same order of magnitude as a multiplication. For example, if $B(n)$ denotes one of the
following total complexities: division of n-bit bigfloat numbers, square root
extraction, or raising to the fixed (rational) power, then $B(n) = 0(M(n))$,
and $M(n) = O(B(n))$.

For references to early results on complexities of power series manipulation and particularly the computations of power series expansions see [K2]
and references in [C4]. Using recurrences on coefficients of power series expansions, the following very simple operational complexity algorithm was
first established in [C4]:

Algorithm 2.1 [C4]. The N-th term in the power series (Puiseux series)
expansions can be computed in $O(N)$ operations with $O(1)$ storage. The
N terms of the expansion can be computed in $O(N)$ operations with $O(N)$
storage.

The Algorithm 2.1 exists for arbitrary algebraic functions and regular
solutions of linear differential equations with polynomial coefficients. Similar

(operational) complexities exist for asymptotic series expansions (normal and subnormal) at irregular singularities of linear differential equations.

Remark. Since we are using only standard operations in Algorithm 2.1 it can be applied to equations over any field. However, over finite fields most linear differential equations do not have regular power series solutions. Those linear differential equations that have sufficiently many linearly independent regular power series solutions over almost all finite fields were conjectured by Grothendieck to have only algebraic function solutions. We proved this conjecture for large classes of linear differential equations, including Lamé equations, equations parametrized by arithmetic groups and others, see [C7].

We derive and prove new algorithms that are the enhancements of 2.1, and simultaneously present fast methods of solution of linear recurrences. In applications to power series expansions, it is enough to look only at linear differential equations, since algebraic functions always satisfy Fuchsian linear differential equations, see proofs in [C4]. (These differential equations were called by Cayley the differential resolvents of algebraic equations.)

Thus, one is looking at regular (Puiseux or power series) expansions of solutions of linear differential equations, either in the matrix form:

$$\frac{dY(x)}{dx} = A(x) \cdot Y(x) \tag{1}$$

for $A(x) \in M_n(\mathbf{C}(\mathbf{x}))$, or in the scalar form:

$$L[\frac{d}{dx}, x] \cdot y = 0 \tag{2}$$

for $L[\frac{d}{dx}, x] + \sum_{i=0}^{n} a_i(x) \frac{d^i}{dx^i}$ with polynomials $a_i(x)$. The typical regular expansion of a solution of (1) or (2) at a regular point $x = x_0$ is represented as

$$Y(x) = \{\sum_{N=0}^{\infty} C_N(x - x_0)^N\} \cdot (x - x_0)^{W_0} \tag{3}$$

for (1) or as

$$y(x) = (x - x_0)^{\nu_0} \cdot \sum_{N=0}^{\infty} c_N(x - x_0)^N \tag{4}$$

for (2). To determine the power series expansion of an arbitrary solution of (1) or (2) at a regular point $x = x_0$, it is enough to determine the expansion (3) or (4) because an arbitrary solution can be represented in terms of the basis of such solutions. (In case of algebraic functions, there is also a problem of initial conditions, discussed in detail in [C4].)

The natural method to determine the coefficients C_N and c_N is to substitute (3) or (4) into (1) or (2), to derive a system of linear equations determining recursively consecutive coefficients. This method was developed by Frobenius for scalar linear differential equations.

For (2), the basis of solutions at a regular point $x = x_0$ can be expressed in terms of regular expansions

$$y(x, \alpha) = (x - x_0)^\alpha \sum_{N=0}^{\infty} y_N(\alpha) \cdot (x - x_0)^N. \tag{5}$$

The coefficients $y_N(\alpha)$ are determined for $n \geq 0$ from the initial conditions at $x = x_0$ and the linear recurrence

$$\sum_{j=0}^{min(N,d)} y_{N-j}(\alpha) \cdot f_j(\alpha + N - j) = 0 \tag{6}$$

$N = 1, 2, 3, \ldots$, with the explicit expression of coefficients $f_j(\beta)$ in terms of a_i in (2) as follows: if $a_j = Q_j(x) \cdot (x - x_0)^j$, $(j = 0, \ldots, n)$, then for $f(x, \alpha) \stackrel{def}{=} \alpha(\alpha - 1)\ldots(\alpha - n + 1) \cdot Q_n(x) + \ldots + \alpha Q_1(x) + Q_0(x)$, we put $f(x, \alpha) = f_0(\alpha) + (x - x_0)f_1(\alpha) + \ldots + (x - x_0)^d f_d(\alpha)$, where d is the bound of degrees for all polynomials $Q_j(x)$. The exponent α in (5) satisfies the indicial equation $f_0(\alpha) = 0$, [C4].

Similarly, we look at regular solutions (3) $Y(x)$ of (1):

$$Y(x) = \{ \sum_{N=0}^{\infty} C_N(x - x_0)^N \} \cdot (x - x_0)^W,$$

where C_0 ($\in M_{n \times n}(\mathbf{C})$) is the initial condition for $Y(x)$ at $x = x_0$, and the (matrix) coefficients C_N of (3) are determined from the matrix linear recurrence of length d (the maximal degree of the rational function in $A(x)$ in (1)). To derive this recurrence, let us consider a case of a regular point $x = x_0$ when $W = 0$. For $A(x)$ from (1) let us put $A(x) = A_0(x)/d(x)$, where $A_0(x) \in M_{n \times n}(\mathbf{C}[x]), d(x) \in \mathbf{C}[x]$. We put $d(x) = \sum_{j=0}^{d} d_j(x - x_0)^j$, where $d_0 \neq 0$, and $A_0(x) = \sum_{j=0}^{d-1} A_j(x - x_0)^j$. Then we have the following recurrence on C_N:

$$C_{N+1} \cdot (N+1) \cdot d_0 = \sum_{i=0}^{min\{N,d-1\}} A_i \cdot C_{N-i} - \sum_{i=0}^{min\{N,d-1\}} d_{i+1}(N-i) \cdot C_{N-i}. \tag{7}$$

Here $d_j = d^{(j)}(x_0)/j!$, $A_i = A^{(i)}(x_0)/i!$.

Thus, the coefficients C_N and c_N in the power series expansions (3)-(4) are always determined by the linear recurrence (7) or (6) of finite length (called the rank of the recurrence) with coefficients that are rational functions of N.From now on we represent these linear recurrences both in the scalar case (6) and in the matrix case (7) as a single matrix first order recurrence (difference equation), by increasing, if necessary, the sizes of the matrices:

$$\mathcal{A}_{N+1} = C(N) \cdot \mathcal{A}_N, \qquad (8)$$

where $\mathcal{A}_N$ is built from C_{N-i} or c_{N-i} and the matrix $C(N)$ is rationally dependent on the index N. The coefficient matrix $C(N)$ can be recognized as a transfer matrix from statistical mechanics or completely integrable systems.

To generate the recurrence (8) from equations (1) or (2) one should use computer algebra systems. E.g. in SCRATCHPAD one can generate the recurrence (6) or (8) starting from an algebraic (polynomial) equation defining an algebraic function $y(x)$.

The matrix recurrence formula (8) justify the Algorithm 2.1. Indeed, to recover C_N or c_N one needs $O(N)$ operations, namely $O(N)$ multiplications and divisions by explicit polynomials of i for $i = 1, \ldots, n$; $O(N)$ nonscalar additions and only $O(1)$ storage. The constant under $O(\cdot)$ depends only on rank and on the order of a linear differential equation (1) or (2).

We present now more efficient algorithms that are based on the fast solution of the matrix linear recurrence (8), that significantly lower both total and operational complexities of computations. Let us look now at the matrix recurrence (8) separately.

$$A_{N+1} = C(N) \cdot A_N \qquad (9)$$

where $C(n)$ is $n \times n$ matrix rationally dependent on N. The solution A_N of (8) with initial conditions $A_{N|N=0} = A_0$ has a symbolic representation

$$A_N = \overleftarrow{\prod_{i=0}^{N-1}} C(i) \cdot A_0, \text{ or}$$

$$ \qquad (10)$$

$$A_N = C(N-1) \cdot C(N-2) \cdot \ldots \cdot C(1) \cdot C(0) \cdot A_0,$$

(the order of the terms in the product is reversed). The fast method of computation of A_N in (8) is the binary-splitting method. This is a well-known method to accelerate the (numerical) solution of linear recurrences. In this method, a binary tree (whenever $N = 2^k$ is a power of 2, or otherwise a more complicated trees associated with addition chain are used) of operations is constructed so that multiplication of terms in (9) proceeds in the way that operands have slow growing sizes. Thus this method is opposite to the obvious method of computation of (9) with consecutive multiplications by $C(i)$, as used in Algorithm 2.1.

With notations of (9) we introduce the following auxiliary variables:

$$\mathcal{A}_{L;K} = \widehat{\prod}_{j=K}^{L-1} C(j) \tag{11}$$

for $L > K$. We put $\mathcal{A}_{K;K} = I_n$.

In these notations A_N is

$$A_N = \mathcal{A}_{N;0} \cdot A_0. \tag{12}$$

There is a chain rule of computations of $\mathcal{A}_{L;K}$ which is the basis of any splitting method including the binary splitting method:

$$\mathcal{A}_{L;K} = \mathcal{A}_{L;M} \cdot \mathcal{A}_{M;K} \ for \ L \geq M \geq K. \tag{13}$$

The chain rule (12) provides a scheme of computations of $\mathcal{A}_{L;K}$ using the binary expansion of indices L and K. This binary method is at its best when N is a power of 2, $N = 2^k$ (otherwise a different addition chain is used).

The algorithm consists of the outer loop over all l from 0 to k and the inner loop over all $(k-l)$-bit (binary) integers. One starts at $l = 0$ with the initialization:

$$\mathcal{A}_{K+1;K} = C(K) \ for \ all \ K \ in \ K = 0, \dots, 2^k = 1. \tag{14}$$

At the l-th step one has determined all $\mathcal{A}_{2^l(K+1);2^l \cdot K}$ for all $(k-l)$-bit integers $K : 0 \leq K \leq 2^{l-k} - 1$. At the step $l + 1$ one uses the rule (12), and obtains:

$$\mathcal{A}_{2^{l+1}(K+1);2^{l+1}K} = \mathcal{A}_{2^l(2K+2);2^l(2K+1)} \cdot \mathcal{A}_{2^l(2K+1);2^{l+1} \cdot K} \tag{15}$$

for $0 \leq K \leq 2^{k-l-1}-1$, i.e. $\mathcal{A}_{2^{l+1}(K+1);2^{l+1} \cdot K}$ are determined for all $(k-l-1)$-bit integers.

Finally, at step k at $l = k$, according to (12), A_N is determined from $\mathcal{A}_{2^k \cdot (K+1); 2^k \cdot K} = \mathcal{A}_{2^k; 0}$ at $K = 0$.

For $l = 1, \ldots, k$, at $l - th$ step of this algorithm one performs in (14) 2^{k-l} matrix multiplications. Binary-splitting method is efficient if the computations are conducted with increased precision, i.e. all bits of information in computations of A_N are preserved. When $C(N)$ is rational in N, the total memory space needed to hold all bits of A_N (as a rational number, i.e. a numerator and a denominator) is $O(n^2 \cdot N \cdot \log_2 N)$. Only in special cases, when the recurrence (8) represents a recurrence associated with a globally nilpotent Fuchsian linear differential equation [C7], the memory requirements are only $O(N)$. It is important to notice that the binary-splitting method of computation of A_N in (13)-(14) requires about the same amount of memory. Indeed, to compute the $l - th$ step of the algorithm only the previous step is needed, with a total amount of memory space of $O(N \log N)$.

The total amount of operations depends on the cost of the multiplication of l-bit numbers–denoted, as above, by $M(l)$. Under the assumptions above, that $C(N)$ is a rational function of N, let

$$C(N) = \frac{C_0(N)}{d(N)} \qquad (16)$$

where $C_0(N) \in M_{n \times n}(\mathbf{Z}[N]), d(N) \in \mathbf{Z}[N]$. let in (15) d be the maximal degree of polynomials in $C_0(N)$ and $d(N)$. With (15) substituted in (9) the solution A_N of (8) can be represented as

$$A_N = \frac{\overleftarrow{\prod}_{i=0}^{N-1} C_0(i)}{\overleftarrow{\prod}_{i=0}^{N-1} d(i)} \, A_0. \qquad (17)$$

We compute $\overleftarrow{\prod}_{i=0}^{N-1} C_0(i)$ using the described above binary-splitting algorithm (13)-(14) for $C_0(N)$ instead of $C(N)$. Let h be the maximum of sizes of coefficients of polynomials in $C_0(N)$ and $d(N)$. Then, as it follows from the iterative scheme (14), all integers $\mathcal{A}_{2^l \cdot (K+1); 2^l \cdot K}$ for $0 \leq K \leq 2^{k-l} - 1$ have sizes bounded by $O((d + h) \cdot 2^l \cdot \log N)$ (for $l \leq k$). Consequently, at every step from $l = 1$ to k the total number of (bit) operations is bounded by $2^{l-k} \cdot M(2^l \log_2 N) \cdot (d + h)$. The total number of bit operations (with $\log_2 N = k$) is $O(n^3 \cdot (d + h) \cdot M(N) \cdot \log^2 N)$. This method gives the numerator in (16).The denominator is computed the same way with the total

number of operations again bounded by $O((d + h) \cdot M(N) \cdot \log^2 N)$.When
necessary, the denominator can be computed faster using the distribution of
prime ideals in the Galois group of polynomial $d(N)$. The (bit) complexity
becomes $O(M(N) \cdot \log N)$.

When $N = 2^k$ we use binary-splitting method as presented above; if N is
arbitrary the corresponding addition chain tree (used for fast computation
of N in $O(\log N)$ additions only) is applied. The consequence of the fast
method (13)-(14) is the following:

Theorem 2.2. Let in the recurrence (8), $C(\cdot)$ be $n \times n$ matrix whose
rational function entries have sizes and degrees bounded by s. Then the
total (bit) complexity of computations of A_N in (9) is bounded by

$$O(M(N) \cdot s \cdot \log^2 N),$$

where $O(\cdot)$ depends on n as $O(\cdot) = O(\mu(n))$, where $\mu(n)$ is the operational
complexity of $n \times n$ matrix multiplication.

Only the term $\log^2 N$ is not the best possible. Apparently, under additional strong arithmetic assumptions it can be improved. This is the case
when all A_N are "nearly integral," i.e. the generating function $Y(x) = \sum_{N=0}^{\infty} A_N x^N$ is a G-function, [C7]. In this case, and also in the case when
$Y(x)$ is a E-function, the $\log^2 N$ term in Theorem 2.2 can be replaced by
$\log N \cdot (\log \log N)^{1+\epsilon}$.

The best results on the upper bounds of the operational complexities of
computations of solutions of the matrix recurrence (8) are not as good as in
Theorem 2.2

Since in operational complexity the count of the sizes of operands are
irrelevant binary-splitting methods or a trivial method of computation of
(9) give the same bound: $O(N \cdot (n^2 \cdot d + n^\mu))$, where $\mu < 2.5$ is the exponent in
the matrix multiplication problem. Thus the complexity count in Algorithm
2.1 is not improved with the binary splitting method.

The linear dependence on N is unsatisfactory. To speed up the computation of (9) one uses a linear acceleration instead of geometric one.
We start with introduction of some auxiliary cost functions. The operational complexity (cost) of multiplication of polynomials of degree d is
$M_p(d) = O(d \log d)$; the cost of evaluation of a polynomial $p(x)$ of degree d
at d consecutive points: $0, 1, \ldots, d - 1$ is $M_e(d) = O(d \log^2 d)$, with $(d \log d)$
essential multiplications, see [K2]. Similarly $M_s(d)$ is the cost of a shift of a
polynomial $p(x)$ to $p(x + h)$ of degree d.

Let us look at new recurrences following from (8):

$$A_{N+k} = C_k(N) \cdot A_N, \tag{18}$$

where for any $k \geq 1$

$$C_k(N) = C(N) \cdots C(N + k - 1). \tag{19}$$

Let us consider, as above, $N = 2^a$ (again similar arguments apply to an arbitrary N):

$$A_{N+2^c} = C_{2^c}(N) \cdot A_N,$$

$$\tag{20}$$

$$C_{2^c}(N) = C_{2^{c-1}}(N + 2^{c-1}) \cdot C_{2^{c-1}}(N).$$

To compute $C_{2k}(N) = C_k(N + k) \cdot C_k(N)$ from $C_k(N)$, as a matrix with entries rational in N, one needs: a) to compute the coefficients of polynomial/rational expansion of $C_k(N + k)$ in (powers of) N—it takes $O(n\, M_s(d_k))$ operations; b) to multiply the rational/polynomial entries of $C_k(N+k)$ and $C_k(N)$–this takes $O(\mu(n) \cdot M_p(d_k))$. Here d_k is the maximum of the degrees of elements of $C_k(N)$.

The scheme (17) for any given k is a version of a general scheme (8), but with such changes: (i) with a different matrix C–the size of it is the same, but the degrees of its polynomial/rational function entries are different; (ii) the length of the recurrence is k times shorter. The maximum d_k of degrees of rational/polynomial entries of $C_k(N)$ is $\leq kd$; say, $d_k = kd$.

We estimate now the total cost of the obvious algorithm of computation of (9) in the case of recurrence (17) (or a binary-splitting version of this algorithm). To evaluate $A_N = A_{t \cdot k}$ in t steps, starting from A_0 following the rule (17) one needs: α) consecutive evaluation of $C_k(0), \ldots, C_k((t-1) \cdot k)$ at $t = N/k$ points. Once the rational function $C_k(\cdot)$ is known, the total operational complexity of α) is $O(n^2 \cdot M_e(d_k) \cdot [\frac{N}{kd_k} + 1])$. Secondly, one needs β) consecutive matrix multiplications $C_k(j \cdot k) \cdot A_{j \cdot k}$ for $j = 0, \ldots, t-1$ with the total cost of $O(\mu(n) \cdot t)$.

The cost of determination of $C_k(\cdot)$ for $k = 2^c$, as follows from the discussion above, is $\sum_{b=0}^{c-1} \{O(\mu(n) \cdot M_p(2^b \cdot d)) + O(n^2 \cdot M_s(2^b \cdot d))\}$, i.e. the cost of determination of $C_k(\cdot)$ is $O(\mu(n) \cdot M_p(kd) + n^2 M_s(kd))$. This result is true not only for $k = 2^c$, but for any k.

Thus, whenever $N \geq k^2 d$ (so that $[N/kd_k] \geq 1$), the total cost of computation of A_N using the scheme (17) is:

$$O(\mu(n) \cdot \{M_p(kd) + \frac{N}{k}\} + n^2\{M_s(kd) + M_e(kd) \cdot \frac{N}{k^2 d}\}).$$

Substituting $M_p(x) = O(x \log x)$, $M_s(x) = O(x \log x)$, $M_e(x) = O(x \log^2 x)$, one derives the following upper bound on the total cost of computations of A_N using the scheme (17) putting $k = [\sqrt{N/d}] + o(\sqrt{N})$:

$$O(\mu(n)\sqrt{Nd} \log N + n^2 \sqrt{Nd} \log^2 N)$$

for $N/d \gg 1$. This bound can be slightly improved, also the cost of a single multiplication has to be multiplied by the corresponding weight (if long numbers are involved). We arrive at the following result:

<u>Theorem 2.3.</u> Let in the recurrence

$$A_{N+1} = C(N) \cdot A_N,$$

$C(\cdot)$ be $n \times n$ matrix with rational coefficients with degrees bounded by d. Then the operational complexity (total number of operations) is bounded by

$$O(\mu(n) \cdot \sqrt{Nd} \log N + n^2 \sqrt{Nd} \log^2 N),$$

with at most $O(\mu(n) \cdot \sqrt{Nd} \log N)$ multiplications.

The operational complexity algorithm of Theorem 2.3 is significantly lower than that of 2.1. The precise running time, of course, has to take into account the precision or the sizes of numbers involved. The results of Theorem 2.3 open the way to compute the coefficients of power series expansions particularly fast, when computations are carried out on bigfloats with fixed precision. In this case we can use asymptotic expansions of solutions of (8) in inverse factorials of N near ∞. Using these Borel and Birkhoff's expansions of A_N near ∞ we get the following bound:

<u>Theorem 2.4.</u> The matrix solution A_N of (8) can be computed within the precision n of leading digits in

$$O(\log^3(nN) \cdot M(n))$$

of total (bit) operations.

To prove Proposition 2.4 one uses $1/N$ expansions of solutions of difference equations, and asymptotic (inverse factorial) series representation of A_N for large N.

Unfortunately, it is not clear, at what N the fast algorithms of Theorems 2.2-2.4 become advantageous over a simple algorithm 2.1. Theoretically, however, the situation is different. E.g. to compute the $N - th$ coefficient of the regular series expansion of a linear differential equation in multipleprecision floating point arithmetic with precision n one needs only

$$O(n) \text{ operations.}$$

E.g. the number of operations needed to compute the $N - th$ coefficient in a fixed precision is independent of N. This phenomenon is very similar to the famous Shamir observation [K2] that on a machine that can perform arithmetic on integers of arbitrary length in one unit of time, the factorial $N!$ can be computed only in $O(\log N)$ steps. On such, "Shamir's computer", the $N - th$ coefficient can be computed completely in $O(\log^2 N)$ steps.

The problem of the minimal complexity of modular computations of recurrence (8) is much more complicated. Any improvement over Theorem 2.3 in computations $\text{mod} M$ for a composite M would give new efficient factorization methods. Indeed, to factor M one has only to compute in modular arithmetic $N!(\text{mod} M)$ for $N = [\sqrt{M}]$, i.e. a solution of a simple recurrence:

$$A_{N+1} = N \cdot A_N \text{ mod } M,$$

and then check for g.c.d of $N!(mod\, M)$ and M.

Then the cost of a single operation mod M is bounded by $O(M(\log M)) = O(\log M \cdot \log \log M \cdot \log \log \log M)$. Thus, Theorem 2.3 implies that we have a method of factorizing M in $O(\sqrt[4]{M} \cdot \log^2 M)$ *bit* operations. This is asymptotically equivalent to two popular factorization methods: Pollard's ρ-method and Shank's quadratic form composition method. Our method, though, is completely deterministic.

One of the most difficult questions in modular computations is the complexity of computations $\text{mod} p$ for a large prime p of coefficients in the expansion of an algebraic function.

First of all, not quite obvious result following from the asymptotic expansions of the difference equation, shows that for a fixed p the computation of the $N - th$ coefficient $c_N(\text{mod} p)$ in the expansion of an arbitrary algebraic function requires only $O(p \cdot \log_p N)$ operations $mod\, p$.

[The same result is true for an arbitrary globally nilpotent equation, i.e. for an equation with a proper lifting of Frobenius.]

We conjecture that one needs much less computations. A reasonable bound for all p and N would be

$$O(\log N \cdot e^{\sqrt{\log p \log \log p}}),$$

and we can prove it for some classes of radical algebraic functions, when $p \pm 1$ is divisible by a high power of 2.

Computations (mod p) of coefficients of power series expansions of algebraic-functions are closely connected with primality testing and factorization see [C8]. One of the most known examples is the problem of computation of the value $P_{(p-1)/2}(\lambda)$ of the Legendra polynomial (mod p). This number determines the order of an elliptic curve $y^2 = x(x-1)(x-\lambda) \bmod p$, and at the same time it is the coefficient at $x^{(p-1)/2}$ of the expansion of the function $(1 - 2\lambda x + x^2)^{-1/2}$ at $x = 0$.

We conclude this chapter with an example of computation of coefficients c_N of power series expansions as in Theorem 2.4 by means of analytic continuation in N. This particular example concerns Legendre polynomials.

In general Nörlund-Birkhoff theorem [B4] of solutions of linear difference equations (8) with polynomial coefficients, one has the expansions of solutions similar to regular expansions and normal/subnormal expansions of solutions of linear differential equations. In this approach one interprets N as a (complex) variable and expands A_N (or c_N in (4)) in factorial (Newton-like) series in N. These expansions have one of the following forms:

$$c_N = \frac{\mu^N \Gamma(x+k)}{\Gamma(x)\Gamma(k)} \cdot \left\{ a_0 + \frac{a_1}{x+k-1} + \frac{a_2}{(x+k-1)(x+k-2)} + \cdots \right.$$

$$\left. \cdots + \frac{a_n}{(x+k-1)\cdots(x+k-n)} + \cdots \right\},$$

$$x = N - r$$

(an expansion at $N = \infty$); and

$$c_N = \frac{\mu^N \Gamma(x+k)}{\Gamma(x)\Gamma(k)} \{ b_0 + b_1(x+k) + b_2(x+k)(x+k+1) + \cdots$$

$$\cdots + b_n(x+k)\cdots(x+k+n-1) + \cdots \},$$

$$x = N - r.$$

These expansions allow us to compute A_N or the coefficients c_N in the power series expansions (and, in fact, within the same method one can compute the value of the truncated power series expansion $\sum_{n=0}^{N} c_n(x-x_0)^N$ at x close to x_0) by means of analytic continuation *in* N. This approach is particularly effective in floating point computations of Theorem 2.4 when one can compute a single A_N (not even for an integral N necessarily) at once within a given precision of computation. Computing A_N or c_N from a factorial expansion allows to increase the precision of computations (by adding more terms to the expansion) without recomputing previous results.

As an example, we look again at computation of Legendre polynomials $P_N(\lambda)$, that are defined as above from the expansion at $x = 0$ of the algebraic function $\alpha(x) = (1 - 2\lambda x + x^2)^{-\frac{1}{2}} = \sum_{N=0}^{\infty} P_N(\lambda)x^N$. Looking on the linear differential equation satisfied by $\alpha(x)$ one deduces a familiar three-term linear recurrence on $P_N(\lambda)$:

$$NP_N(\lambda) - (2N-1)\lambda P_{N-1}(\lambda) + (N-1)P_{N-2}(\lambda) = 0.$$

Expanding $P_N(\lambda)$ near $N = \infty$ in inverse factorial series we get the following asymptotic series expansions: Let $\theta = (\sqrt{\lambda^2 - 1} - \lambda)/(2\sqrt{\lambda^2 - 1})$; then

$$P_N(\lambda) = \{(\lambda - \sqrt{\lambda^2 - 1})^N \cdot \theta^{\frac{1}{2}} \cdot \frac{\Gamma(-\frac{1}{2} - N)}{\Gamma(-N)\Gamma(\frac{1}{2})} \times$$

$$\times \sum_{n=0}^{\infty} \frac{(\frac{1}{2})_n^2}{n!} \cdot \frac{\theta^n}{(\frac{3}{2} + N)(\frac{3}{2} + N + 1)\cdots(\frac{3}{2} + N + n - 1)}\} +$$

$$+ \{(\lambda - \sqrt{\lambda^2 - 1})^{-N} \cdot \theta^{-\frac{1}{2}} \cdot \frac{\Gamma(\frac{1}{2} + N)}{\Gamma(1 + N)\Gamma(\frac{1}{2})} \times$$

$$\times \sum_{n=0}^{\infty} \frac{(\frac{1}{2})_n^2}{n!} \cdot \frac{\theta^n}{(\frac{1}{2} - N)(\frac{1}{2} - N + 1)\cdots(\frac{1}{2} - N + n - 1)}\},$$

where $(\frac{1}{2})_n = \frac{1}{2} \cdot (\frac{1}{2} + 1)\cdots(\frac{1}{2} + n - 1)$, and N is an arbitrary complex number such that $N + \frac{1}{2}$ is not an integer.

3   Evaluation and Analytic Continuation of Solutions of Linear Differential Equations.

The methods of §2 based on fast computation of solutions of matrix linear recurrences (8) of §2 allow us to construct new algorithms of low total complexity for evaluation of solutions of linear differential equations with an arbitrary precision. Such algorithms of low total complexity were till now unknown for general special functions. Low complexity algorithms were known only for algebraic function computations (standard Newton method), and for elementary functions.

These low operational complexity algorithms for computation of the logarithmic function based on Landen's transformation for elliptic functions, or the Gauss arithmetic-geometry mean iteration, were proposed by Salamin and improved by Brent [B5- B6]. These low operational complexity algorithms are translated into low total (bit) complexity algorithms for computation of values of the logarithmic function. In particular, n digits of π can be computed in $O(M(n)\log n)$ short (bit) operations. These particular methods are used in all recent computation of π.

The computational cost of analytic continuation of solutions of linear differential equations is very similar to the computational cost of evaluation of coefficients of power series expansions (and, in fact, they both can be computed together). E.g. for Fuchsian linear differential equations we have similar to the basic Algorithm 2.1 the following simple

Algorithm 3.1 [C5]. In order to compute the value of analytic continuation of a solution $y(x)$ of a linear differential equation with given initial conditions at $x = x_0$ one needs at most

$$O(M \cdot log(\parallel x \parallel +1))$$

operations, where $\parallel x \parallel$ is the distance from the base point $x = x_0$ to the point $x = x_1$ of evaluation as counted along a given path γ on the Riemann surface of $y(x)$

We describe now entirely new algorithms of low total complexity for computation of values of solutions of differential equations, neither related to any rapidly convergent analytic transformations, nor to any low operational complexity methods. These methods differ in bit complexity only by factor $\log^2 n$ or $\log^3 n$ from the bit complexity $M(n)$ of algebraic computations.

Let us look again at an arbitrary linear differential equation with rational (polynomial) coefficients, either in the scalar form

$$a_n y^{(n)} + a_{n-1} y^{(n-1)} + \cdots + a_1 y' + a_0 y = 0, \tag{1}$$

or, in the general matrix form,

$$\frac{d}{dx} Y(x) = A(x) \cdot Y(x), \tag{2}$$

where $a_i \in C(x)$, and $A(x) \in M_n(C(x))$. We are interested in the evaluation of solutions of (1) and (2) with arbitrary precision using the method of (formal) power series expansions of §2. A solution of (1) or (2) is determined by its initial conditions at $x = x_0$. If $x = x_0$ is not a singular point of (1) or (2), then the solution $y = y(x)$ of (1) or $Y(x)$ of (2) is uniquely determined by its initial conditions $y(x_0), \ldots, y^{(n-1)}(x_0)$ in the case of (1), or $n \times n$, matrix $Y(x_0)$, in the case of (2), respectively. If $x = x_0$ is a regular singular point of (1) or (2), then the initial conditions take the form of a few leading terms in the expansion of $y(x)$ or $Y(x)$ in powers of $(x - x_0)^\alpha$ for various local exponents α. Finally, if $x = x_0$ is an irregular singularity of (1) or (2), one can interpret as initial conditions at $x = x_0$ few first term in the asymptotic expansions that are linear combinations of formal power series from $\mathbf{C}[[x - x_0]]$ times functions like $exp(Q((x - x_0)^{-\alpha}))$ for $Q(x) \in \mathbf{C}[x]$. In all these cases, having specified initial conditions for $y(x)$ or $Y(x)$ at $x = x_0$, we want to evaluate within a given precision l $y(x)$ or $Y(x)$ at another point $x = x_1$. For all practical purposes we assume that values of x_0 and x_1 are given correctly with the precision of l bits, or as rational or algebraic numbers of sizes less than l. To determine values at $x = x_1$ from those at $x = x_0$ one has to specify a path γ from x_0 to x_1 on the Riemann surface (or its universal covering) of $y(x)$ or $Y(x)$, see [C5]. But first we look at the most important case when $x = x_1$ lies within the disc of convergence of power series expansions defining $y(x)$ or $Y(x)$. In the case of a regular point $x = x_0$ this is simply the disc of convergence of $y(x)$ or $Y(x)$. If $x = x_0$ is a regular or irregular singularity, and the expansion at $x = x_0$ looks like $y(x) = \phi_0(x) \cdot y_0(x)$, where $y_0(x) \in \mathbf{C}[[x - x_0]]$, and $\phi_0(x) = (x - x_0)^\alpha$ or $\phi_0(x) = exp(Q((x - x_0)^{-\alpha}))$, then $x = x_1$ should lie within the disc of convergence of $y_0(x)$.

In the cases of (1) and (2), when the coefficients are rational functions and the set S of singularities of (1) or (2) is a finite set, the radius of convergence of $y(x)$ or $Y(x)$, if nonzero, is bounded from below by the distance from the point $x = x_0$ to the nearest point in S. We consider now only the case of nonzero radius of convergence (though the general case can be treated in the same framework using the generalized Borel transform). For this reason we again consider, as in §2, the case when $x = x_0$ is a regular or a regular singular point only.

Again, as in §2, the general regular solution of (1) or (2) can be written in terms of the basis of regular solutions

$$y(x, \alpha) = (x - x_0)^\alpha \sum_{N=0}^{\infty} y_N(\alpha) \cdot (x - x_0)^N \tag{3}$$

or

$$Y(x) = \left\{ \sum_{N=0}^{\infty} C_N (x - x_0)^N \right\} \cdot (x - x_0)^W \tag{4}$$

with the recurrences on $y_N(\alpha)$ or C_N being as in (6)-(7) of §2:

$$\sum_{j=0}^{min\{N,d\}} y_{N-j}(\alpha) \cdot f_j(\alpha + N - j) = 0 \qquad (5)$$

or

$$C_{N+1} \cdot (N+1) \cdot d_0 = \sum_{i=0}^{min\{N,d-1\}} A_i \cdot C_{N-i} - \sum_{i=0}^{min\{N,d-1\}} d_{i+1} \cdot (N-i) \cdot C_{N-i}. \qquad (6)$$

In applications, in order to evaluate the power series expansions (3) or (4), we should look at recurrences satisfied by $C_N(x - x_0)^N$ instead of C_N. E.g., the recurrence for $C_N(x - x_0)^N$ has the form

$$C_{N+1}(x - x_0)^{N+1} = \{ \sum_{i=0}^{min\{N,d-1\}} \alpha_{i;N} C_{N-i}(x - x_0)^{N-i} \}/\delta_0, \qquad (7)$$

where $\delta_0 = (N + 1) \cdot d_0$, $\alpha_{i;N} = (x - x_0) \cdot [A_i(x - x_0)^i - d_{i+1}(N - i)(x - x_0)^i]$: $i = 0, \dots, min\{N, d - 1\}$. This recurrence (like (5) or (6)) can be written in the matrix form. We write this matrix recurrence that computes simultaneously d consecutive coefficients $C_{N-i}(x - x_0)^{N-i}$: $i = 0, \dots, d-1$ ($i \leq min\{N, d - 1\}$), and simultaneously, the $N - th$ partial sum $Y_N(x) = \sum_{i=0}^{N-1} C_i(x - x_0)^i$ of $Y(x)$. This new recurrence follows from (7), if to add one more formula:

$$Y_{N+1}(x) = Y_N(x) + C_N(x - x_0)^N. \qquad (8)$$

To represent the matrix recurrence, we introduce a $n(d + 1) \times n$ matrix $\mathcal{Y}_N = (Y_n, C_N \cdot (x - x_0)^N, \dots, C_{N-(d-1)} \cdot (x - x_0)^{N-d+1})^t$. From (7)-(8) we deduce a matrix recurrence:

$$\mathcal{Y}_{N+1} = C(N) \cdot \mathcal{Y}_N, \qquad (9)$$

where $C(N)$ is a $n(d+1) \times n(d+1)$ matrix consisting of blocks of $n \times n$ matrices: $C(N) = (B_{ij}(N))_{i,j=1}^{d+1}$. Here $B_{1,1}(N) = 1$, $B_{1,2}(N) = 1$, $B_{i,j}(N) = 0$ for $j = 3, \dots, d + 1$; $B_{k,l}(N) = \delta_{k-1,l}$ for $k \geq 3$, $B_{2,1}(N) = 0$, and

$$B_{2,j}(N) = \alpha_{j-2;N}/\delta_0 : j = 2, \dots, d+1,$$

in the notation of (7). Similarly, in the case of equation (1), we get a matrix recurrence

$$\mathcal{Y}_{N+1} = C(N) \cdot \mathcal{Y}(N),$$

for $(d+1) \times (d+1)$ matrix $C(N)$ and $\mathcal{Y}_N = (y_N(x), y_N(\alpha) \cdot (x-x_0)^N, \ldots$
$\ldots, y_{N-d+1}(\alpha)(x-x_0)^{N-d+1})^t$, for $\mathcal{Y}_N(x) = (x-x_0)^\alpha \cdot \sum_{i=0}^{N-1} y_i(\alpha) \cdot (x-x_0)^i$.

To derive (9) we need to know the original coefficients of (1) or (2) and to know α or W. In all important cases α is a fixed rational number (and W is a diagonal matrix with rational number entries). To compute coefficients in (4) or (6) one needs to determine the coefficients of the translated polynomials in (1) or (2) after the translation $x \rightarrow x + x_0$. Thus the total number of operations to compute the coefficients matrix $C(N)$ in (9) is $O(n^2 d \log d)$ in the case of (2), and is $O(n \, d \log d)$ in the case of (1) (with $d \geq n$ in the case of (1)). The matrix $C(N)$ in (9) can be represented as $C(N) = C_0(N)/d_0(N)$, where $C_0(N)$ and $d_0(N)$ are polynomial in N, x_0 and x.

In the case of (2) (with $W = 0$), $C(N)$ is an $n(d+1) \times n(d+1)$ matrix, with $C_0(N)$, $d_0(N)$ linear in N, polynomial in x_0, and $x - x_0$ of degrees at most d in x_0, and with $C_0(N)$ of degree d in $(x-x_0)$ ($d_0(N)$ is independent of x). More precisely, $C_0(N)$, as a polynomial in $x - x_0$ and x_0 has a total degree of at most d. In the case of the equation (1), $C_0(N)$ and $d_0(N)$ are polynomial of degree n in N; they are polynomials in $x - x_0$ and x_0. $d_0(N)$ is polynomial in x_0 of degree at most d, and $C_0(N)$ is polynomial in $x - x_0$ and x_0 of total degree at most d.

The growth of the coefficients in the power series expansions $y(x)$ or $Y(x)$ can be estimated e.g. from the recurrences (5) or (6) (or (9)) according to the Poincaré-Perron theorem that the asymptotics of solutions of recurrences are determined by roots of the limit characteristic equation of a linear recurrence with constant coefficients that one deduces from (5) or (6) as $N \rightarrow \infty$. An explicit form of (5) or (6) shows that these roots of the limit characteristic equation are $1/(s-x_0)$ for $s \in S$ (the set of singularities of (5) or (6)), when x_0 is a regular point. More precisely, an asymptotic analysis of recurrences (5) or (6) (or (9)) gives the following leading term of the asymptotics of coefficients $y_N(\alpha)$ or C_N:

$$|y_N(\alpha)| \leq \gamma_1 \cdot N^\nu \cdot dist^*(x_0, S)^{-N}, \qquad (10)$$

$$\| C_N \| \leq \gamma_2 \cdot N^\mu \cdot dist^*(x_0, S)^{-N}. \qquad (11)$$

Here $dist^*(x_0, S) = min\{|s - x_0| : s \in S, x_0 \neq S\}$, and $\| \cdot \|$ is a co-norm of $n \times n$ matrices. Asymptotic bounds (10) or (11) hold for arbitrary equations (1) or (2) with any initial conditions at a regular point $x = x_0$. In general one can complement (10) and (11) with a similar lower bound for $N \geq N_0$, but with different constants γ_1 and γ_2. Bounds (10) and (11) show that whenever $\delta > 0$, and $|x_1 - x_0|/dist^*(x_0, S) < 1 - \delta$, one obtains the value of $y(x_1)$ or $Y(x_1)$ from the matrix $\mathcal{Y}_N$ at $x = x_1$ because $y_N(x_1)$ or $Y_N(x_1)$ converges to $y(x_1)$ or $Y(x_1)$, respectively, as a geometric progression in N. To evaluate $y(x_1)$ or $Y(x_1)$ with the precision l one needs the values of $\mathcal{Y}_N$ with the precision $l + O(\log l)$ for $N = O(-l/\log(|x_1 - x_0|/dist^*(x_0, S)))$, if $|x_1 - x_0|/dist^*(x_0, S) < 1 - \delta$ (for fixed $\delta > 0$). As above, let us assume that for a fixed $\delta > 0$, $|x_1 - x_0|/dist^*(x_0, S) < 1 - \delta$. As it was shown above, to compute the $n \times n$ matrix with precision $l + O(\log l)$ one can compute $\mathcal{Y}_N$ for $N = -O(l/\log(|x_1 - x_0|/dist^*(x_0, S)))$ from (9) with initial conditions $Y(x)|_{x=x_0} = I_n$ (i.e. $C_0 = I_n$.)

To estimate the bit complexity of computations of $\mathcal{Y}_N$ one needs to know the amount of bit information in the representation of x_0 and x_1. We can represent x_0 and x_1 as rational numbers p/q of logarithmic size, size $(p/q) = \log(\max\{1, |p|, |q|\})$, or as b-bit binary (floating point) numbers $\underbrace{b_1 b_2 b_3 \ldots b_b}_{b} \times 2^{EXP}$ (for $EXP = O(b)$). The matrix of coefficients $C(N)$ in (9) depends polynomially on x_0 and $x_1 - x_0$. Convenient parameters can be x_0 and $x_1 - x_0$ or $\frac{x_1 - x_0}{x_0}$ (if $x_0 \neq 0$). Let us look at x_0, $x_1 - x_0$ written with a common denominator: $x_0 = \frac{X_0}{D}$, $x_1 - x_0 = \frac{X_{10}}{D}$ (e.g. $D = 2^b$ in the binary representation). We define then $b = \log_2 \max\{l, |D|, |X_0|, |X_{10}|\}$ as the (logarithmic) size of x_0, $x_1 - x_0$. Then $C(N) = C_0(N)/d_0(N)$ where $C_0(N)$, $d_0(N)$ are linear in N with coefficients that are $0(d \cdot b)$-bit integers. Let $e = -\log_2(|x_1 - x_0|/dist^*(x_0, S)) > 0$ be the measure of closeness of x_1 to x_0. Then we need to compute $\mathcal{Y}_N$ for $N = O(l/e)$ with the recision $l + O(\log l)$.

To compute $\mathcal{Y}_N$ from the matrix recurrence (9) we use the fast binary-splitting technique of Theorem 2.2. This way we arrive at the following *local* evaluation of the solution $Y(x)$ of (2), where we specify all the dependencies on the sizes of coefficients involved.

<u>Theorem 3.2.</u> Let (2) be a fixed matrix linear differential equation, where all elements of $A(x)$ are rational functions of the total degree at most d, and coefficients of these functions are rational numbers of sizes at most k (or arbitrary complex numbers represented by their binary approximations as k-bit binary floats). Let $x = x_0$ be a regular or regular singular point of (2) represented by a rational number of size of at most b, or by a b-bit binary floating point number. Let $Y(x)$ be a solution of (2) with fixed initial conditions at $x = x_0$, and let x_1 be another binary b-bit number, lying within the disc of convergence of $Y(x)$. Then to compute $Y(x)$ at

$x = x_1$ with precision of l leading digits one needs at most

$$c_1 \cdot \mu(n) \cdot \frac{db + k}{e} \cdot M(l) \cdot \log^2 l + O(l)$$

bit operations. Here $e = -\log_2(|x_1 - x_0|/dist^*(x_0, S)) > 0$, $\mu(n)$ is the number of operations needed for $n \times n$ matrix multiplications, $M(l)$ ($= l \cdot \log(l) \cdot \log\log(l)$) is the number of operations for l-bit multiplication, and c_1 is an absolute constant. The constant in $O(l)$ depends on initial conditions of $Y(x)$.

Theorem 3.2 is the basis of bit-burst method of fast evaluation and of analytic continuation of any solution of a linear differential equation to any point on its Riemann surface. According to this method the evaluation of any branch of any solution with the precision of l leading bits (digits) requires at worst any $O(M(l) \cdot \log^3 l \cdot (1 + o(l)))$ total bit operations.

For this we use the method of analytic continuation, following [C5].

Let $Y(x; x_0)$ be an $n \times n$ matrix solution of (2) normalized at $x = x_0$:

$$Y(x; x_0)|_{x=x_0} = I_n,$$

for a unit matrix $I_n = (\delta_{ij})_{i,j=1}^n$. The basic rule of analytic continuation is the superposition formula, see [C5], according to which

$$Y(x; x_1) \cdot Y(x_1; x_0) = Y(x; x_0) \qquad (12)$$

for any three points x_0, x_1, x in $\mathbf{CP}^1$. The superposition formula gives the following simple chain rule of analytic continuation of an arbitrary solution of (2) along any path γ in $\mathbf{CP}^1$. Let us assume that γ is not passing through any of the singularities of (2). Let x_0 be the initial point of γ and x_{fin} be its end-point (they can coincide). Then, by choosing $m + 2$ vertices $x_0, x_1, \ldots, x_m, x_{m+1} = x_{fin}$ on γ, we can replace the process of analytical continuation of a normalized solution $Y(x; x_0)$ along γ from x_0 to x_{fin} by the process of successive solution of (2) with new initial conditions:

$$Y(x_{fin}; x_0) \, (\text{the analytic continuation of } Y(x; x_0) \text{ from } x_0$$

$$(13)$$

$$\text{to } x_{fin} \text{ along } \gamma) \underset{\gamma}{\rightarrow} Y(x_{fin}; x_m) \cdots Y(x_2; x_1) Y(x_1; x_0).$$

In order to apply this chain rule, one has to be sure that each of the factors, $Y(x_{i+1}; x_i)$ is defined nonambiguously. For this it is sufficient to assume that x_{i+1} lies in the disk with the center x_i and the radius of this disk is smaller than the radius of convergence of $Y(x; x_i)$ e.g. whenever for a fixed $\delta > 0$, $|x_{i+1} - x_i|/dist^*(x_i, S) < 1 - \delta$.

From the point of view of minimal complexity, the problem is to determine the minimal number of $m + 2$ points x_i and their positions, for which the number of operations necessary to complete the analytic continuation within a given precision is minimal.

As it turned out, see [C5], the minimal number of $m + 2$ points <u>does not depend on the chosen precision</u> but depends only on the positions of singularities of a linear differential equation and on the path γ. Moreover, this number is always bounded (for a given set of singularities) by the function of logarithm of the total length of γ.

Let us fix a path γ in $\mathbf{CP}^1 \backslash S$. There are two natural formulations of optimal analytic continuation along γ described above.

I) One fixes γ having initial point x_{in} and its end point x_{fin}. We want to find a number m and points $x_0 = x_{in}, x_1, \ldots, x_m, x_{m+1} = x_{fin}$ lying on γ such that the computation of power series expansion of a given solution at x_i and their evaluations at $x = x_{i+1}$ consecutively for $i = 0, \ldots, m$ requires the minimal number of operations for a given precision of calculations.

II) Since analytic continuation along γ depends only on the homotopy class of γ in $\pi^1(\mathbf{CP}^1 \backslash S)$, one can ask in I) to find a polygon $\Delta = \overrightarrow{x_0 x_1}$ $\overrightarrow{x_1 x_2} \cdots \overrightarrow{x_m x_{m+1}}$ with vertices $x_0, \ldots, x_{m+1}$ equivalent to the path γ in $\pi^1(\mathbf{CP}^1 \backslash S)$, and for which the process of consecutive computation of power series expansions of a given solution at x_i and their evaluations at x_{i+1} for $i = 0, \ldots, m$ requires the minimal number of operations for a given precision of calculations.

As above, one needs at most $O(M/log(dist(x_0, S)/|x_1 - x_0|))$ terms in the local power series expansion of $Y(x)$ at $x = x_0$ to evaluate $Y(x)$ at $x = x_1$ within precision M.

Consequently, for the fixed precision M, the Problem I is reduced to the following extremal problem:

<u>Problem I'</u>. For a given path γ, find the minimum:

$$\underset{m}{min} \cdot min\{I(x_0, \ldots, x_{m+1}) : x_0 = x_{in}, x_{m+1}$$

$$= x_{fin}, x_i \in \gamma, |x_{i+1} - x_i| < dist^*(x_i, S)\}$$

$$\text{with} I(x_0, \ldots, x_{m+1}) = \sum_{i=0}^{m} 1/log(\frac{|x_{i+1} - x_i|}{dist^*(x_i, S)}).$$

Problem I ' can be easily solved for paths γ that lie in the domains, where $dist^*(x, S) = |x - a_i|$ for a fixed a_i. Two crucial examples are the circle around the singularity and a straight line moving away from a given singularity. In both cases we found m, x_i and the value of the minimum, see [C5]:

Proposition 3.3. Let γ be a straight path from x_{in} to x_{fin}, where x_{in} is not a singularity and for all points on γ, there is a single $a_i \in S$, which is the closest element of S. Without loss of generality we can assume $x_{in} = 1$, $a_i = 0$ and x_{fin} is real > 0. Then the solution to problem I' has the following form:

$$x_i = x_{fin}^{i/(m+1)} : i = 0, \ldots, m + 1$$

(so that $x_0 = x_{in}$, $x_{m+1} = x_{fin}$). The number m is defined so that $m + 1$ is (non-negative) integer closest to $\ln x_{fin} / \ln \beta$. Here $\beta = 1.318...$ is the solution of a transcendental equation

$$\frac{\ln(\beta - 1)}{\ln \beta} + \frac{\beta}{\beta - 1} = 0.$$

Proposition 3.4. Let γ be a closed circular path encompassing a single singularity as a center. Let us assume that the radius of γ be smaller than a half distance from a given singularity to any other one. The in Problem I' the minimal number $m+1$ of distinct points of γ (we assume that $x_{in} = x_{fin}$) is 17, and these points form a regular 17-gon inscribed into γ.

The global recipes above for the analytic continuation of $Y(x)$ along γ are combined with better local methods of evaluation of $Y(x)$ from $x = x_0$ to $x = x_1$. For this we are using Theorem 3.2 to evaluate $Y(x)$ from $x = x_0$ to $x = x_1$ making several steps between x_0 and x_1, again using the chain rule (13), releasing consecutive blocks of x_1 in bursts. We call this method "bit-burst" method. In this approach one, in order to evaluate $Y(x)$ at a bigfloat $x = \sum_{N=0}^{\infty} b_N \cdot B^{-N}$, evaluates $Y(x)$ consecutively at $x_i = x_i + \sum_{N=2^i}^{2^{i+1}} b_N \cdot B^{-N}$ by analytic continuation of $Y(x)$ (rule (13)): "adding more bits of a number, but computing with the same accuracy". In this approach we are matching the distance from the evaluation point to the point of expansion with the size of the point of evaluation. In particular, we move initial conditions from $x = x_0$ to a nearby point $x = x_0'$ of bounded size, and then evaluate $Y(x)$ at $x = x_1$ starting from $x = x_0'$ in bit-bursts.

This way we arrive at the following general theorem that gives an upper bound on the total (bit) complexity:

<u>Theorem 3.5.</u> Let (2) be a given linear differential equation with rational function coefficients, and $Y(x)$ be its arbitrary (regular) solution with initial conditions at $x = x_{in}$, where x_{in} is an K-bit number. Given a path γ from x_{in} to an K-bit number x_{fin} (on the Riemann surface of $Y(x)$) of length L, one can evaluate $Y(x)|_{x=x_{fin}}$ at $x = x_{fin}$ with the full K-bit precision at most

$$O(M(K)(\log^3 K + \log L))$$

bit operations.

The bit-burst method in the general form of Theorem 3.5 is not the best possible: one would like to see $\log^3 K$ replaced by $\log K$ always. Sometimes the complexity can be lowered:

I. If x_{in} and x_{fin} are fixed rational numbers, then the computation of $Y(x)$ with the full K bits of precision has total (bit) complexity at most $0(M(K)(\log^2 K + \log L))$.

II. If the differential equation (1) or (2) possesses special arithmetic properties, bit-complexity can be lowered. E.g. if the (1)-(2) possesses a solution which is either an E-function or a G-function, then the general bit bound of Theorem 3.5 can be lowered to

$$O(M(K) \cdot (\log^2 K + \log L)).$$

If, further, like in x_{in} and x_{fin} are fixed rational numbers, and $Y(x)$ is built from E- or G-functions, then K significant digits of $Y(x_0)$ can be computed in

$$O(M(K) \cdot (\log K + \log L))$$

bit operations.

This last bound is unsurpassed by any other algorithms even for elementary functions, like the exponent, where low operational complexity algorithms are well known, see [B5-B6].

Our original interest in the development of power series evaluation facilities was purely transcendental. We wanted to have the ability to compute the monodromy group of linear differential equations with an arbitrary precision to check various hypotheses of transcendence, algebraicity and the existence of algebraic relations among elements of the monodromy matrices. The "constants" appearing as elements of monodromy matrices encompass classical constants of geometry and analysis. Among them there are periods of algebraic varieties including values of the Euler Γ- and B-functions (at rational points) and other integrals of elementary and algebraic functions over closed paths.

How to compute the monodromy group of linear differential equations? Existing methods include Poincaré-Lappo-Danilevsky methods of polylogarithmic series representations of monodromy matrices (or multiple path integral representation) and Poincaré-vonKoch methods of infinite determinants, see reviews in [C5]. We choose the power series method based on the direct analytic continuation of a fundamental system of solutions because of its low complexity. This method follows the definition of monodromy and had been, in a sense, considered by L. Fuchs. In this method one starts with a fundamental system of solutions

$$\vec{Y} = (y_1(x), \ldots, y_n(x))$$

of (1) [given by their initial conditions] and analytically continues it along a closed path with no singularities on it. The system Y analytically continued along a path γ undergoes a linear transformation

$$Y \underset{\gamma}{\mapsto} Y \cdot M(\gamma).$$

The set of all matrices $M(\gamma)$ is a monodromy group of (1). Typically, one is interested only in the conjugacy class of this group or in its Zariski closure.

Our monodromy finder package RIEMANN [C5] allows us to determine a monodromy group of an arbitrary linear differential equation with rational function coefficients (or of an arbitrary algebraic function). (As initial data one has to provide the coefficients of a differential equation and (approximate) positions of singularities.)

Typically the monodromy group is computed by analytic continuation of regular expansions from one singularity to another through the connection formuas. Our package, though designed for computer algebra systems, had been implemented in FORTRAN (for fixed precision computations).

Areas of applications include:

1) multiple precision computation of Abelian integrals and their periods. These computations are used then in the transcendental solution of the problem of reduction of Abelian integrals. In order to determine the reducibility of Abelian integrals to the lower genra (e.g. when an Abelian integral is an elementary function) one looks at the $\mathbb{Z}$-relations between

periods. Multiple precision computations of periods together with multi-dimensional continued fraction methods (Jacobi-Perron, Ferguson-Forcade or Lenstra methods) provides an effective tool in the solution of reduction problem.

2) Another application of monodromy computations is to the solution of the direct and inverse Galois problem, when one wants to find a Galois group of a given algebraic function field (differential equation) or wants to construct a field with a given Galois group. RIEMANN package is designed mainly for the direct Galois problem, but we found it extremely convenient to use for solution of the inverse Galois problem when the number of parameters is not large and one can numerically invert the function (monodromy matrix output) generated by RIEMANN. Alternative methods in the inverse Galois problem, that we are using, include the solution of the Riemann boundary value problem. Interesting areas of future research should include the explicit construction of algebraic function fields over $\mathbf{Q}$ with a given Galois group, whose existence follows from Belyi-Matzat-Thompson theorems. Large simple groups are the most interesting objects.

3) Uniformization theory seems to be an attractive proving ground for application of monodromy packages. For us the crucial problem was the question of arithmetic nature of parameters in the solution to the uniformization problem, including: a) algebraic equations (i.e. their coefficients) defining Riemann surfaces to be uniformized; b) invariants of discrete groups that uniformize these Riemann surfaces (Fricke parameters and more sophisticated parametrizations of Teichmüller spaces); and finally, the most notorious group of parameters c) assessory parameter that uniquely determine the differential equation of the second order, ratio of solutions of which determines the inverse to the uniformizing function. See §6-7 for details.

4   Continued Fraction Expansions.

Power series expansions are usually blamed for inferior rate of convergence (which is true), and, in practice and theory, their convergence rate is usually accelerated, or they are substituted by rational (Padé) approximations. In number theory particularly, rational approximations are crucial because they provide a unique tool for study of diophantine approximations to numbers that are values of well approximated functions. Especially interesting classes of functions, that one wants to approximate rationally, are once more solutions of linear differential equations or algebraic functions.

Fast algorithms of computation of power series allow for relatively fast computation of rational (Padé) approximations to these power series. The number of operations needed to compute the N-th (diagonal) Padé approximation this way can be estimated as

$$O(N \log^2 N).$$

Is there a faster way to compute Padé approximations and continued fraction expansions of, say, algebraic functions, without computing first their power series expansions?

The answer is yes, at least for a large class of functions. Let us look at the continued fraction expansion of a function $y(x)$ at $x = \infty$.

$$y(x) = a_0(x) + \cfrac{1}{a_1(x) + \cfrac{1}{a_2(x) + \ldots}}$$

with polynomials $a_i(x) : i = 0, 1, 2, \ldots$. Here we denote by $[h(x)]_\infty$ a singular part of a function $h(x)$ at $x = \infty$:

$$h(x) = [h(x)]_\infty + h_{-1}/x + h_{-2}/x^2 + \cdots,$$

and the algorithm of the continued fraction expansion of $f(x)$ is the following: $f_0(x) = f(x)$, $a_i(x) = [f_i(x)]_\infty$, $f_{i+1}(x) = (f_i(x) - a_i(x))^{-1} : i \geq 0$.

<u>Algorithm 4.1.</u> Let $y(x)$ be an algebraic function satisfying quadratic or cubic equations over a rational function field. Then in order to compute N-th elements $a_N(x)$ in the continued fraction expansion of $y(x)$ one needs at most $O(\log N)$ operations. The constant in $O(\cdot)$ depends only on the genus of a function field to which $y(x)$ belongs and on the representation of $y(x)$ in terms of the basis of this field.

This algorithm holds also for all radical functions $y(x) = \prod_{i=1}^{k} P_i(x)^{k_i/n_i}$ The key to the algorithm is the representation of partial fractions in the continued fraction expansion of $y(z)$ in terms of θ-functions, see [C5]. We conjectured that Algorithm 4.1 can be constructed for an arbitrary algebraic function.

How can solutions of linear differential equations be efficiently Padé approximated? One can, apparently, construct Padé-Hermite approximations to solutions of linear differential equations and their derivatives without first computing their power series expansions. For this one needs $O(N)$ operations in the sequential algorithm and $O(\log N)$ steps in the parallel algorithm to get the N-th diagonal Padé-Hermite approximation, for the description of the method, see [C9].

5   Parallel Algorithms.

Complexities of power series computations can be investigated from the point of view of parallel (vector) implementation. The parallel (vector) methods are important because they seem to be the only way to address large jobs. Algorithms that compute power series coefficients and values of functions by means of linear recurrences are particularly well-suited for various vector and parallel implementations. For computations in the parallel environment one has to take into account the following parameters: a) the number of parallel steps (the depth of the circuit); b) the total number of bit (boolean) operations; and c) the number of (micro) processors.

For a problem with I or O of $O(n)$ bits, e.g. in the evaluation with the full n-bit precision of the value of an algebraic or transcendental function, one can hope at best for the depth of $O(\log n)$. With this depth one would like also to have parallel algorithms of total circuitry (number of processors) to be optimal as well: on the order of $O(M(n) \cdot \log n)$.

There are serious obstacles in generation of such fast parallel algorithms from the serial algorithms with the same bit complexity. Iterative algorithms, that are used to construct low bit (and operational) complexity serial algorithms, do not provide depth $O(\log n)$. These iterative algorithms, particularly the arithmetic-geometric mean, and also the Newton iterational method, can get the depth at best $O(\log^2 n)$.

Popular iterative methods of computations of elementary functions give depth $O(log^2 n)$, though with only $O(M(n))$ of total circuitry for elementary function computations. Only recently new algorithms of Reif, Bini, Schonhage and Pan, made it possible to compute elementary functions in $O(\log n \cdot \log \log n)$ depth circuits. Unfortunately, the total circuitry significantly increases: the typical number of processor becomes $n^{O(1)}$.

The methods that we propose, bit-burst algorithms for computations of arbitrary linear differential equations, have always depth $O(\log n \log \log n)$ even though, in general, the total bit operation count is $O(M(n) \cdot \log^3 n)$. In fact, for E- and G-functions the total bit operation count with the same depth is only $O(M(n) \cdot \log^2 n)$. Also, the full n-bit precision computations of values of nonlinear differential equations (whose right hand side is built from "known" functions, that can be themselves solutions of differential equations) can be complete in only $O(\log n)$ steps with the total circuitry of $n^{0(1)}$ processors.

6 Computations of the Accessory Parameter in Lamé Equations

Efficient algorithms of §3 allow for fast high precision computations of (invariants of) the monodromy group of linear differential equations. These algorithms can be used in particular in the solution of the uniformization problem for Riemann surfaces in terms of the classical language of accessory parameters in linear differential equations for which the monodromy group of these differential equations are representable by real matrices, see [K2] and [P2]. Our interest in the uniformization problem as have been already mentioned in §3, is purely transcendental: we want to know the arithmetic nature (algebraicity, transcendence, measures of diophantine approximations) of numbers arising in the explicit analytic uniformization of Riemann surfaces defined over $\overline{\mathbf{Q}}$. Among numbers whose algebraic or transcendence properties we are interested to examine are the invariants of the uniformizing Fuchsian groups (typically expressed through Fricke parameters), and the values of the accessory parameter in the auxiliary differential equation.

One should exclude (as usual) rational and elliptic curves. In the elliptic curve case the transcendence problem is settled by Siegel-Gelfond-Schneider results. The next crucial case is the uniformization of a punctured tori (equivalently, 4 punctures on a Riemann sphere), when the Teichmüller space is nontrivial. Our initial effort in the applications of RIEMANN package were directed to the multiple precision computations of uniformizing parameters of a punctured tori defined over $\mathbf{Q}$ and $\overline{\mathbf{Q}}$.

In the case of a punctured tori, the modular invariant $k^2 = k^2(\tau)$, is the fourth singularity together with $0, 1, \infty$. The uniformization problem in this case is equivalent to the uniformization problem for the Riemann sphere with punctures at $0, 1, k^2$ and ∞. For the punctured tori case the uniformization problem can be studied using a linear differential equation with singularities at $0, 1, k^2$ and ∞ with the local exponents at $0, 1$ and k^2 as $(0, \frac{1}{2})$, and at ∞ as $(\frac{1}{4}, \frac{1}{4})$. This equation is a particular case of a Lamé equation with the parameter $n = \frac{1}{2}$, depending on one more constant known as an accessory parameter. This equation can be represented as follows [C5], [K3]:

$$\left\{\frac{d^2}{dx^2} + \frac{1}{2}\left(\frac{1}{x} + \frac{1}{x-1} + \frac{1}{x-k^2}\right)\frac{d}{dx} + \frac{\frac{1}{16}x + B}{x(x-1)(x-k^2)}\right\}y = 0 \qquad (1)$$

The parameter B is called an accessory parameter. The Fuchsian group that uniformize the punctured tori corresponding to the elliptic curve

$$y^2 = x(x-1)(x-k^2) \tag{2}$$

coincides with the monodromy group of a linear differential equation (1), *provided that* the monodromy group of (1) consists of real matrices. This reality condition determines uniquely the accessory parameter B. (Fuchsian parameter, in this case.)

Our main interest, as we explained, was to try to establish any relationship between the algebraicity of the fourth singularity k^2—the modulus of the elliptic curve and the arithmetic nature of B and the Fuchsian group that corresponds to this (Fuchsian) value of the accessory parameter B.

Specifically, we are interested in the arithmetic nature of B and the invariants of the monodromy group of (1) for algebraic k^2, at least for those k^2 that have number-theoretic significance (like singular moduli $k^2 = k^2(\tau)$ corresponding to the quadratic imaginary τ). Conversely, we were interested in the possibilities that k^2 and B in (1) can be algebraic, when the monodromy group of (1) is Fuchsian, representable by real 2×2 matrices with algebraic entries. There are 4 cases, when the monodromy group of (1) is an arithmetic one (a congruence subgroup). In these 4 cases the accessory parameter can be computed explicitly and is an algebraic (rational) number. The 4 corresponding elliptic curves (2) are also defined over $\mathbf{Q}$ and are the following ones:

$$y^2 = x^3 - 1, \; y^2 = x(x^2 - 1),$$

$$\tag{3}$$

$$y^2 = x(x^2 + 11x - 1), \; y^2 = x(8x^2 + 7x - 1).$$

The first two cases above had been known to Poincaré, the cases three and four lead to the Lamé equation arising in the Apéry work on irrationality of $\zeta(2)$ and $\zeta(3)$, [C6].

Before we started our numerical experiments, it seemed to us that the condition of the Fuchsian uniformization of (1) should guarantee the algebraicity of the (punctured) tori uniformized by the Fuchsian subgroup of $SL_2(\overline{\mathbf{Q}} \cap \mathbf{R})$. That is why we had implemented in the SCRATCHPAD environment a multiprecision package that determines the accessory parameters B in (1) and the corresponding uniformizing Fuchsian group starting from the punctured tori with the modulus k^2 (and vice versa). The corresponding values of the Fuchsian accessory parameter B and the Fricke parameters defining the corresponding Fuchsian group were checked for the possibility of algebraicity using multidimensional continued fraction methods. Unfortunately, our search for algebraic Fuchsian accessory parameters failed to find any new elliptic curve, other than those in (3), defined over $\mathbf{Q}$ or $\overline{\mathbf{Q}}$, for which the corresponding punctured tori has an algebraic value of an accessory parameter, or for which the Fuchsian group is represented by matrices with algebraic entries only.

Our computations were a part of a more general series of uniformization calculations based on the fast methods of computation of elements of monodromy matrices of arbitrary linear differential equations, described in §3. There are other methods of monodromy computations. These are: the Poincaré-Lappo-Danilevsky method [L9] of expansion of a monodromy group in series of periods of polylogarithmic functions and the Koch infinite determinant method [K4]. The Koch's method of infinite determinants is closely connected with the method of Hill's determinants, representing the characteristic numbers of solutions of equations with periodic coefficients. The relationship between the process of finding of Floquet exponents for equations with periodic coefficients and the determination of the invariants of the monodromy group was described in the classical literature, [I1]. More recently this relationship was presented by Magnus [M5] who used the Fourier expansion of the elliptic functions to represent the invariants of the monodromy group of a Lamé equation in terms of Hill determinants. This method was used by L. Keen, H.E. Rauch and A.T. Vasques in their studies [K3] of uniformization of punctured tori. These Hill determinants correspond to the infinite order recurrences depending on transcendental parameters $q = e^{2\pi i \tau}$, $K = K(k^2)$ and $E = E(k^2)$ of the fourth regular singularity $k^2 = k^2(\tau)$ of (1).

In our paper [C5, §13] we described a variety of different approximation methods of representation of elements of the monodromy group of (1) and of general Lamé and Heun equations using continued fraction expansions and three-term linear recurrences. These recurrences are based on Hermite-Ince-Darwin transformations [C5] of the Lamé equation to its trigonometric form and on the auxiliary third order linear differential equations. These transformations are valid for an arbitrary (generalized) Lamé equation. If one denotes the local multiplicities of the Lamé equation at ∞ by $-n/2$ and $(n+1)/2$, then the algebraic form of the Lamé equation is the following:

$$\frac{d^2y}{dx^2} + \frac{1}{2}\left(\frac{1}{x} + \frac{1}{x-1} + \frac{1}{x-k^2}\right)\frac{dy}{dx} + \frac{hk^{-2} - n(n+1)x}{4x(x-1)(x-k^2)}y = 0. \quad (4)$$

The transcendental (Hermite's) form of the Lamé equation (4) corresponds to the change of variables $x = (sn(z, k))^2$, where $sn(z, k)$ is the Jacobi sn-function. Under this transformation, one obtains from

$$\frac{d^2y}{dx^2} + \{h - n(n + 1)k^2sn^2(z, k)\}y = 0. \tag{5}$$

The monodromy group of (4) is generated by 3 matrices: M_0, M_1 and M_2 corresponding to an arbitrary choice of two linearly independent solutions of (4). The invariants of the monodromy group of (4) are generated by 3 Fricke parameters: $x = tr(M_0M_1)$, $y = tr(M_0M_2)$, $z = tr(M_1M_2)$. There exists a single algebraic relations connecting x, y, z. In the most interesting case of $n = -\frac{1}{2}$ (the punctured tori case) the relation is

$$x^2 + y^2 + z^2 = xyz. \tag{6}$$

The uniformization problem of punctured tori with a given invariant k^2 is equivalent to the problem of determination of the value of the accessory parameter in (1) or (4) such that the Fricke parameters x, y, z in (6) of the monodromy group of (1), (4) are all real. The reality of x, y, z means that the monodromy group (1), (4) is a Fuchsian group. Recent interest in the accessory parameter solution to the uniformization problem arose after Polyakov-Zamolodchikov work on conformally invariant field theories. Using the theory of Liouville equation, Takhatadjan and Zograf established a representation of the Weyl metrics on Teichmüller spaces in terms of explicit functions of acessory parameters furnishing the Fuchsian uniformation, [T2].

According to [T2] the Weyl metric for the Teichmüller space of a Riemann sphere with 4 punctures at 0, 1, a, ∞ has in the a coordinate representation of the form $m(a, a*)dada*$, where $m(a, a*) = 1/2\pi\partial_{a*}(S)$, $S = (1/2 - h)/(a(a - 1))$, see (4). On Fig. 1 and Fig 2 one can see landscapes of $h = h(a)$ and of $m = m(a, a*)$, respectively, in the a-complex plane obtained on a VAX780 after a few hours of computation of Fuchsian accessory parameters and numerical differentiation. The size of the grid is 51 × 51; computations were in FORTRAN double complex precision.

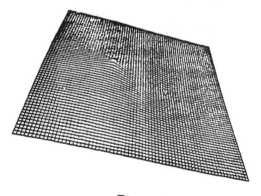

Figure 1

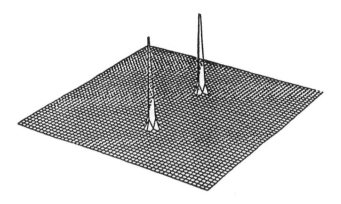

Figure 2

Another application of the fixed precision computations (in FORTRAN) of accessory parameters and the monodromy group of linear differential equations include the problem of conformal mappings of arbitrary circular (curvilinear) polygon onto the upper half-plane (unit circle). In the Schwarz-Christoffel formula for such mappings unknown parameters are the accessory parameters of the second order linear differential equation, like that of (4). In this and other cases we are solving the inverse monodromy problem, thus reducing a linear (boundary value) problem to the inversion of a nonlinear one. When the number of (accessory) parameters is moderate, this approach gives a good convergence.

In the multi-precision computations of algebraicity of B and x, y, z in (1) or (4) for an algebraic k^2 we were checking the possible algebraicity for degree up to 200 and with algebraic relations between powers of a given number with integral coefficients not exceeding 10^{10}.

Our experiments, particularly for curves with complex multiplication, did not reveal new algebraicity cases in addition to those in (3). Floating point computations were supplemented by the p-adic analysis aimed at discovering of new globally nilpotent equations (see definitions in [C7]) of the form (1). As a result of our computer experiments (SCRATCHPAD environment), we conjecture that no new equations of the form (1) defined over **Q** exists, which are either globally nilpotent or give a Fuchsian uniformization of the punctured tori defined over **Q** (with $B \in$ **Q**). Arithmetic problems here seem to be closely connected with the classification of all linear differential equations that come from geometry, or have solutions with nearly integral power series expansions [C7]. In the case of (1) (for curves in (3)), and for other hyperelliptic equations the only such arithmetico-geometric equations correspond to the blow-up of the hypergeometric equations or uniformizations by arithmetic groups.

7   The Uniformization Problem. The General Case and the Whittaker Conjecture.

In this chapter we report on the applications of computations of solutions of linear differential equations and of the monodromy groups of these equations to the classical uniformization problem of Riemann surfaces.

Let us review briefly classical, pre-Koebe, description of the uniformization theory of Riemann surfaces in terms of Schwarzians, Fuchsian linear differential equations, their monodromy groups and accessory parameters on which these groups depends (in that order). This description was the basis of Klein's [K2] and Poincaré's [P2] work on the uniformization problem, where the continuity method was employed for the proof of the existence theorem.

The continuity method is still not rehabilitated, and the contemporary theory of uniformization and conformal (quasiconformal) mappings is trying to avoid it. Nevertheless, the relationship between the uniformization of Riemann surfaces of genus $g \geq 2$ and linear differential equations of the second order, depending on $3g - 3$ complex parameters (accessory parameters), holds; it is simply hard to find these parameters explicitly; even their existence is, in general, a corollary of complex function theory. This relationship can be explored numerically (or analytically, when possible) to arrive to an acceptable explicit solution of the uniformization problem, when one tries to find a Fuchsian group, its fundamental domain and the uniformizing functions starting from a Riemann surface given by an explicit algebraic equation. One can call this uniformization problem a direct uniformization problem, while the inverse uniformization problem is a problem of reconstruction of a Riemann surface from the Fuchsian group uniformizing it, given by appropriate Fricke parameters (where available). Though there are several numerical methods aimed at the solution of these problems (name but a few: Myrberg method [M2] based on the Koebe theorem, various integral equations methods aimed at the solution of the inverse problem, see [H2] etc.), one of the best methods one can think of requires the determination of accessory parameters. When the uniformization problem is solved in terms of the accessory parameters, one has to determine $6g - 6$ real parameters on which the second order Fuchsian equation depends linearly, for which the monodromy group of this equation is represented by real 2×2 matrices (so that this group is Fuchsian). Whenever one has an efficient algorithm that computes the monodromy group of a second order equation, this algorithm becomes a subroutine in the iterative (Newton-type) program that tries to find that particular choice of acccessory parameters for which traces of products of all monodromy matrices have zero (or very close to zero, within a given precision) imaginary parts. This iterative program can be viewed, e.g. as a descent method to find a minima of squares of these imaginary parts. The global minima is unique and is achieved at Fuchsian accessory parameters.s

We have already reported in this paper and elsewhere [C5] on our progress in the fast computation of solutions of linear differential equations and their analytic continuations. Fast computations of monodromy group allow an application of descent method to find the Fuchsian accessory parameters when their number is not too large. In practice it means $g \leq 6$ if you want to keep high (double) precision and one is using traditional numeric FOR-TRAN packages. In reality $g \leq 4$ is a more reasonable constraint because of certain difficulties with the determination of proper cuts on an arbitrary Riemann surface and the determination of normalized differentials on the surface. This last restriction will be overcome in the future development of computer algebra systems, particularly in the SCRATCHPAD to come.

Earlier in this paper our attention was focused on Riemann surfaces of genuses 0 and 1 with several prescribed branch points (including the notorious case of once-punctured tori). We generalized our programs to include the genus $g \geq 2$ (particularly, $g = 2$ and 3). The case of hyperelliptic curves of an arbitrary genus $g \geq 2$ was studied in greater detail both because of the special role of hyperelliptic curves and because the number of accessory parameters is smaller (only $2g - 1$ accessory parameters), and because the Fuchsian equation depending on them has polynomial coefficients. We did not expect to find in this case any clues to possible closed form expressions of accessory parameters in terms of known algebraic or transcendental quantities. These pessimistic expectations had been confirmed. In particular, a nice, in its simplicity, conjecture of Whittaker [W2] on the possible simply expression of accessory parameters in the hyperelliptic case, turned out to be, in general, incorrect. Moreover, this conjecture proved since 1929 in many special cases, seems to be correct only in these cases.

To be more specific, we describe the role of accessory parameters in the explicit solution of the uniformization problem, following the exposition in [H4].

Let us consider an algebraic curve given by an equation $P(x, y) = 0$ for an irreducible polynomial $P(x, y)$. One can interpret this curve as a ramified covering of a Riemann sphere $\mathbf{CP}^1$ by looking at y as a multivalued function of x.

Let us assume that this algebraic curve has a genus $g \geq 2$, and let us denote the corresponding compact Riemann surface by Γ.

If $\hat{\Gamma}$ denotes the universal covering surface of Γ, then the existence theorem of the uniformization theory implies the conformal equivalence of $\hat{\Gamma}$ to the upper-half plane H. [If $\hat{\Gamma}$ would have genus $g = 0$, $\hat{\Gamma}$ would be conformally equivalent to the whole plane, while for $g = 1$ it would have been a torus.] This conformal equivalence with H determines the uniformization of Γ:

$$\phi : H \to \Gamma.$$

The group of cover transformations of Γ is

$$G = \{T : \phi(T \circ z) = \phi(z) \; for \; all \; z \in H\}.$$

Here G is a discrete group of Möbius transformations over $\mathbf{R}$, i.e. a Fuchsian group. Moreover, one can assume that G is a discrete subgroup of $SL_2(\mathbf{R})$.

Graphically one forms a tesselation of H starting from a fundamental polygon F of Γ, and looking at $\cup_{\sigma \in G} F^\sigma$. Thus one represents Γ as a quotient space

$$\Gamma = H/G,$$

(by identifying the sides of polygon F).

So far, we had mainly topology and the existence results (though an explicit determination of the polygon F is already a nontrivial problem, because even if the basis of homology on Γ is chosen, then to find F one needs to compute the periods of the integrals of the first kind on Γ). Let us pass now to a simple calculus.

It is easier to look at the inverse map

$$z = \phi^{-1}$$

than on ϕ. The function z is multi-valued on Γ and its monodromy is related to the action of $G : z \mapsto \sigma(z)$ for $\sigma \in G$. Since G is a group of Möbius transformations, and the Schwarzian derivative $\{\cdot, x\}$ (here $\{f, x\} = (\frac{f''}{f'})' - \frac{1}{2}(\frac{f''}{f'})^2$) is invariant under the fractional transformations, the function $\{z, x\}$ is a single-valued function on Γ.

This means that the function

$$\{z, x\} = 2R(x, y) \; on \; \Gamma \tag{7}$$

is a rational function on Γ. The classical relationship between the Schwarzian and the second order differential equations states that a change of variables

$$y_1 = (\frac{dz}{dx})^{-\frac{1}{2}}, y_2 = z(\frac{dz}{dx})^{-\frac{1}{2}}, z = y_2/y_1$$

reduces (1) to a second order linear differential equation on y_1 and y_2:

$$(\frac{d}{dx^2} + R(x,y))y = 0. \tag{8}$$

Whenever z is locally one-to-one, $R(x,y)$ can be determined as

$$R(x,y) = R_0(x,y) + Q(x,y),$$

where $Q(x,y)dx^2$ is a regular quadratic differential and $R_0(x,y)$ is a second derivative of an Abelian integral of the third kind (an explicit function that can be determined, see [H4], whenever Γ is known algebraically). If one can determine explicitly the basis $\{Q_j(x,y) : j = 1, \ldots, 3g-3\}$ of regular quadratic differentials on Γ (and this can be done whenever one has a program that computes a basis for the function space from the Riemann-Roch theorem), then an equation (2) can be represented as

$$(\frac{d^2}{dx^2} + R_0(x,y) + \sum_{i=1}^{3g-3} c_j \cdot Q_j(x,y))y = 0 \tag{9}$$

for $3g-3$ complex accessory parameters $c_j (j = 1, \ldots, 2g-2)$. For an analytic continuation of a chosen basis (y_1, y_2) of (3) around the loop γ of $\pi_1(\Gamma; \mathcal{O})$, one has 2×2 monodromy matrix M_γ:

$$(y_1, y_2)^t \underset{\gamma}{\mapsto} M_\gamma(y_1, y_2)^t.$$

When the accessory parameters $c_j (j = 1, \ldots, 3g-3)$ in (3) are <u>Fuchsian</u>, i.e. (3) arises as above from the uniformization of Γ by the Fuchsian group G, $y_2/y_1 = z = \phi^{-1}$ for $\phi : H \to \Gamma$,–then the monodromy group $\{M_\gamma\}$ of (3) is <u>simply</u> G, at least up to conjugation in $SL_2(\mathbf{C})$.

Consequently for a given Γ with the properly chosen R_0 and Q_j in (3) there exist a unique selection of parameters $c_j : j = 1, \ldots, 3g-3$ $(6g-6$ real parameters) for which the corresponding monodromy group of (3) is Fuchsian (i.e. is a subgroup of $SL_2(\mathbf{R})$).

This description is a bit different for a hyperelliptic curve. In this case simple transformations allow us to reduce (2) to an equation with coefficients polynomial in x. To be more explicit let us consider a general hyperelliptic equation of genus g:

$$y^2 = (x - e_1)\ldots(x - e_{2g+2}) \tag{10}$$

(the fact that genus is g means that all $e_i : i = 2,\ldots,2g + 2$ are distinct). Simple arguments (Schottky in the general case and Whittaker [W1] 1898 for the hyperelliptic curves), show that the Fuchsian group G of an equation (4) is a subgroup of index 2 of a group G^* of self-inverse transformations, i.e. elliptic transformations of period 2. The group G can be easily determined from G^*. But G^* has genus 0, and thus its automorphic function can be expressed as a rational function of a single automorphic function. Consequently, in equation (2) corresponding to the hyperelliptic curves (4) R is a rational function of x only, with singularities only at e_i. Analyzing the behavior of the uniformizing variable at ∞ one is able to get even the following clear description of (2), Whittaker, [W3]. The uniformizing variable z of (4) is represented as a ratio of two linearly independent solutions of the following second order linear differential equation

$$(\frac{d^2}{dx^2} + R(x))y = 0,$$

where

$$R(x) = \tfrac{3}{16} \cdot$$

$$\{\sum_{i=1}^{2g+2} \frac{1}{(x - e_i)^2} - \frac{2(g + 1)x^{2g} - 2g \sum_{i=1}^{2g+2} e_i \cdot x^{2g-1} + \sum_{i=0}^{2g-2} C_i \cdot x^{2g-2-i}}{(x - e_1)\ldots(x - e_{2g+2})}\}. \tag{11}$$

Here $2g - 1$ parameters $C_0,\ldots,C_{2g-2}$ are accessory parameters. The Fuchsian accessory parameters C_i are determined uniquely by the condition that the monodromy group of (5) is Fuchsian, i.e. is represented by real 2×2 matrices.

Since Poincaré [P2] many researchers applied the transformations of Riemann surfaces (the action of modular subgroups on Teichmüller spacers) to gain the information on the accessory parameters. In particular, in cases when Riemann surfaces are uniquely determined as fixed points of some of these transformations, one can find the explicit expressions of accessory parameters. Poincaré [P2] used these arguments in two famous cases of a punctured tori with a fundamental domain being a square or a rombus.

Whittaker [W2] studying the birational transformations of (4) found explicit expressions for uniformization of a special case of (4) when

$$y^2 = x^5 + 1 \qquad (12)$$

(here $g = 2$ and $e_6 = \infty$). The group G^* in this case is conjugate to the group generated by 5 transformations

$$\sigma_j^* : x \mapsto \frac{\alpha x - \epsilon_j}{\epsilon_j^* x - \alpha} : j = 0, \ldots, 4$$

for $\epsilon_j = exp((4j+1)\pi i/10) :\cdot j = 0, \ldots, 4$ and $\alpha = \sqrt{\frac{\sqrt{5}+1}{2}}$. The group G itself is generated by $\sigma_0\sigma_1, \ldots, \sigma_0\sigma_4$, and the corresponding linear differential equation in this case is a lifting of a particular Gauss hypergeometric equation.

Whittaker called automorphic functions in the case (6) the hyperlemniscate functions, and, on the basis of (6) and other similar examples, had been led to a conjecture describing $R(x)$ in (5) explicitly:

Conjecture 7.1 [W2]. Let $P(x) = (x - e_1) \ldots (x - e_{2g+2})$. Then the Fuchsian accessory parameters $C_0, \ldots, C_{2g-2}$ in (5) are polynomial in e_i, and we have the following explicit expression for $R(x)$ in (5):

$$R(x) = \frac{3}{16} \cdot \{(\frac{P'(x)}{P(x)})^2 - \frac{2g+2}{2g+1} \cdot \frac{P''(x)}{P(x)}\}. \qquad (13)$$

Whittaker students and collaborators studied this conjecture (Mursy [M3-M4], J.M. Whittaker [W3], Dalzell [D2], Dhar [D3] and others). Among their results are: a) the verification of the Conjecture in cases $y^2 = x^{2n+1} + 1$ and $y^p = x^n - 1$ for $p|n$. In these case G^* is a subgroup of one of the Schwarz's triangle subgroups, and the differential equation (5) of the form (7) is a lifting of a hypergeometric equation; b) if accessory parameters C_i are indeed polynomials in singularities e_j, then the conjecture is true. Recent results on the Whittaker conjecture belong to Rankin [R2-R3] who showed the truth of the Whittaker conjecture for a large number of equations associated with one of the finite groups on the Schwarz's list, when (5) is related to the 3-singularity equation.

We checked the Conjecture 4.1 numerically for random curves of genus g between 2 and 6, it is incorrect.

Various numerical experiments suggest even more gloomy state of Conjecture 7.1. We conjectured that Conjecture 7.1 is valid only for thos hyperelliptic curves (4) with a given $e_1, \ldots, e_{2g+2}$ for which there exists an algebraic transformation $x = r(x')$ which reduces Gauss hypergeometric equation to the form (5).

For a fixed g there are only finitely many transformations for which the reduction of Gauss hypergeometric function to an equation of the form (5) is possible.

Consequently, if our "conjecture" holds, for a given g there would be only finitely many hyperelliptic equations (4) for which the accessory parameters C_i in (5) are expressed as polynomials of e_j, as in (7). Still, there are many such equations. In addition to examples presented above here are some others, determined in [R2-R3]:

$$y^2 = x(x^{10} + 11x^5 - 1);$$

$$3y^2 - 2y^3 = 4x^3 - 3x^4;$$

$$y^m = x^p(x^9 - 1)^r$$

$$y^m = \frac{(x^8 + 14x^4 + 1)^3}{x^4(x^4 - 1)^4}.$$

The crucial question now is not the global algebraicity of accessory parameters as functions of coefficients of an equation defining a Riemann surface—this is disproved in the hyperelliptic case. Rather, one is interested in the arithmetic nature of uniformization parameters, as in the following

Problem. 7.2. Let us consider a linear differential equation (5) for the hyperelliptic equation (4), that gives a Fuchsian uniformization (or the corresponding equation (2) for the general algebraic curve). Let us assume that this algebraic curve is defined over $\overline{\mathbf{Q}}$ (or even $\mathbf{Q}$). Are the accessory parameters algebraic? If they are not algebraic, are they algebraically independent? Conversely, if one starts with a Fuchsian group G in $SL_2(\mathbf{R})$ that uniformizes a hyperelliptic curve Γ, and such that $G \subset SL_2(\overline{\mathbf{Q}})$ (i.e. G is represented by 2×2 matrices with algebraic, or even integral algebraic, entries), does one get a hyperelliptic curve defined over $\mathbf{Q}$? Are the corresponding accessory parameters algebraic?

The algebricity of accessory parameters is known only in cases when the answer to the Whittaker conjecture is positive, or when the uniformizing group G is an arithmetic Fuchsian group (typically arising from the quaternion algebras over totally real fields, see Swinnerton-Dyer [SW1]).

Our multiprecision computations of the accessory parameters for hyperelliptic curve (4), run in the SCRATCHPAD environment, seem to indicate the transcendence of accessory parameters for curves not uniformized by arithmetic groups, or groups reducible to triangle ones. We determined the absence of any nontrivial algebraic relations (via multidimension Perron's analogue of the continued fraction expansion or Lenstra's lattice algorithm) of moderate degrees with moderate size integer coefficients. It would be surprising, if algebraic accessory parameters arise in cases different from arithmetic subgroups or Schwarz triangle groups (not all of these groups are arithmetic).

One can ask, see [C5], a more general question on the transcendence of the elements of the monodromy matrices of an arbitrary Fuchsian linear differential equation. Even in the case of the second order Fuchsian linear differential equation very little is known on the transcendence even of a single element of the matrix. Among the second order equations the transcendence of elements of the monodromy matrices had been studied only in case when a monodromy group of this equation is <u>commutative</u>. [Hint: if a monodromy group is commutative, the solutions of the equation can be parametrized by means of θ-functions, and the elements of the monodromy group are expressed in terms of the combinations of periods of Abelian integrals of the third kind. In this case one can use traditional methods of the transcendental number theory to obtain some information on the arithmetic nature of these elements.]

Recently we were able to prove new transcendental results for monodromy groups of Fuchsian differential equations. These results are based on methods that stem from a uniformization of these Fuchsian equations by special subgroups of $SL_2(\mathbf{C})$, using the uniformization of algebraic curves by special function in H having only 3 branch points [C6]. The only condition we impose is the existence of a G-function solution for this differential equation. We remind that a G- function is a function having an expansion at some algebraic point x_0, $f(x) = \sum_{N=0}^{\infty} c_N (x - x_0)^N$ with algebraic c_N such that the common denominator of $\{c_0, \ldots, c_N\}$ is bounded by the geometric progression in N. Then we have:

<u>Theorem 7.3.</u> Let a Fuchsian differential equation $L[\frac{d}{dx}, x]y = 0$ of order n be satisfied by a (transcendental) G-function $f(x)$ (that does not satisfy

any equation over $\overline{\mathbf{Q}}(x)$ of order less than n). If $L[\frac{d}{dx}, x] \in \overline{\mathbf{Q}}(x)[\frac{d}{dx}]$, and $\mathcal{M}$ is a monodromy group of L, corresponding to a choice of a fundamental system of solutions $y_1, \ldots, y_n$ of $Ly = 0$ with algebraic initial conditions $y_i^{(i)}(x_1)$ for an algebraic x_1. Then at least one element of one matrix from $\mathcal{M}$ is transcendental provided that the base point x_1 of $L[\cdot]y = 0$ is nonsingular.

The last assumption in Theorem 7.3 is crucial otherwise a monodromy group of $L[\cdot]y = 0$ might be represented by matrices in $GL_n(\overline{\mathbf{Q}})$.

Example 7.4. The Gauss hypergeometric function $_2F_1(\frac{1}{2}, \frac{1}{2}; 1; x)$ at $x = 0$. This is a G-function, satisfying a Legendre linear differential equation, whose monodromy group is $\Gamma(2)$, if you choose a basis on $K(k), K'(k)$. [Here $x = k^2$.]

8 Formal Groups and Their Applications.

The study of power series expansions over $\mathbf{Z}$, $\mathbf{Z}_p$ and $\mathbf{F}_p$ is of particular importance in many number-theoretic and combinatorial problems. The nearly integral power series expansions (mentioned in §§2-3) are important in many problems of diophantine geometry. Among these expansions the class of Siegel's G-functions [C7] plays a special role. According to Siegel, a function $f(x)$ with the expansion $f(x) = \sum_{n=0}^{\infty} a_n x^n$ is called a G-function if $f(x)$ satisfies a linear differential equation over $\overline{\mathbf{Q}}(x)$, if all coefficients a_n are algebraic numbers, and if, for all n, the sizes $|\overline{a_n}|$ of the algebraic numbers a_n and common denominators $den\{a_0, \ldots, a_n\}$ are bounded by geometric progressions in n:

$$|\overline{a_n}| \le C_1^n, \; den\{a_0, \ldots, a_n\} \le C_2^n : n \ge 1$$

with constants $C_1 > 1$, $C_2 > 1$ depending only on $f(x)$.

Among G-functions, in addition to algebraic functions, one finds generalized hypergeometric functions $_{p+1}F_p(^{a_1,\ldots,a_{p+1}}_{b_1,\ldots,b_p}|x)$ with rational a_i, b_j, and various periods of algebraic varieties (solutions of Picard-Fuchs equations) depending on a parameter x. Following the celebrated Siegel's conjecture on the structure of E-functions one can speculate that all G-functions "arise from geometry" (Dwork), i.e. are reducible to integrals of algebraic functions, [C7].

In order to prove a variety of new number-theoretic results on the irrationality and the linear independence of values of G-functions at algebraic points one had to prove that linear differential equations satisfied by *G*-functions are subject to severe arithmetic constraints. Using methods of Padé approximations we were able to prove that the existence of a nontrivial *G*-function solution of a linear differential equation implies a very good *p*-adic convergence of all solutions of this equation. One of our results [C7] can be formulated as

Theorem 8.1. Let $L[y] = 0$ be a linear differential equation of order n over $\overline{\mathbf{Q}}(x)$ satisfied by a *G*-function $y(x)$, which does not satisfy any linear differential equation over $\overline{\mathbf{Q}}(x)$ of order $< n$. Then all solutions of the equation $L[y] = 0$ with algebraic initial conditions at an algebraic point $x = x_0$ have *G*-function expansions at this point.

In particular, a linear differential equation $L[y] = 0$ is globally nilpotent (in the sense of Katz, see [C7]), which means that

$$(\frac{d}{dx})^{pn} \equiv 0 \bmod (p, L[\cdot])$$

for almost all (density one) p. The last condition is called the "*p*-curvature of L is nilpotent" (Katz). A stronger condition

$$(\frac{d}{dx})^{p} \equiv 0 \bmod (p, L[\cdot]),$$

is called "*p*-curvature is zero", and is the subject of the Grothendieck conjecture:

The Grothendieck Conjecture: Let $L[y] = 0$ be a linear differential equation of order n over $\mathbf{Q}(x)$. If $L[y] \bmod p = 0$ has n linearly independent solutions in $\mathbf{F}_p[x]$ or, equivalently, $L[y]$ has a "zero *p*-curvature" for almost all p, then all solutions of $L[y] = 0$ are algebraic functions.

According to this conjecture, see [C7], strong integrality properties of power series expansions of all solutions of a given linear differential equation imply that all these solutions are algebraic functions.

As we have shown in [C7], [C6], [C11], it is possible to prove a variety of positive results whenever the assumptions of near integrality of coefficients of power series expansions— nonarchimedean conditions—are coupled with assumptions on the analytic continuation of an expanded function in the complex plane (its Riemann surface)—an archimedean condition. An early example of such a theorem is the Borel-Polya theorem, according to which power series expansions with integral coefficients, meromorphic in the domain of conformal radius > 1 represent rational functions. Padé approximations methods allowed us to generalize considerably this result to functions uniformized by various cases of meromorphic functions [C7], [C11]. Roughly speaking, one of our results [C11] means that n functions $f_1(\bar{x}), \ldots f_n(\bar{x})$ in g variables $\bar{x} = (x_1, \ldots, x_g)$, $n \geq g + 1$, having "nearly integral" power series expansions at $\bar{x} = \bar{0}$, and being uniformized near $\bar{x} = \bar{0}$ by meromorphic functions in $\mathbb{C}^g$ of finite order of growth, are *algebraically dependent*. The condition of "near integrality" means that denominators of coefficients in the expansions grow slower than factorials. We present one example of our results from [C11]:

<u>Theorem 8.2.</u> Let $n \geq g + 1$ functions $f_1(\bar{x}), \ldots, f_n(\bar{x})$ be uniformized near $\bar{x} = \bar{0}$ by n meromorphic functions $U_1(\bar{u}), \ldots, U_n(\bar{u})$ in $\mathbb{C}^g$ of finite order $\leq \rho$ of growth under a nonsingular transformation $\bar{x} \rightleftharpoons \bar{u}$ (near $\bar{x} = \bar{0}$). Let $f_i(\bar{x}) = \sum_{\bar{m} \in \mathbf{Z}^g} a_{\bar{m}}^i \bar{x}^{\bar{m}}$ be expansions at $\bar{x} = \bar{0}$ such that $a_{\bar{m}}^{(i)} \in K$ for an algebraic field K of degree d over $\mathbf{Q}$. Put $f_1(\bar{x})^{k_1} \ldots f_n(\bar{x})^{k_n} = \sum_{\bar{m}} a_{\bar{m}; \bar{k}} \bar{x}^{\bar{m}}$ for $k_i \geq 0$ and denote

$$\Delta_M = den\{a_{\bar{m}; \bar{k}} : \| \bar{m} \| < M, \| \bar{k} \| < M\};$$

$$\xi = \limsup_{M \to \infty} \frac{\log |\Delta_M|}{M \log M},$$

$$\sigma = \overline{\limsup_{M \to \infty} \{|a_{\bar{m}}^{(i)}|^{1/M} : \| \bar{m} \| < M\}}.$$

Then, if $\sigma < \infty$ and $\xi < \frac{(1-g/n)}{d\rho}$, the functions $f_1(\bar{x}), \ldots, f_n(\bar{x})$ are algebraically dependent over K. [Here $\| \bar{m} \| = m_1 + \ldots + m_g$, etc.]

From these general theorems we deduced positive solutions to the Grothendieck conjecture, whenever all solutions of a given linear differential equation over $\overline{\mathbf{Q}}(x)$ are uniformized by meromorphic functions of finite order of growth. This is the case of the Lamé equations with integral n. Other equations we have proved the Grothendieck conjecture include all rank one equations over elliptic curves and other curves of positive genus, and all linear differential equations describing the so called finite-band potentials, see [C7].

These results on the Grothendieck-Katz conjecture can be applied to a group of interesting questions proposed by Matthews, including the following amusing problem on indefinite integration of algebraic functions. Let Γ be a curve over $\mathbf{Q}$ and D be any derivation of a function field $\mathbf{Q}(\Gamma)$. Let $f \in \mathbf{Q}(\Gamma)$ be such that for almost all p, we can find g in $\mathbf{F}_p(\Gamma \bmod p)$ such that, $\bmod p$, $f = Dg$. Is it true then, that

$$f = Dg$$

for some g in $\mathbf{Q}(\Gamma)$? (If an integral is locally algebraic from the same field, is it globally algebraic?) The answer is "yes".

How about other linear differential equations? Most of them are not uniformized by meromorphic functions (e.g. Lamé equations with nonintegral n). Again, the answer seems to depend on the combination of nonarchimedean and archimedean conditions imposed on solutions of a linear differential equation. For example, if solutions of a linear differential equation can be uniformized by an arithmetic group, one can prove a positive answer to the Grothendieck conjecture. Moreover a much weaker condition: that the monodromy group of a linear differential equation of order n is up to conjugation a subgroup of $GL_n(\overline{\mathbf{Q}})$—implies a positive answer to the Grothendieck conjecture, provided that this linear differential equation satisfies the assumptions of the Grothendieck conjecture [C6].

These results on algebraicity properties of functions satisfying archimedean and nonarchimedean analytic continuation conditions can be applied to formal groups associated with elliptic curves. Let us look at a plane cubic model of an elliptic curve in E: $y^2 + a_1 xy + a_3 y = x^3 + a_2 x^2 + a_4 x + a_6$. The functions x and y on E are parametrized by Weierstrass elliptic functions:

$$x_E = \mathcal{P}(u) - (a_1^2 + 4a_2)/12,$$

$$2y_E = \mathcal{P}'(u) - a_1 x - a_3;$$

$\mathcal{P}'(u)^2 = 4\mathcal{P}(u)^3 - g_2 \mathcal{P}(u) - g_3$. An invariant differential on E is $\omega = dx/(2y + a_1 x + a_3) = du$. The local parameter z near the origin is $z = -x/y$ (Tate). Then the expansions $\omega = \sum_{n=1}^{\infty} b_n z^{n-1} dz$ of ω with $b_1 = 1$, $b_i \in \mathbf{Z}[a_1, \ldots a_6]$ gives rise to the eliptic logarithm of E:

$$l_E(z)(= u) = \sum_{n=1}^{\infty} \frac{b_n}{n} z^n.$$

If E is defined over $\mathbf{Z}$, i.e. $a_i \in \mathbf{Z}$, then the formal group law $F_E(X, Y) = l_E^{-1}(l_E(X) + l_E(Y))$ is defined over $\mathbf{Z}$. Honda's criterion for isomorphism of one-dimensional group laws over $\mathbf{F}_p$ can be interpreted, in case of elliptic curves, as a nonarchimedean convergence criterion. Combination of this

Chudnovsky and Chudnovsky

nonarchimedean convergence with the uniformization by the Weierstrass elliptic functions can be used to prove isogeny theorems that are particular cases of the Tate conjecture. For example, let E_1 and E_2 be two elliptic curves defined over $\mathbf{Q}$ that have isomorphic group laws over $\mathbf{F}_p$ (or equivalently, the same number of solutions $\bmod p$) for almost all p. Then E_1 and E_2 are isogenous over $\mathbf{Q}$, [C11].

Indeed, according to the Honda criterion (see [C11]), the two formal group laws $F_{E_1}(X,Y)$ and $F_{E2}(X,Y)$ are strictly isomorphic over $\mathbf{Z}_p$ for almost all p. This means that the power series $f(z) = L_{E_1}^{-1}(L_{E_2}(z)), (= z + 0(z^2))$ has p-integral coefficients for almost all p. We can apply Theorem 8.2 (with $g = 1$, $n = 2$) to two functions z and $f(z)$. These functions have nearly integral expansions and are uniformized by the Weierstrass elliptic functions

$$z = z_{E_1}(u), f(z) = z_{E_2}(u),$$

where

$$z_{E_i}(u) = -x_{E_i}(u)/y_{E_i}(u) : i = 1, 2.$$

Theorem 8.1 implies that the function $f(z)$ is algebraic over $\mathbf{Q}(z)$, i.e. that the elliptic functions $\mathcal{P}_1(u)$ and $\mathcal{P}_2(u)$ are algebraically dependent over $\mathbf{Q}$. This means that E_1 and E_2 are isogeneous over $\mathbf{Q}$. This result is effective, for bounds see [C11]. Of course, more general results on the Tate conjecture belong to Faltings [F2].

Formal groups over $\mathbf{F}_p$, particularly formal completion of elliptic curves, can be used for primality testing and factorization.

It is well known that the necessary test of primality of p is based on Fermat's Little Theorem $x^p \equiv x(\bmod p)$ for any $x \bmod p$. This test, say, for $x = 3$, is also an efficient "practically sufficient" test of primality of a prime suspect p, provided p has no small factors [K2]. Though the inverse to the Little Fermat Theorem is incorrect, there are various ways to obtain sufficient conditions of primality from the structure of the multiplicative group of $(\mathbf{Z}/n\mathbf{Z})^*$. The famous $n-1$ primality criterion (Pocklington, 1914; Selfridge, see [K2]) states: if for every prime divisor p of $n-1$ there exists $x_p \bmod n$ such that $x_p^{n-1} \equiv 1 \bmod n$, but $x_p^{(n-1)/p} \not\equiv 1 \bmod n$, then n is prime.

To see how these and other criteria of primality are derived one can reformulate the Little Fermat Theorem in terms of a Frobenius acting on an algebraic (formal) group over $\mathbf{F}_q$ [C8].

<u>Little Fermat Theorem.</u> Let an algebraic group A be defined over $\mathbf{F}_q$

(e.g. A is a reduction of an algebraic group scheme defined over $\mathbf{Z}[1/N]$ for $(N, p) = 1$), and let $P(x)$ be a characteristic polynomial of a Frobenius automorphism F_q of A. Then for an arbitrary X of A, the sequence of multiples

$$X_m = [m]_A(X) : m \in \mathbf{Z}$$

in the group law of A has a *rank of apparition of p*, and this rank, $\tau(p)$, divides $P(1)$. This means that

$$X_m = \bar{0} \ in \ A/\mathbf{F}_q$$

if and only if

$$\tau(p)|m$$

for $\tau(p)|P(1)$.

In these notations the Lucas sequence [K2] correspond to the multiplicative formal group law

$$F_{\sqrt{D}}(x, y) = x + y + \sqrt{D} \cdot x \cdot y.$$

Similar to $n \pm 1$ criteria of primality we have presented in [C8] sufficient (and necessary) criteria of primality based on the divisibility of orders of elliptic curves (and CM-Abelian varieties) A defined over $\mathbf{Z}/N\mathbf{Z}$. E.g. let $N = \sqcap_{i=1}^{k} p_i^{\alpha_i}$, where p_i are (distinct) divisors of $n (> 3)$, and let E be an elliptic curve defined over $\mathbf{Z}/N\mathbf{Z}$ (over $\mathbf{F}_{p_i}$ for all p_{i_k}). If one denotes by $\phi_E(\cdot)$ the analog of Euler function: $\phi_E(N) = \sqcap_{i=1}^{k} p_i^{\alpha_i - 1}(p_i + 1 - a_{p_i}(E))$, where $a_{p_i}(E)$ are traces of Frobenius of $E \bmod p_i$, then $\phi_E(N)$ coincides with the order of $E(\bmod N)$ if and only if N is prime.

Goldwasser et al. suggested a (probabilistic) primality test, based on elliptic curves, where to prove the primality of a suspect prime n, one had to find (by random search) an elliptic curve $\bmod n$, whose order is $2q$, where q is a (suspect) prime. Then one checks the primality of q, etc. The polynomiality of the running time of this algorithm depends on conjectures of distribution of primes. Moreover, this algorithm, and similar ones (see Lenstra's talk at this Conference) are hard to implement for large n, because they depend on fast methods of computation of numbers of points $(\bmod n = p)$ on elliptic curves over $\mathbf{F}_p$. Though there is Schoof's algorithm that allows to compute this number in $O(\log^6 p)$ steps, its realization is unrealistic for p having a hundred digits. It is much more efficient to start with elliptic curves and Abelian varieties with complex multiplication, where one has explicit formulas for the traces of Frobenius. Atkin constructed an algorithm that uses modular equations that can work with primality proving for 200-400 digit long numbers, see Lenstra's talk. The best results, though, are achieved using CM-varieties.

Other applications of algebraic curves over $\mathbf{F}_p$ to new low complexity algorithms include new interpolation algorithms from [C12] that reduces the problem of polynomial multiplication to the interpolation on algebraic curves of positive genus. Using special curves over $\mathbf{F}_p$ with a large number of rational points over $\mathbf{F}_p$ (in terms of genus of a curve), we had constructed new algorithms of multiplication in $\mathbf{F}_p[t]$. These are bilinear algorithms that have multiplicative complexity (rank) *linear* in the degrees of multiplied polynomials, see [C12]. An important problem is the extension of the results of [C12] to $\mathbf{Z}$ or $\mathbf{Z}[1/2]$.

We complete this chapter with one application of computer algebra environment to formal groups arising from algebraic topology. This is the computation of an explicit expression of the characteristic power series of the elliptic genera in terms of modular functions of level two, conducted by us in the summer of 1985 as an answer to the question of Landweber-Stong [L8], see [C10].

We adopt the standard definitions of [H5]. By a genus one means a ring homomorphism $\phi : \Omega_*^{SO} \to \Lambda$ from the oriented bordism ring to a commutative $\mathbf{Q}$-algebra with unit. The "universal" elliptic genus $\rho_t : \Omega_*^{SO} \to \mathbf{Q}[[t]]$, vanishing on all $[CP(\xi^{2m})]$, arises from a multiplicative characteristic class $\rho_t = \sum_{k \geq 0} \rho_k t^k$ with $\rho_k = \pi_k + \sum_{0 < |\omega| < k} a_\omega^k s_\omega^\pi$. One studies ρ_t by looking at the characteristic power series

$$f_t(y) = 1 + yt + \sum_{k \geq 2} p_k(y)t^k, \qquad (14)$$

where $p_k(y) = \sum_{i=1}^{k-1} a_{(i)}^k y^i \in \mathbf{Q}[y], a_{(1)}^k = 0 : k \geq 2$.

At the request of Landweber and Stong we studied the question of integrality of the coefficients of power series expansions of $f_t(y)$. In the course of the study of this question in SCRATCHPAD environment we derived a closed form expression for $f_t(y)$ in terms of modular functions. For this [C10] one looks at an elliptic curve $y^2 = 1 - 2\delta z^2 + \epsilon z^4$, where $\delta = \delta(t)$ and $\epsilon = \epsilon(t)$ are power series from $\mathbf{Q}[[t]]$, and at associated Jacobi's elliptic function $a \cdot sn(\frac{x}{a}|k^2)$, where a and the modulus k^2 are connected with δ and ϵ as follows:

$$a^2 \delta = (1 + k^2)/2, \quad a^4 \epsilon = k^2. \qquad (15)$$

The relationship of $f_t(y)$ with $a \cdot sn(\frac{x}{a}|k^2)$ is the following [L8], [C10]:

$$\frac{1}{asn(\frac{x}{a}|k^2)} = \frac{1}{2\sinh(\frac{x}{2})} f_t(e^x + e^{-x} - 2). \tag{16}$$

From (3) one deduces the fundamental relations between a and quarter-periods K and K':

$$\frac{2\pi i}{a} = 2K + 4mK + 2niK' \tag{17}$$

(for rational integers m and n). In [C10] we determined the properties of relations (4) under the action of the modular group in the upper half plane of $\tau = i\frac{K'}{K}$. The knowledge of the leading term in the expansion of δ and ϵ in powers of t : $\delta \equiv -\frac{1}{8} + 3t, \epsilon = -t + 7t^2 + \dots$ allows us to identify m and n in (4). Namely, we show that $1/a$ is represented as a power series in $\lambda = k^2$. Since $Y = 1/a$ satisfy, in view of (4), the Legendre's linear differential equation $[\lambda(1-\lambda)\frac{d^2}{d\lambda^2} + (1-2\lambda)\frac{d}{d\lambda} - \frac{1}{4}]Y = 0$, one can identify $1/a$ with K $(= \frac{\pi}{2}2F_1(\frac{1}{2},\frac{1}{2};1,\lambda)$—the basis of solutions regular at $\lambda = 0)$. From the modular structure of $1/a$ and its leading term in the expansion at $\lambda = 0$ we derive a single transcendental relation

$$\frac{\pi i}{a} = K. \tag{18}$$

The relations (5) and (3) uniquely determine $f_t(y)$ as a function of $q = e^{\pi i \tau}$:

$$f_t(y) = \prod_{n=1}^{\infty} \{\frac{1 - yq^{2n-1}/(1 - q^{2n-1})^2}{1 - yq^{2n}/(1 - q^{2n})^2}\}. \tag{19}$$

The parameter t is related to q in the following way $t = -q + O(q^2)$. In fact, as it is noted in [C10], the most natural choice of t is $t = -q$, which makes coefficients in the expansion of $f_t(y)$ in (6) into modular functions of τ.

We also considered in [C10] normalization of $f_t(y)$ under the transformations $t \to t' = t + \sum_{k \geq 2} b_k t^k$. The most important normalization is that of $a_{(i)}^k = 0$ for $i > [\frac{k+1}{2}]$ in (1). In this case t becomes a modular function of q and can be represented in terms of ϑ-constants as follows:

$$t = \frac{1}{24}\{1 - \vartheta_2(0)^4 - \vartheta_3(0)^3\}. \tag{20}$$

As a consequence of the formulas, all coefficients in the expansion of $f_t(y)$ are rational integers.

Recently our formulas (6-7) were applied to some physical problems and reinterpreted in a geometric way by Landweber and Witten [W4]. In this interpretation one starts with characteristic classes Θ_t defined as follows. For vector bundle E one put:

$$\lambda_t(E) = \sum_{n \geq 0} \lambda^n(E)t^n, \; S_t(E) = \sum_{n \geq 0} S^n(E)t^n,$$

where $\lambda^n(E)$, $S^n(E)$ are exterior and symmetric powers of E. Then Witten defines:

$$\Theta_t(E) = \bigotimes_{n=1}^{\infty} [\lambda_{t^{2n-1}}(E) \otimes S_{t^{2n}}(E)].$$

With this characteristic class one obtains a genus $\Theta_t : \Omega_*^{SO} \to \mathbf{Q}[[t]]$ in a natural way

$$\Theta_t(M^{4n}) = \hat{A}(M)ch\frac{\Theta_t(TM)}{\Theta_t(1)^{4n}}[M^{4n}].$$

9 Algebraic Complexities of Polynomial Root Finding. A Review.

A classical problem of algebra is that of resolution of polynomial equations. In its purely algebraic form it involves the computation of the Galois group of a given polynomial (say, with rational integral or rational function coefficients) and its complete resolution (not solution). One can look at the problem of root finding as on the factorization problem: given a polynomial $P(x)$ with coefficients from $K[x]$ for a field K we want to decompose $P(x)$ into the product of irreducible polynomial factors. [This problem is even more difficult, if one replaces K by a ring A.] One has to distinguish two different classes of cases:

I) K is an algebraic number field or, more precisely, $K = \mathbf{Q}$. In this case one wants to decompose $P(x) \in \mathbf{Z}[x]$ into products of irreducible factors having integral rational coefficients. Computer algebra systems usually deal with factorization of polynomials in one or many variables over the rational number field, finite fields and their finite extensions. By now the factorization algorithms for these fields are well developed, and these algorithm are implemented in all modern computer algebra systems. We refer to the review of Kaltofen [K1] for the historical development.

II) K is an algebraically closed field, or, more precisely, $K = \mathbf{C}$ (though various fields containing $\mathbf{Q}_p$ can also be considered). In this case one looks at all roots of $P(x)$ in K (with the proper multiplicities). Unless the Galois field of $P(x)$ is solvable, all roots of $P(x)$ cannot, in general, be determined by arithmetic operations $(+, -, *, \div)$ and root extraction (i.e. $\sqrt[n]{\cdot}$) in finitely many steps. In this case, one can determine roots of $P(x)$ only with a relative precision ϵ. The precision ϵ is $\epsilon = 10^{-M}$, where M is a number of significant decimal digits that are correct.

In both factorization Problems I and II, the main question is the cost of computations. It is usually formulated as an (algebraic) complexity problem. Crucial parameters are the input data (or, rather, the size of inpute data). These parameters, in the dense case, when $P(x)$ has integral rational coefficients, are:

a) the degree n of $P(x)$;

b) the height H of $P(x)$ (the maximum of absolute values of its coefficients);

c) in the root finding problem in the complex plane, the precision ϵ or $M = -log_{10}\epsilon$, is another parameter.

In the case of sparse polynomials the input parameters are described by the sequence of exponents and coefficients (see §10).

The main problem in the algebraic complexity framework is the polynomiality, which we are going to discuss in general. However, we have to mention the essential parameters in which the problem has to be polynomial:

$$n, \, log \, H \, \text{ and } \, log \, | \, \frac{1}{\epsilon} | (\text{or even } log \, M)$$

In parallel computations the crucial quantity is the depth–the number of parallel steps and the number of processors. The polynomiality, which one would like to expect, is the polynomiality in terms of the parameters above, in the number of unit cost operations. The number of parallel steps would be, however, logarithmic in these parameters, while the number of parallel processors is allowed to be polynomial in these parameters.

This model of compexity reflects computational rather than bit (boolean) costs. That is why one has to take into account the price per unit operation. If it is a single arithmetic operation (multiplication) performed on an M-digit float or an M-digit integer, the price is bounded by $M(M)$. This concerns the input data with the size of $log \, H$, and results of intermediate computations, performed with the precision M.

The problem I on the factorization of polynomial with rational integer coefficients is well studied, particularly using reduction $mod \, p$ methods and the examination of short vectors in lattices. Important theoretical developments belong to Lenstra et al. [L1] and Kaltofen [K1] (in the parallel case).

None of the traditional number-theoretic methods can be used in Problem II. Here the greatest contribution has been made by classical mathematicians.

There is a constant interest since Gauss's time in the development of theoretical methods of complex root finding. It is not that easy, though, to implement the theoretical method in an automatic polynomial root-finding problem for polynomials of relatively large degree. Concerning the definition of "large degree," it was not so long ago that Lanczos [L2] remarked that the degree of a polynomial "rarely exceeds 6", and G. Birkhoff [B1] in 1973 listed as an outstanding problem, "finding the roots of polynomial equations of degree up to 100."

Even in computer algebra systems one needs polynomial root finding. E.g. in our RIEMANN package we required means of separating the singularities of linear differential equations. This particular problem and studies of multisoliton interaction were the starting points in our work on sequential and parallel algorithms for complex root finding.

Let us review briefly the existing methods of polynomial root finding in the complex plane.

An important class of methods constitute the class of local or iterative methods, based on the construction, starting from an initial approximation, of a denumerable sequence of (complex) numbers that converge to a zero of $P(x)$. This sequence z_m is defined iteratively as:

$$z_{m+1} = \phi(z_m, z_{m-1}, \ldots, z_{m-k}) : m \geq k.$$

Some of these methods are called globally convergent (in Traub's [T1] terminology), because they converge starting from an arbitrary initial point. These methods may require, however, high order iterations.

Among the most widely used local methods are

9.1 The Newton method.

For an equation $f(z) = 0$, the Newton iterations are:

$$z_{m+1} = z_m - \frac{f(z_m)}{f'(z_m)}.$$

As it is well known, if z_0 is chosen very close to a simple root of $f(z) = 0$, the sequence z_n converges to this root quadratically. The first qualitative study of the Newton method had been undertaken in the polynomial case by Smale and Shub-Smale see [S1],[SS1]. Smale [S1] showed that for a large class of polynomials (of measure $\geq 1 - \delta$ in an appropriate measure on the space of polynomials), the Newton method converges after at most $O(n^{0(1)}\delta^{-0(1)})$ iterations (starting at $z_0 = 0$):

$$z_{i+1} = z_i - h\frac{P(z_i)}{P'(z_i)}, h = h(n, \delta),$$

and one arrives at z_m, starting from which the Newton method already converges quadratically to a zero of $P(x)$. Smale and Smale-Shub later improved this result for the modified Newton method. In this method, for "almost all" polynomials it takes "on average" $O(n + log \mid \frac{1}{\epsilon} \mid)$ iterations

$$z_{m+1} = z_m + \frac{\omega_m - P(z_m)}{P'(z_m)}$$

(starting from a "random"$z_0, |z_0| = 3$) to get an "approximate" zero z_k:

$$\mid P(z_k) \mid < \epsilon.$$

Unfortunately, Newton-like methods suffer from the presence of multiple or close zeros, and in order to guarantee the convergence, zeroes are to be separated. E.g. in the case of clustered zeroes Newton-like methods are inefficient.

Among other iterative methods one can name the method of secants, Müller's method, Laguerre method and Euler-Newton iterations of higher order [H1]. The Euler-Newton method is actually related to our current work because it tries to study iteratively the inverse function $x = x(t)$ of $P(x) = t$. The (ordinary) Newton method looks just on the leading term of the expansion of x.

9.2 Laguerre method.

Highly recommended method (iterations of the third order), believed to be globally convergent. If $x_1 = P'(x)/P(x), x_2 = (P'(x)^2 - P(x)P''(x))/P^2(x)$, then the Laguerre iterations are:

$$x' = x - \frac{n}{x_1 \pm \sqrt{(n - 1)(nx_2 - x_1^2)}}$$

(choose the sign $\pm$ to minimize the addition to x).

More elaborate iterative methods (including the descent method) follow the pattern of behavior of $P(x), P'(x), \ldots$ etc. to establish the multiplicity of the root and to isolate it from its close neighbors and critical points. The most successful procedure of such kind is the Jenkins-Traub algorithm [JT1] adopted in the most standard packages. Iterative methods are reasonably fast, when they converge. But even the global iterative methods have serious handicaps. First of all, they typically produce one root at a time, and so such are not well suited for parallelization. Attempts to start at several initial points at once, to make iterative methods parallel, are usually unsuccessful because clustered roots cannot be found this way. To separate one root from another one has to divide the polynomial by the found linear factor–the procedure of deflation. Masters of numerical analysis gave various recipes of deflation (start from the root with the smallest absolute value,... etc.) but the loss of the accuracy occurs at every deflation step. The only remedy is the increase in the precision of initial computations. Deflation becomes particularly unpleasant in the fixed precision computations, like in double precision FORTRAN, after 30-50 deflations. (Sometimes one cannot "polish" a root after the deflation.) One of the best automatic packages based on a clever deflation was developed by Y.F. Chang [C1], whom we thank very much for providing access to his programs.

More attractive are global methods, when all or a large group of zeroes are determined at once. This seems to be the strategy in recent theoretical studies of root finding. One of the first global methods can be attributed to Bernoulli. We try to put the global methods together under the general umbrella of Padé approximationsn to (rational) functions, whose poles are precisely the roots of a given polynomial.

Let $f(x) = \sum_{N=0}^{\infty} c_N x^N$ be regular at $x = 0$ an $[L/M]$ Padé approximation to $f(x)$ is a rational function $P_L(x)Q_M(x)$ for polynomials $P_L(x), Q_M(x)$ of degrees at most L and M, respectively, such that $\mathrm{ord}_{x=0}(Q_M(x)f(x) - P_L(x)) \geq L + M + 1$. An $[L/M]$ Padé approximation always exists and is unique see [H2]. Padé approximations are used to find the singularities of $f(x)$. A particularly simple recipe is provided by classical de Montessus de Ballore theorem

<u>Theorem (de Montessus) 9.1.</u> Let $f(x)$ be meromorphic in the disk $|x| \leq R$ with m poles at distinct points $z_1, \ldots, z_m$ with

$$0 < |z_1| \leq \ldots \leq |z_m| < R.$$

Let the total multiplicity of poles be $M = \sum_{i=1}^{m} mult(z_i)$. Then

$$f(x) = \lim_{L \to \infty} [L/M]$$

uniformly on any compact subset of $\{x : |x| \leq R, x \neq z_i : i = 1, \ldots, m\}$.

This result and its generalizations provide the cornerstone in a variety of global methods of root finding. To apply these methods one usually looks at an expansion of a rational function

$$f(x) = \frac{Q(x)}{P(x)} \text{ for some polynomial } Q(x),$$

$deg(Q) < deg(P)$. Typically one takes

$$f(x) = \frac{P'(x)}{P(x)} \text{ or } f(x) = \frac{1}{P(x)}$$

and changes the variables x to x^{-1}. Even the case of $M = 1$ in the de Montessus theorem is important. This case is related to the definition of the radius of convergence of series $f(x) = \sum_{N=0}^{\infty} c_N x^N$. If $f(x)$ is a rational function, then its radius of convergence is the distance to the nearest singularity and is

$$\limsup_{N \to \infty} |c_N|^{\frac{1}{N}}.$$

This observation is the basis of Bernoulli method: if $P(x) = a_0 \cdot \prod_{i=1}^{n} (x - x_i)$, and $s_N = \sum_{i=1}^{n} x_i^N$, then:

$$\frac{P'(x)}{P(x)} = \sum_{N=0}^{\infty} \frac{s_N}{x^{N+1}}.$$

<u>Corollary 9.2.</u> If $|x_1| \leq \ldots \leq |x_{n-1}| < |x_n|$, then

$$\lim_{N \to \infty} \frac{s_{N+1}}{s_N} = x_n.$$

In this Bernoulli method, s_i for $i = 0, \ldots, n$ are evaluated from Newton's formulas for symmetric functions, and higher s_N can be determined from the $n - th$ order recurrence with constant coefficients:

$$a_0 s_N + a_1 s_{N-1} + \ldots + a_n s_{N-n} = 0 : N \geq n$$

for

$$P(x) = a_0 x^n + a_1 x^{n-1} + \ldots + a_n.$$

If there is no single dominant root, or there are several roots with an absolute value close to that of the dominant one, an immediate application of the Bernoulli method is impossible. In this case the estimate of the radius of convergence of $P'(x)/P(x)$ at $x = \infty$ gives us an absolute value of a dominant root:

$$|x_n| = \limsup_{N \to \infty} |s_N|^{\frac{1}{N}}.$$

A very good precise bound of this kind was proved by Turan [T2]:

Proposition (Turan) 9.3. Let $|x_1| \leq \ldots \leq |x_n|$ and $s_N = \sum_{i=1}^n x_i^N$. Then for any $N \geq 1$:

$$5^{-\frac{1}{N}} < \frac{max\{|s_{Ni/n}|^{1/(Ni)} : i = 1, \ldots, n\}}{|x_n|} \leq 1.$$

There are several well-known methods to recover x_n from its absolute value. They include: i) a slight shift $x \to x + h_0$ in $P(x)$ (for $h_0 \leq min \frac{|x_i - x_j|}{2}$); ii) a shift $x \mapsto x + \epsilon$ for $\epsilon \to 0$; iii) a transformation of a circle $|x| = |x_n|$ into the real axis; iv) a subresultant methods looking for quadratic factors $x^2 + Px + \rho^2$ of $P(x)$, $\rho = |x_n|$, having two roots that are two roots of $P(x)$ of the same absolute value. The last method is particularly attractive for real polynomials $P(x)$ and a complex x_n.

Instead of determining roots one by one, one can try to determine them all at once, or in large groups. The Padé approximants $[L/M]$ to $P'(x)/P(x)$ for $M > 1$ can do this, provided that:

$$|x_1| \leq \ldots \leq |x_{n-M}| < |x_{n-M+1}| \leq \ldots \leq |x_n|.$$

One of the best methods to do the group root finding is the qd-method of Rutishauser popularized by Henrici [H2]. Another variation of the same method involves the direct computation of Hankel determinants

$$H_{N,m} = \begin{vmatrix} s_{N-m+1} & \cdots & s_N \\ s_{N-m+2} & \cdots & s_{N+1} \\ \cdots & \cdots & \cdots \\ s_N & \cdots & s_{N+m-1} \end{vmatrix}.$$

In fact, the nondegeneracy of these determinants is a sufficient condition of the convergence of qd-method.

To bound the algebraic complexity of this group of global methods, one needs fast methods to compute the coefficients s_N of the expansion of $P'(x)/P(x)$. The usual linear recurrence requires $O(nN)$ operations in the straightforward implementation. This number can be significantly reduced if one realizes that $s_0, \ldots, s_N$ can be determined at once by looking at the polynomial quotient of

$$x^N \cdot P'(x) \bmod P(x).$$

This algorithm can be accelerated further if one computes $x^N \bmod P(x)$ for large N using successive squarings $x^{2N} \bmod P(x) = (x^N \bmod P(x))^2 \bmod P(x)$ etc. This method of successive squarings gets

$$s_2, s_4, s_8, s_{16}, \ldots.$$

The fast method of successive squaring was proposed long ago as a means to separate further already separated roots. It had been suggested at various times by Dandelin [D1] (1826), Lobachevsky [L3], (1834), and Graeffe [G1] (1827), who presented it in the algorithmic form. We prefer to call it Lobachevsky in honor of D.H. Lehmer. The essence of the method is the following:

Starting from $P(x) = a_0 \prod_{i=1}^{n} (x - x_i) = a_0 x^n + \ldots + a_{n-1} x + a_n$, we put

$$P^{<2>}(x) \stackrel{\text{def}}{=} P(-\sqrt{x})P(\sqrt{x}) = a_0^2 \prod_{i=1}^{n} (x - x_i^2).$$

and, in general, $P^{<m>}(x) = a_0^{2^m} \prod_{i=1}^{n} (x - x_i^{2^m})$. There is a simple law to generate the coefficients of $P^{<m>}(x) = a_0^{<m>} x^n + \ldots + a_n^{<m>}$:

$$a_\alpha^{<m+1>} = a_\alpha^{<m>2} + 2 \sum_{i \geq 1} (-1)^i a_{\alpha-i}^{<m>} a_{\alpha+i}^{<m>}.$$

Simple analysis shows that not only the (absolute value of a) dominant root is determined this way, but all the roots as well, provided that their absolute values are well separated. E.g. if

$$|x_1| \leq \ldots \leq |x_n|$$

and

$$|x_{k-1}| < |x_k| < |x_{k+1}|,$$

then

$$|x_k| = \lim_{m \to \infty} |a_{n-k+1}^{<m>}/a_{n-k}^{<m>}|.$$

Again, one can recover the value of x_k from $|x_k|$ by means of one of the methods i)-iv) above.

The attractiveness of Lobachevsky method is the fast computation of coefficients of $P^{<m>}(x)$. The use of the sequential FFT allows us to do this in $O(mn \, log \, n)$ unit cost operations. In practice this number is much larger because of increase in the sizes of coefficients (the scaling leads to the sacrifice of the FFT method). Parallel implementation of Lobachevsky offers even better possibilities. They were exploited recently by V. Pan [P1]. To prove good complexity bounds Pan required the knowledge of the root separation. E.g. Pan showed that for polynomials with

$$|1 - |\frac{x_i}{x_j}|| \geq n^{-q} \, for \, i \neq j,$$

all roots of $P(x) = 0$ can be determined with precision ϵ in $O(n \, log^2 n + nlogn \, log|\frac{1}{\epsilon}|)$ unit cost operations, $O(log^2 n + log \, n \, log \, |\frac{1}{\epsilon}|)$ parallel steps with n processors.

If no assumptions on $1 - |x_i/x_j|$ are made, Pan's methods require $O(n^2 log^2 n + nlogn \, log|\frac{1}{\epsilon}|)$ parallel steps with n processors.

Other methods of root finding include homotopy methods and variety of proximity tests, that determine the domains in the plane where the roots are present, and isolate them. One of the most popular is the Schur-Cohn-Lehmer test for circular domains. In the case of polynomials with real roots only, the classical method of Sturm sequences and the change of sign-rule provide the best tool for the isolation of real roots. A new idea in the proximity tests (generalizing H. Weyl's method) belongs to Schonhage, who suggested to improve the proximity test into a method of partial factorization of polynomials. Namely, if one knows that there are k roots $x_1, \ldots, x_k$ of $P(x) = 0$ in the domain D with the smooth boundary ∂D (without roots on ∂D), then symmetric functions $\sum_{i=1}^{k} x_i^N (N = 0, \ldots, k)$ can be determined from the contour integral

$$\int_{\partial D} x^N \frac{P'(x)}{P(x)} dx.$$

This gives a factor $P_1(x) = \prod_{i=1}^{k} (x - x_i)$ of $P(x)$. For a detailed description of Schonhage methods see [S2].

This method was recently used by Ben-Or, Feig, Kosen and Tiwari [BOFKT] in the new parallel method of root finding for polynomials $P(x) \in \mathbf{Z}[x]$ with real roots only.

Remark. So far NC hypothesis for parallel root finding was not proved

(NC means a solution in polylogarithmic time using the polynomial number of processors).

The most important deficiency of the described global methods: the number of unit cost operations is polynomial in $M = log_{10}(\frac{1}{\epsilon})$, while it is natural to assume the number of operations to be polynomial in $log\,M$. Such fast convergence is achieved only in local Newton-like methods.

10 Algebraic Complexities of Polynomial Root Finding for Sparse Polynomials.

One calls a (univariant) polynomial sparse, if it has a number of nonzero coefficients significantly less than its degree.

[In general, one can consider a different notion of "sparsity" of a polynomial according to which a polynomial is "sparse", if the definition (e.g. a program describing it) of this polynomial is of low algebraic complexity. This might be, say, a case of a polynomial defined as a characteristic polynomial of a sparse matrix.]

Sparse polynomials, particularly trinomials and quadrinomials, are extremely interesting from many points of view. Let us mention now just one problem of significant interest to the Galois theory:

which Galois groups G can be realized as Galois groups of trinomials $ax^n + bx^m + c = 0$ (for $a, b, c \in \mathbf{Z}$ or for $a, b, c \in \mathbf{Z}[\mathbf{x}]$)?

There are obvious restriction on G, stemming from the fact that a trinomial cannot have more than three real roots. It is not known what are the sufficient conditions for G to be realized as a Galois group of a trinomial (quadrinomial).

Further relationships with multi-soliton theory will be discussed later.

The generic Galois group of a trinomial of degree n is S_n, so roots of a trinomial cannot be determined in closed form for $n \geq 5$. Nevertheless one can express the roots of trinomials in a closed form as infinite series with binomial coefficients. Such an expression can be found in Ramanujan's writing which receives attention every now and then.

Ramanujan normalizes the trinomial equation in the following form

$$aqx^p + x^q = 1, \, a > 0 \text{ and } 0 < q < p.$$

For $n > 0$ and a particular root of a trinomial $aqx^p + x^q = 1$, Ramanujan

[R1,Ch.3 of his Second Notebook] proves the expansion:

$$x^n = \frac{n}{q} \cdot \sum_{N=0}^{\infty} (-qa)^N \cdot \frac{\Gamma(\frac{n+pN}{q})}{N!\Gamma(\frac{n+pN}{q} - N + 1)}. \tag{21}$$

This expansion converges when $|a| \leq p^{-p/q} \cdot (p-q)^{(p-q)/q}$.

This expansion of Ramanujan is usually attributed to Lambert [L6] (1758). The usual derivation of this formula is based on Lagrange inversion theorem following Lagrange's original derivation [L5] (1770). References to this and other similar Ramanujan's formulas can be found in Brendt, Evans, Wilson, [B2].

"Ramanujan's" formula can be also derived directly from the trinomial equation using the transformation of contour integrals into Barn's type integrals of Γ-functions. This derivation had been achieved by Hj. Mellin in his memoir [M1]. Mellin's Γ-function representation of all roots of a normalized trinomial equation

$$x^n + a_1 x^p - 1 = 0, \; p < n; \; q = n - p \tag{22}$$

is the following:

$$x_j^{\mu} = \epsilon_j^{-\mu} \cdot \frac{\mu}{n} \cdot \frac{1}{2\pi i} \int_{c-i\infty}^{c+i\infty} \epsilon_j^{-zp} \cdot \frac{\Gamma(z)\Gamma(\frac{\mu-pz}{n})}{\Gamma(1 + \frac{\mu+qz}{n})} a_1^{-z} dz \tag{23}$$

for n roots of unity: $\epsilon_j^n = 1 : j = 1, \ldots, n$ and $0 < c < \frac{\mu}{p}$. This immediately implies the "Ramanujan" formulas:

$$x_j^{\mu} = \epsilon_j^{\mu} \cdot \frac{\mu}{n} \sum_{N=0}^{\infty} \frac{(-\epsilon_j^p)^N \Gamma(\frac{\mu+pN}{n})}{\Gamma(N+1)\Gamma(1 + \frac{\mu-qN}{n})} a_1^N : j = 1, \ldots, n, \tag{24}$$

convergent for $|a_1| < n/\sqrt[n]{p^p q^q}, q = n - p$.

Mellin developed Γ-function representations for roofs of arbitrary algebraic equations, see [M1].

These integral representations are particularly convenient for sparse polynomials, where they also can be converted into multivariate power series expansions.

Let us look at an arbitrary $k + 2$-nomial

$$x^n + a_1 x^{n_1} + a_2 x^{n_2} + \ldots + a_k x^{n-k} - 1 = 0, \tag{25}$$

$n > n_1 > n_2 > \ldots > n_k > 0$. Then we have the following expansion of all n roots x_j of this equation, or of their $\mu - th$ powers x_j^μ:

$$\epsilon_j^\mu \cdot x_j^\mu = 1 + \mu \sum_{N=1}^\infty \frac{(-1)^N}{n^N} \cdot \sum_{j_1+\ldots+j_k=N} A_{j_1,\ldots,j_k} \cdot a_1^{j_1} \ldots a_k^{j_k}, \qquad (26)$$

where one puts:

$$A_{j_1,\ldots,j_k} = \frac{\prod_{i=1}^{N-1}(\mu - n_i + j_1 n_1 + \ldots + j_k n_k)}{j_1! \ldots j_k!}.$$

The series (6) converge always whenever

$$max(|a_i| : i = 1, \ldots, k)$$

$$< min(n/(k \sqrt[n]{n_1^{n_1}(n-n_1)^{n-n_1}}), n/(k \sqrt[n]{n_k^{n_k}(n-n_k)^{n-n_k}})).$$

In principle, the power series formulas like Ramanujan (1) or Mellin's (6) can be used directly for the root finding of sparse polynomials. The immediate advantage of power series formulas: the simplicity of generation of coefficients of power series expansions–they are generated by two-term linear recurrences. This procedure is particularly well suited for parallel implementation, because the coefficients can be computed in parallel and roots can be evaluated in parallel as well. Unfortunately, simple formulas like (1) and (6) suffer from a serious handicap: series converge only in a small disk around the cyclotomic polynomials. It is much more practical to use our methods of analytic continuation of power series solutions of linear differential equations satisfied by algebraic functions. These linear differential equations were called by Harley [H3] (1862) or by Cayley the differential resolvents. These differential resolvents can be represented in terms of equations on multi-dimensional generalized hypergeometric functions. These equations can be integrated and analytically continued everywhere. The explicit expression for the resolvents can be deduced from power series representations of (1) or (6). The most general expression for a system of Fuchsian linear differential equations on x_i^μ, as functions of $a_1, \ldots, a_k$, can be deduced from the power series representation of (6) for x_j^μ.

Namely, the (algebraic) functions x_j^μ for roots x_j of (5) satisfy the following system of k Fuchsian linear differential equations on $\partial/\partial a_1, \ldots, \partial/\partial a_k$:

$$\{(-1)^{n_i} \cdot n^n \cdot \frac{\partial^n}{\partial a_i^n} - \sqcap_{m=0}^{n_i-1}(n_1 a_1 \frac{\partial}{\partial a_1} + \ldots + n_k a_k \frac{\partial}{\partial a_k} + \mu + nm)$$

$$\times \sqcap_{m=0}^{n_i'-1}(n_1' a_1 \frac{\partial}{\partial a_1} + \ldots + n_k' a_k \frac{\partial}{\partial a_k} - \mu + nm)\}x_j^\mu = 0; n_i' \qquad (27)$$

$$= n - n_i : i = 1,\ldots, k.$$

This system (7) of linear differential equations can be integrated starting from $a_1 = a_2 = \ldots = a_k = 0$ with initial conditions $x_j = \epsilon_j$ for $n - th$ roots of unity $\epsilon_j^n = 1 : j = 1,\ldots, n$. Alternatively integration can start at ∞, or at any other point in k- space, where x_j are known.

We want to remark that for trinomials the differential equation (7) can be reduced to the ordinary generalized hypergeometric equation, Birkeland [B3]. Namely, for

$$x^n + ax^p - 1 = 0$$

and $t = (-1)^p p^p q^q a^n / n^n, q = n - p$ one has the representation:

$$x^\mu = 1 + \frac{\mu}{n} \sum_{r=0}^{n-1} a^r F_r(t),$$

where

$$F_r(t) =$$

$$F\left(\begin{array}{c} \frac{r}{n} + \frac{\mu}{pn}, \ldots, \frac{r}{n} + \frac{\mu+n(p-1)}{pn}, \frac{r}{n} - \frac{\mu}{qn}, \ldots, \frac{r}{n} - \frac{\mu-n(q-1)}{qn} \\ \frac{r+1}{n}, \frac{r+2}{n}, \ldots, \frac{n-1}{n}, \frac{n+1}{n}, \ldots, \frac{n+r}{n} \end{array} \middle| t \right) : \qquad (28)$$

$$r = 0,\ldots, n - 1.$$

The methods of analytic continuation of algebraic functions x_j from $a_i = 0$ can be easily implemented, and its only nontrivial part concern the regular (Puiseux) power series expansions of x_j in the neighborhood of singularities of differential resolvent equations (7), see [C4]. This leads to an iterative algorithm, where to compute the roots of $k + 2$-nomial one has to precompute the roots of $k + 1$-nomials (that give the singularities of the former branches of algebraic functions), etc. E.g. to compute the roots of trinomials, one needs to precompute the roots of unity, etc. This method is perfect for parallel implementation, since on each level of iteration all roots are computed independently. The storage requirements are determined only by the number of roots one wants to see. Moreover, this iterative method was implemented in vector and array hardware, e.g. on IBM 3090-VF or on CRAYs, because most of the operations are vectorizable loops.

The complexity bound for the computation of roots of sparse polynomials using the analytic continuation of solutions of differential resolvents is the following.

<u>Theorem 10.1.</u> One can compute all n roots of a k-nomial of degree n with the precision M (of leading digits) in $O(k^2 log^2 M)$ parallel steps on n processors. If $k \ll n$, then one can compute $m (\leq n)$ roots of a k-nomial of degree n with the precision of M (digits) in $O(k \, log^2 M)$ parallel steps on $O(m)$ processors.

One should note that the price of a unit operation depends not only on M, but also on the size of the exponents (i.e. n), though logarithmically so.

A few words are in order concerning the applications of these numerical methods. One of the applications involves soliton interactions in the theory of the Korteweg-de Vries (KdV) like equations. In this theory of isospectral deformation equations it is well known, and relatively easy to prove, that the interaction of real (confined) solitons separates asymptotically as time $t \to \pm\infty$ into an elastic interaction with the proper shift of phases, see [L7]. What is much less known, is the multisoliton interection in the intermediate range of time t. This intersection is crucial in many problems including: 1) the study of S-matrices of competely integrable systems using interaction of test, soliton, or kink-like particles; 2) the study of interaction between real solutions corresponding to the continuous spectrum; and, 3) the complex picture of multisoliton and finite-band potential interaction. The last problem arises in the study of the blow-up of solutions of the Boussinesq and KdV equations.

The multisoliton interaction can be better understood by analysis of the singularities of multisoliton solutions $u(x,t)$ of the KdV equation. The famous Hirota's expression of the N-soliton is the following:

$$u(x,t) = 2\frac{d^2}{dx^2} log F(x,t). \tag{29}$$

Here the KdV equation on $u = u(x,t)$ is: $u_t + 6uu_x + u_{xxx} = 0$ and $F(x,t)$ is the following exponential polynomial. (Below x_i^0 are initial positions and v_i are the speeds of solitons; $i = 1, \ldots, N$):

$$F(x,t) =$$

$$\sum_{\substack{\mu_1=0,1 \\ \vdots \\ \mu_N=0,1}} \Pi_{i<j}^N \left(\frac{v_i - v_j}{v_i + v_j}\right)^{2\mu_i\mu_j} \cdot \Pi_{i=1}^N exp(x_i^0 + v_i x + v_i^3 t)^{\mu_i}. \tag{30}$$

The roots of $F(x, t)$ in complex x-plane are precisely the poles of the N-soliton (9). These roots can be looked at as zeroes of a sparse exponential polynomial or, if you will, after a proper logarithmic transformation, as roots of the sparse polynomial with arbitrary exponents. The precise locations of roots and their motion as t changes from $-\infty$ to ∞ (competely depend on the arithmetic nature of ratios v_i/v_j. If all these ratios are rational, then poles of N-soliton are periodic with roots of the sparse polynomial determining their position. The motion of singularities is patterned according to the structure of the monodromy group of the corresponding algebraic equation. However, for a general choice of irrational (particularly Liouville-like) v_i/v_j, the position of singularities does not have any immediate symmetry, though one can discover some patterns by matching best diophantine approximations to v_i/v_j. To analyze these patterns and, simultaneously, to describe the S-matrix of the singularity interaction, is to observe the position of roots of the corresponding sparse exponential polynomials by approximating them with k-nomials of large degrees, considering good rational approximations to v_i/v_j.

The analogy with particle interaction (in KdV theory) is now well-known for singularities of meromorphic solutions, see [C2]. According to this interpretation, the singularities interact as particles with the potential $1/x^2$ as the time evolves. This interpretation can be extended to the description of interaction of zeroes of sparse polynomial as functions of polynomial coefficients. From such interpretation it follows that the roots of a k-nomial in general lie on at most $k - 1$ circles with a definitive distribution functions for (relatively) large exponents of a polynomial. These distribution functions depend on the coefficients of polynomials according to certain dynamical systems. One can describe this interaction as an interaction between $k - 1$ strings with self-interaction potential. This situation is similar to the one-dimensional kinetic theory of gases for very special potential cf. with [C3]. A variety of numerical experiments were performed. They involved the computation of roots of trinomial and quadrinomial polynomials and exponential polynomials with variable coefficients, which were displayed as continuous curves. Our programs were run on different computers starting with PCAT with the live screen output to IBM 3090-VF and CRAY II (at

Minnesota Supercomputer Center, courtesy NSF grant). On CRAY II one can determine in real time roots of a trinomial/quadrinomial of degree of up to 1,000,000.

A warning should be given here to those who will try high degree computations in the traditional FORTRAN environment. The crucial osbtacle in the complex root finding for large degrees of polynomials is the need for multiple precision computations. For the discussion of the precision requirements see Schonhage [S2]. These requirements make it necessary to combine the fast root finding methods with fast bignum packages available currently only in the framework of computer algebra systems. Unfortunately high precision requirements make the programming of the root finding methods awkward in any vector or parallel environment. Because of these constraints our fast polynomial root finding algorithms for <u>dense</u> polynomials involved degrees only in thousands. The largest "random" dense polynomial we completely solved in double precision on a single processor of CRAY II had degree 15,000.

The target of our large degree polynomial root finding programs was the analysis of the distribution of complex roots of a real (or complex) polynomial with random coefficients. Specifically we looked at normally distributed random coefficients, though various other distributions were analyzed as well. The distribution of real roots of a random polynomial of degree n (there are $O(log n)$ of them) was described by Hardy-Littlewood-Kac. Much less is known about the distribution of the complex roots in addition to the obvious statement that "most of the roots are uniformly distributed around the unit circle." [One can see easily that a root cannot have a large absolute value because there will be no cancellation between monomials. Similarly, roots cannot have a small absolute value.]

We refer to [BR] for a review of known theoretical results

We have conducted extensive computations of complex roots for large series of random polynomials (normal, uniform, uniform $\{0,1\}$, and other distribution of coefficients) of degrees varying from 500 to 15,000. While degrees up to 1,000 are easy to handle on a PCAT with an accelerator board (in our case it was DSI-020), degrees higher than 1,000 required a vector facility. We used the vector facility of IBM 3090-VF for degrees up to 7,000, and CRAY II for degrees 7,000-10,000 with the maximum of 15,000. The limit on degrees is a consequence of precision constraints, and an increase in precision significantly inhibits the vectorization and slows down the computations. We present some of the outputs of the pictures of complex roots of random polynomials with normal distribution of coefficients, with parts of the picture blown up for detail.

These computations were based on our new parallel algorithms of root finding for dense polynomials.

The crucial problem in the construction of the truly polynomial time root finding methods is the ability to deal with possible clustering of roots. Multiple roots themselves can be separated, particularly when $P(x) \in \mathbf{Z}[\mathbf{x}]$, by looking at the GCD of $P(x)$ and $P'(x)$, which can be accomplished relatively easily. The only lower bound on the closeness of different roots of $P(x)$ that one has, is the lower bound on the discriminant $\Delta(P)$ of $P(x) : \Delta(P) \geq 1$, which translates into the separation bound $|x_i - x_j| \geq 0(1)n^{-0(n)}H^{-n}$ for $x_i \neq x_j$.

Without this bound, otherwise fast methods like Lobachevsky's become exponential. We propose a new (probabilistic) method based on the analytic continuation of algebraic functions through their singular points. This method is rather involved, and uses in an indirect way, the homotopy method, the Euler-Newton method and the Lobachevsky method. Our method is based on the deformation of a polynomial $P(x) \in \mathbf{Z}[\mathbf{x}]$ $deg P = n$, into a bundle $P(x, t)$ with the initial position $P(x, 0) = P_0(x)$ for a random polynomial $P_0(x)$, and with the final position $P(x, 1) = P(x)$.

We then study the algebraic functions $x_i = x_i(t)$ that are branches of an algebraic function defined by the equation

$$P(x, t) = 0.$$

One analytically continues these functions $x_i(t)$ simultaneously from $t = 0$ to $t = 1$ using the Fuchsian (resolvent) differential equation on x:

$$L[\frac{d}{dt}, t]x_i = 0.$$

In the analytic continuation of $x_i(t)$, steps are chosen according to the radius of the convergence of the local power series expansion of $x_i(t)$ at $t = t_0$. The singularity of $L[.] = 0$, nearest to $t = t_0$, is determined using Lobachevsky's-like method applied to the polynomial $r(t) \stackrel{\text{def}}{=} discriminant_x P(x, t)$. This method is fast even in scalar or vector modes. The theoretical description of the complexity of this method is the following:

<u>Theorem 10.2.</u> If $P(x) \in \mathbf{Z}[\mathbf{x}]$ has degree n, then all n zeroes of $P(x)$ can be found with the precision M of leading digits ($M \gg 1$) in $0(M \log n)^{0(1)}$ parallel steps on $n^{0(1)}$ processors. For a generic polynomial $P(x)$ of degree

n, all n complex zeroes of $P(x)$ can be found with the precision of M leading digits ($M \gg 1$) in $O(\log M \log n)^{0(1)}$ parallel steps on $n^{0(1)}$ processors.

It is possible that one can omit the condition for $P(x)$ to be generic.

New possibilities of polynomial time algorithms of root finding are also open by the introduction of multidimensional θ- functions in connection with our work on Padé approximations to algebraic functions see §4. These methods are particularly effective for polynomials with real roots only.

Plot Shows Complex Zeros of the Polynomial

The Degree of the Polynomial Is 1001

This is a random polynomial. Seed is -1

This is a Whole Root Picture

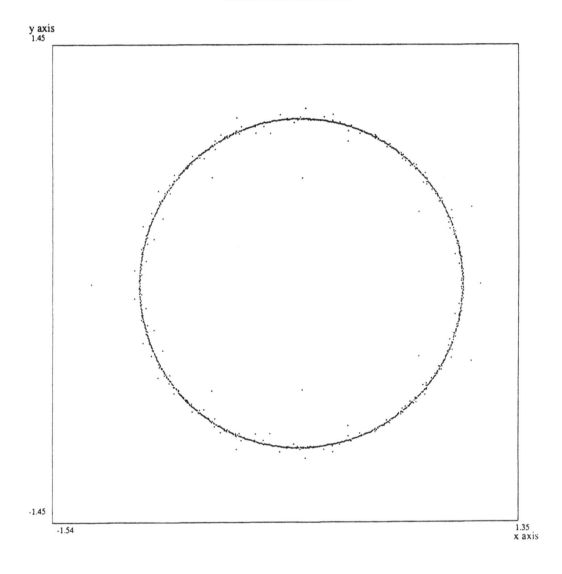

y axis
1.45

-1.45

-1.54

1.35
x axis

Plot Shows Complex Zeros of the Polynomial

The Degree of the Polynomial Is 1001

This is a random polynomial. Seed is -2

This is a Whole Root Picture

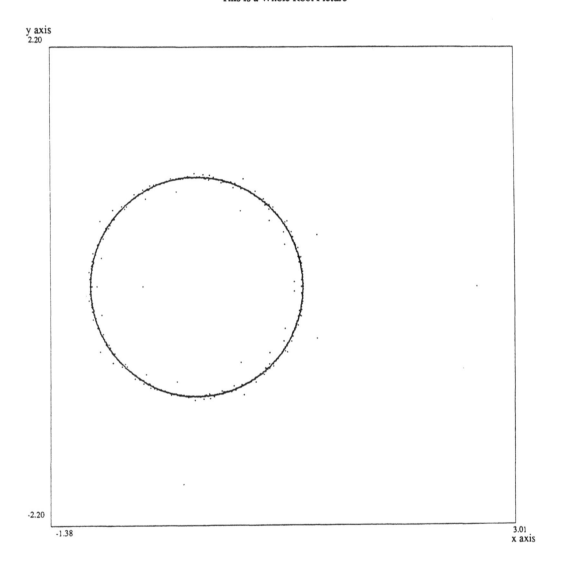

Plot Shows Complex Zeros of the Polynomial

The Degree of the Polynomial Is 1001

This is a random polynomial. Seed is -3

This is a Whole Root Picture

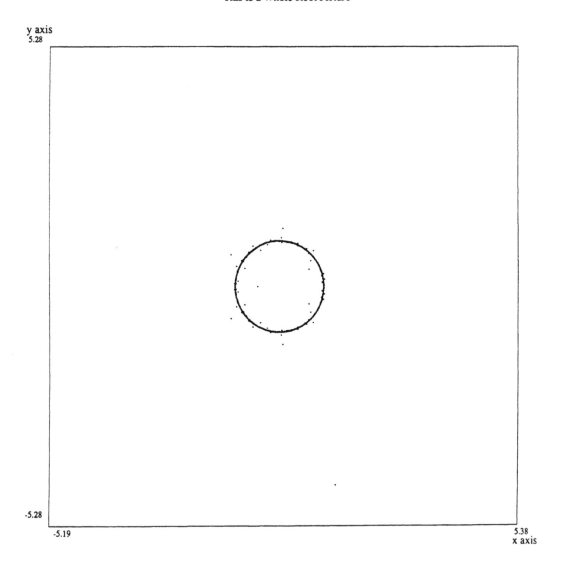

y axis
5.28

-5.28

-5.19

5.38
x axis

Plot Shows Complex Zeros of the Polynomial

The Degree of the Polynomial Is 1001

This is a random polynomial. Seed is -4

This is a Whole Root Picture

y axis
1.32

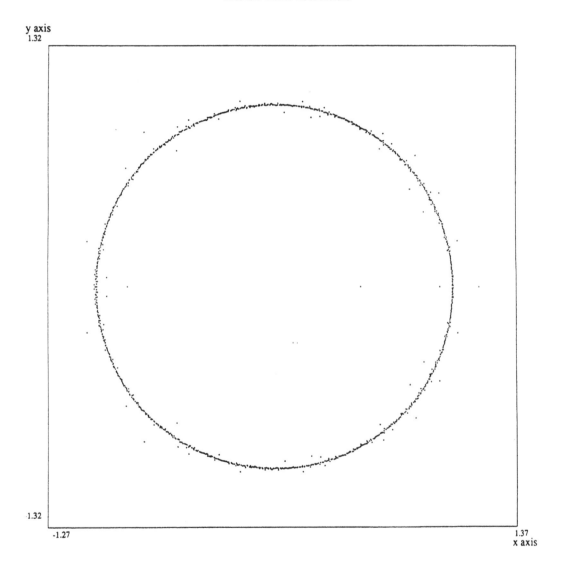

-1.32

-1.27   1.37
                                                                                             x axis

Plot Shows Complex Zeros of the Polynomial

The Degree of the Polynomial Is 1001

This is a random polynomial. Seed is -5

This is a Whole Root Picture

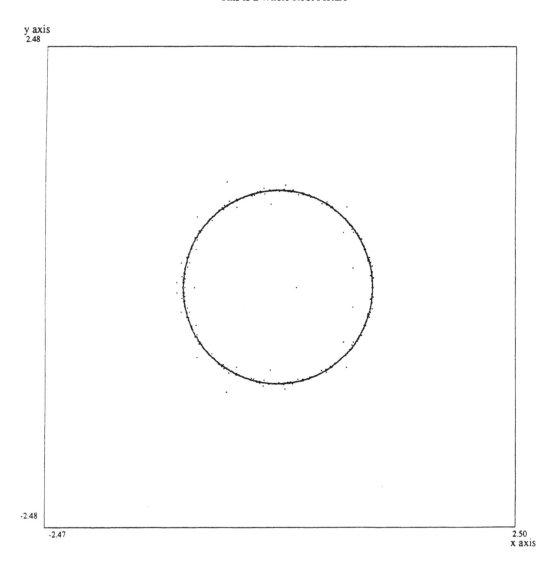

y axis
2.48

-2.48

-2.47

2.50
x axis

Plot Shows Complex Zeros of the Polynomial

The Degree of the Polynomial Is 1001

This is a random polynomial. Seed is -6

This is a Whole Root Picture

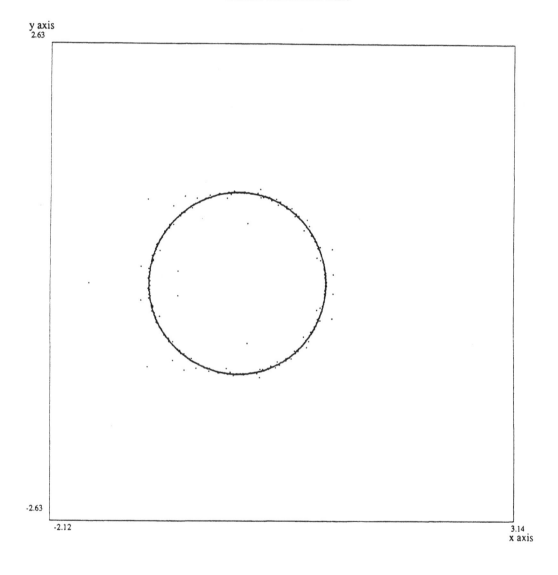

y axis
2.63

-2.63

-2.12   3.14
                                                             x axis

Plot Shows Complex Zeros of the Polynomial

The Degree of the Polynomial Is 1001

This is a random polynomial. Seed is -7

This is a Whole Root Picture

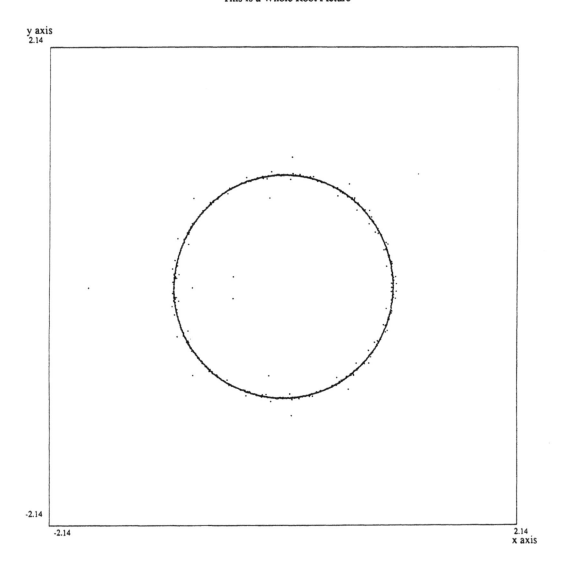

y axis
2.14

-2.14

-2.14   2.14
                                                                          x axis

Plot Shows Complex Zeros of the Polynomial

The Degree of the Polynomial Is 1001

This is a random polynomial. Seed is -8

This is a Whole Root Picture

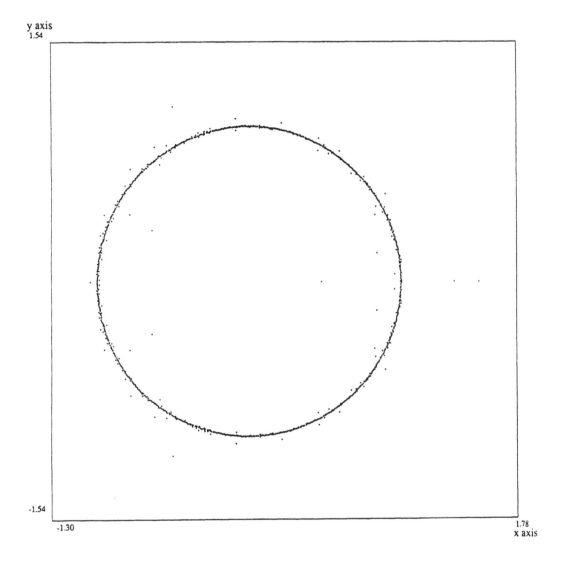

Plot Shows Complex Zeros of the Polynomial

The Degree of the Polynomial Is 1001

This is a random polynomial. Seed is 9

This is a Whole Root Picture

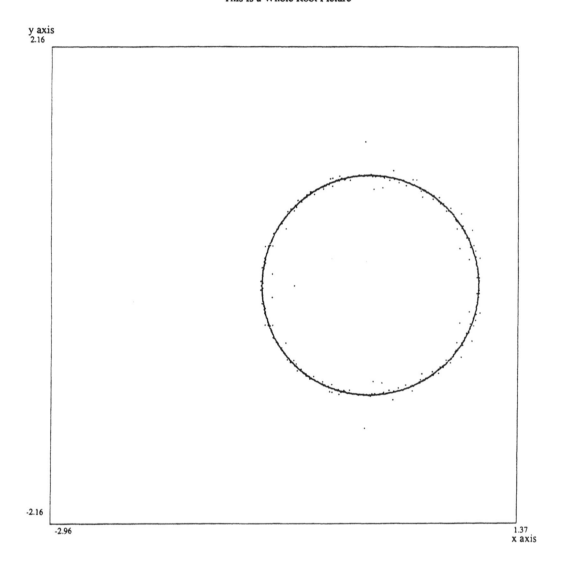

Plot Shows Complex Zeros of the Polynomial

The Degree of the Polynomial Is 5000

This is a random polynomial. Seed is -2

This is a Whole Root Picture

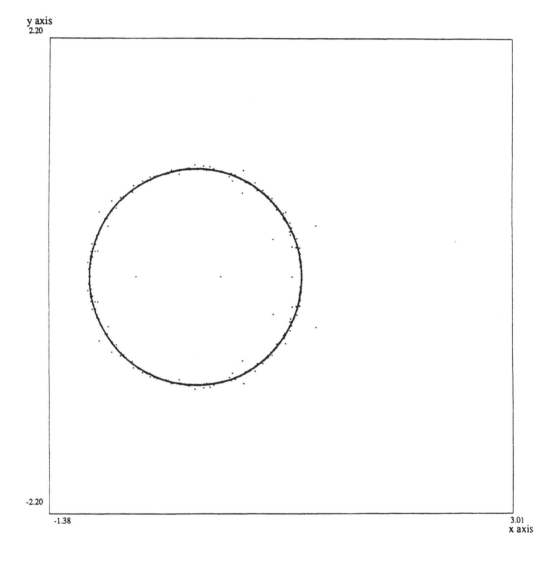

Plot Shows Complex Zeros of the Polynomial

The Degree of the Polynomial Is 5000

This is a random polynomial. Seed is -2

This is a Window of the Whole Root Picture

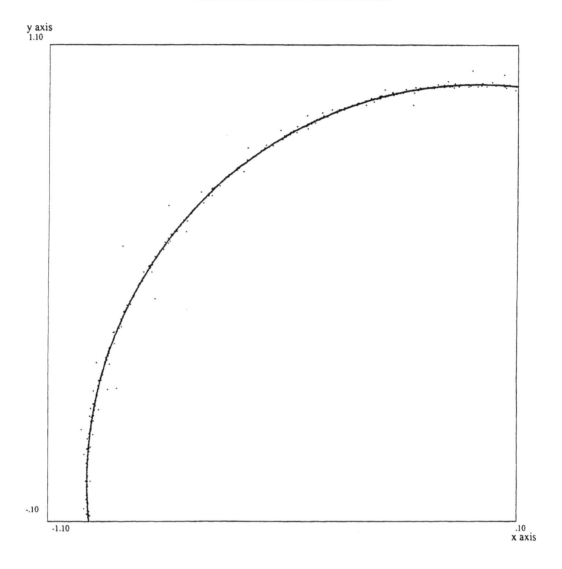

Plot Shows Complex Zeros of the Polynomial

The Degree of the Polynomial Is 5000

This is a random polynomial. Seed is -2

This is a Window of the Whole Root Picture

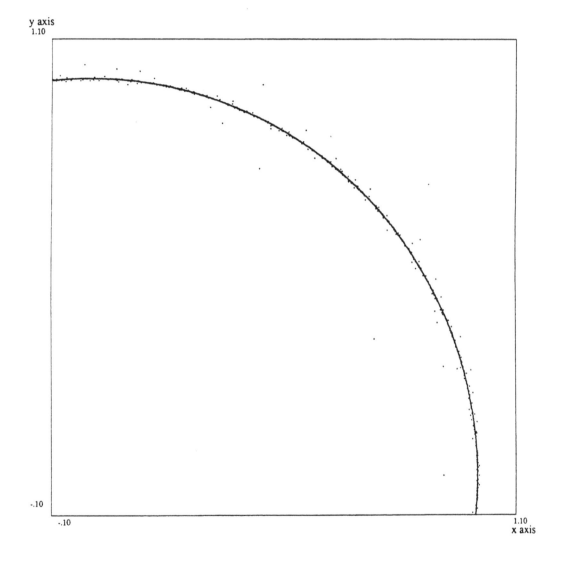

Plot Shows Complex Zeros of the Polynomial

The Degree of the Polynomial Is 5000

This is a random polynomial. Seed is -2

This is a Window of the Whole Root Picture

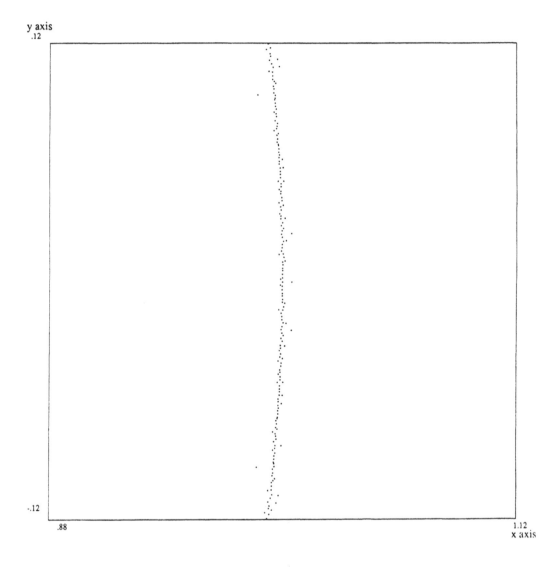

y axis
.12

-.12

.88

1.12
x axis

Plot Shows Complex Zeros of the Polynomial

The Degree of the Polynomial Is 5000

This is a random polynomial. Seed is -2

This is a Window of the Whole Root Picture

y axis
1.12

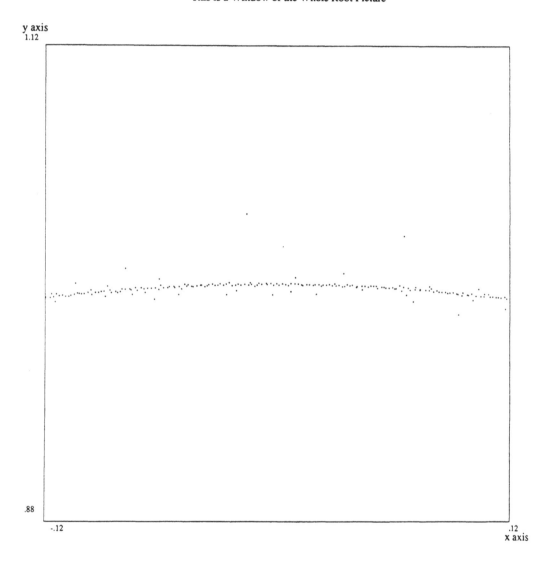

.88

-.12

.12
x axis

Plot Shows Complex Zeros of the Polynomial

The Degree of the Polynomial Is 5000

This is a random polynomial. Seed is -2

This is a Window of the Whole Root Picture

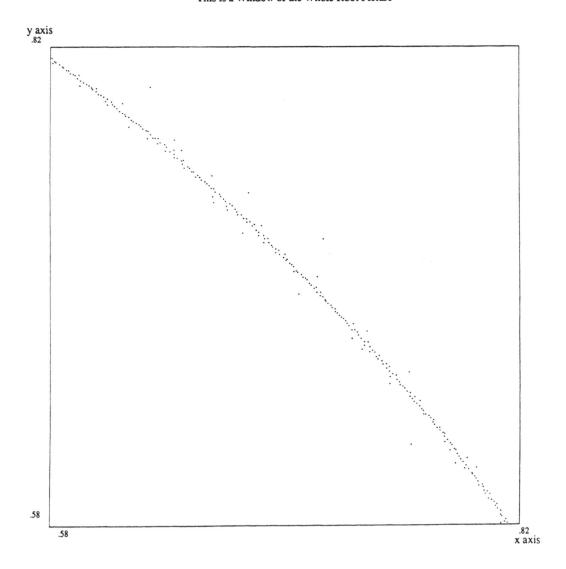

Plot Shows Complex Zeros of the Polynomial

The Degree of the Polynomial Is 5000

This is a random polynomial. Seed is -3

This is a Whole Root Picture

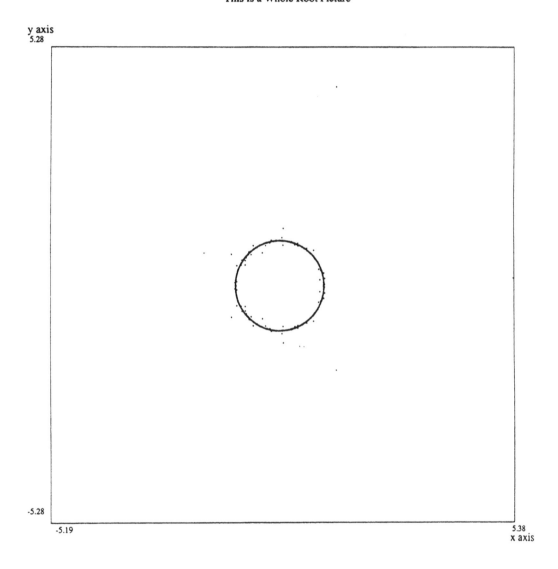

Plot Shows Complex Zeros of the Polynomial

The Degree of the Polynomial Is 5000

This is a random polynomial. Seed is -3

This is a Window of the Whole Root Picture

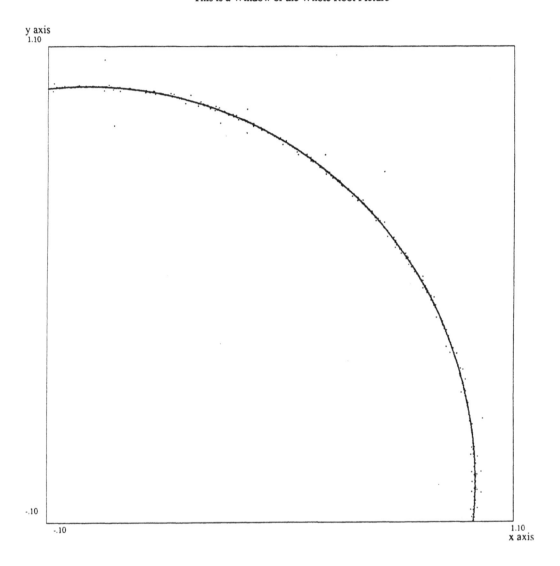

Plot Shows Complex Zeros of the Polynomial

The Degree of the Polynomial Is 5000

This is a random polynomial. Seed is -3

This is a Window of the Whole Root Picture

y axis
1.10

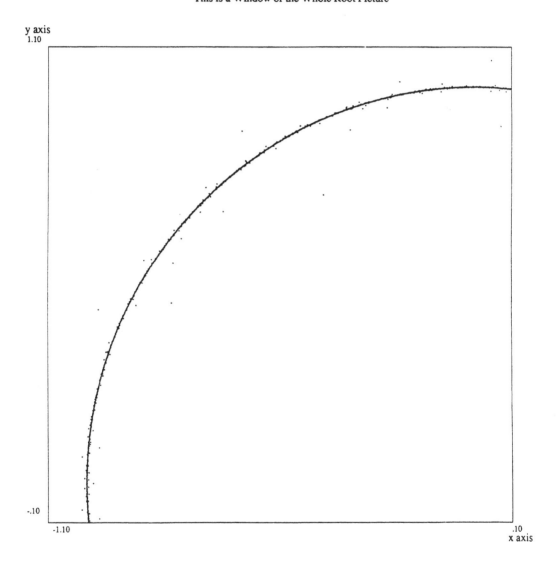

-.10

-1.10

.10
x axis

Plot Shows Complex Zeros of the Polynomial

The Degree of the Polynomial Is 5000

This is a random polynomial. Seed is -3

This is a Window of the Whole Root Picture

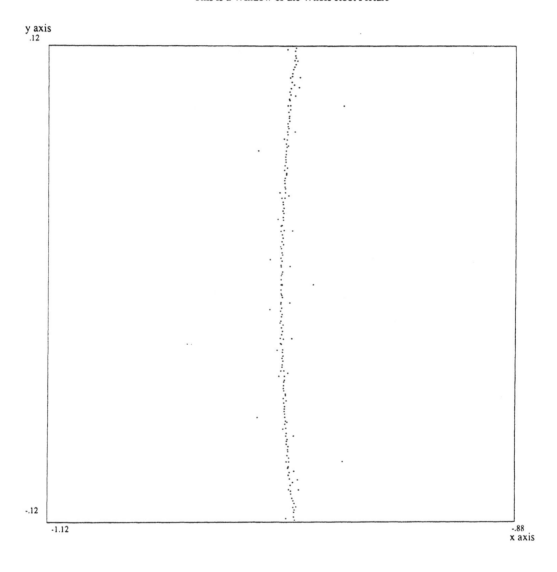

Plot Shows Complex Zeros of the Polynomial

The Degree of the Polynomial Is 5000

This is a random polynomial. Seed is -3

This is a Window of the Whole Root Picture

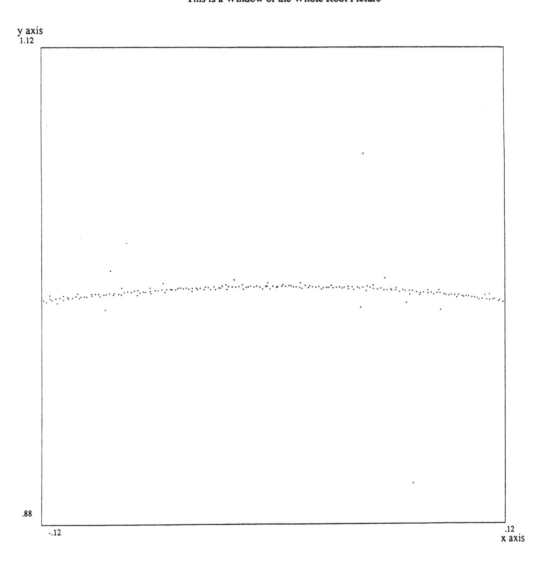

Plot Shows Complex Zeros of the Polynomial

The Degree of the Polynomial Is 5000

This is a random polynomial. Seed is -3

This is a Window of the Whole Root Picture

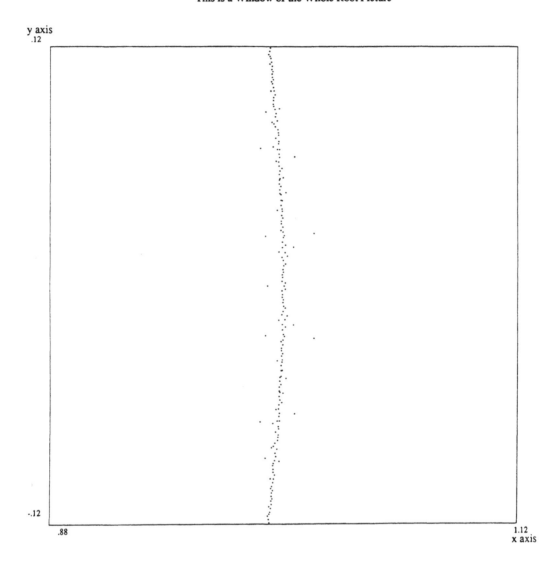

y axis
.12

-.12

.88   1.12
                                                                             x axis

Plot Shows Complex Zeros of the Polynomial

The Degree of the Polynomial Is 5000

This is a random polynomial. Seed is -9

This is a Whole Root Picture

y axis
2.16

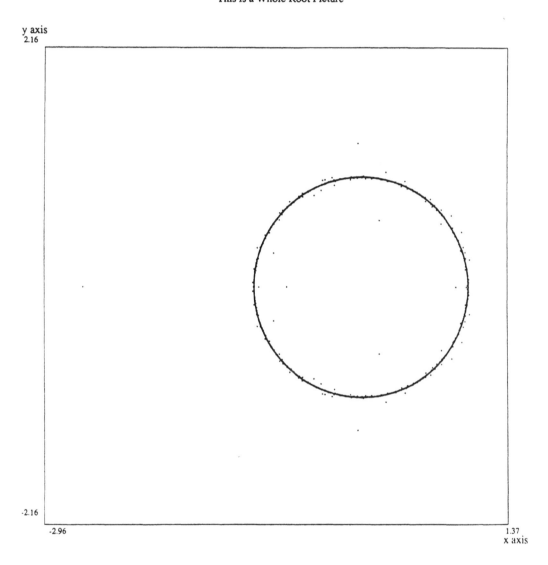

-2.16

-2.96

1.37
x axis

Plot Shows Complex Zeros of the Polynomial

The Degree of the Polynomial Is 5000

This is a random polynomial. Seed is -9

This is a Window of the Whole Root Picture

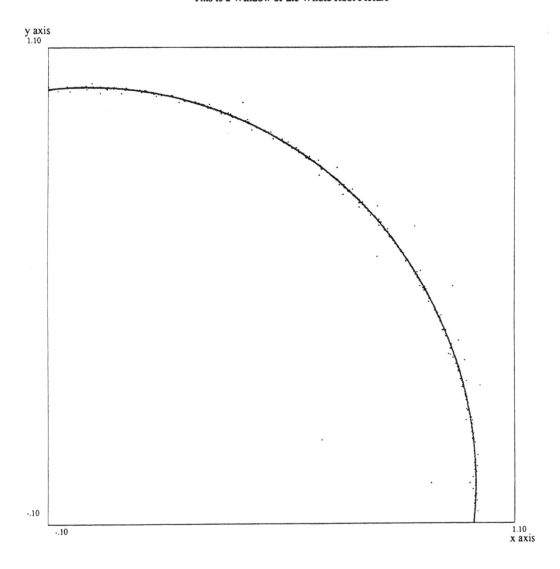

Plot Shows Complex Zeros of the Polynomial

The Degree of the Polynomial Is 5000

This is a random polynomial. Seed is -9

This is a Window of the Whole Root Picture

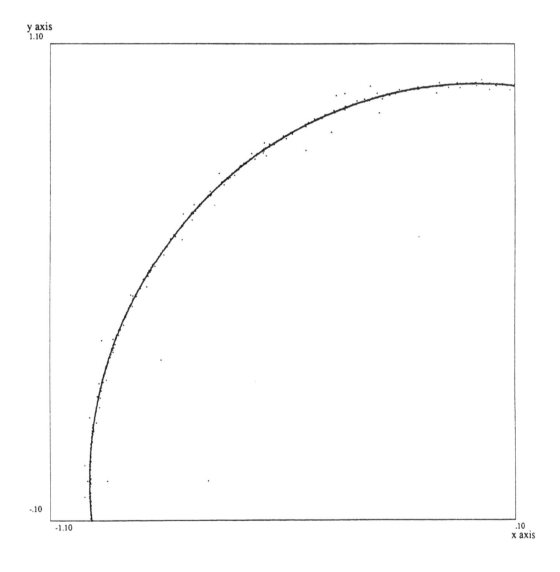

y axis
1.10

-.10

-1.10

.10
x axis

Plot Shows Complex Zeros of the Polynomial

The Degree of the Polynomial Is 5000

This is a random polynomial. Seed is -9

This is a Window of the Whole Root Picture

y axis
.12

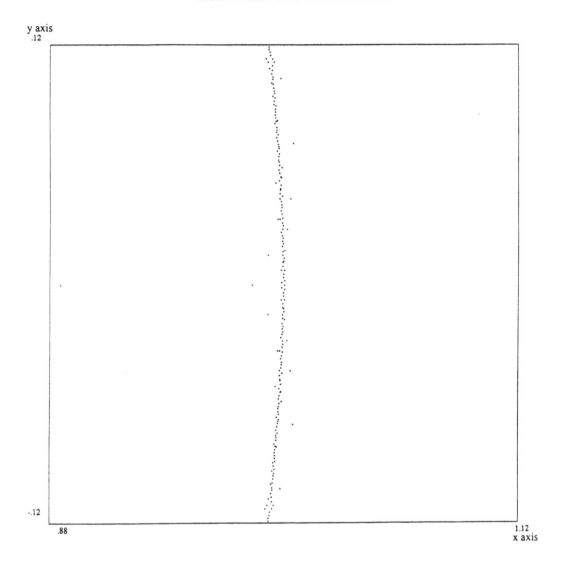

-.12

.88   1.12
                                                                                       x axis

Plot Shows Complex Zeros of the Polynomial

The Degree of the Polynomial Is 5000

This is a random polynomial. Seed is -9

This is a Window of the Whole Root Picture

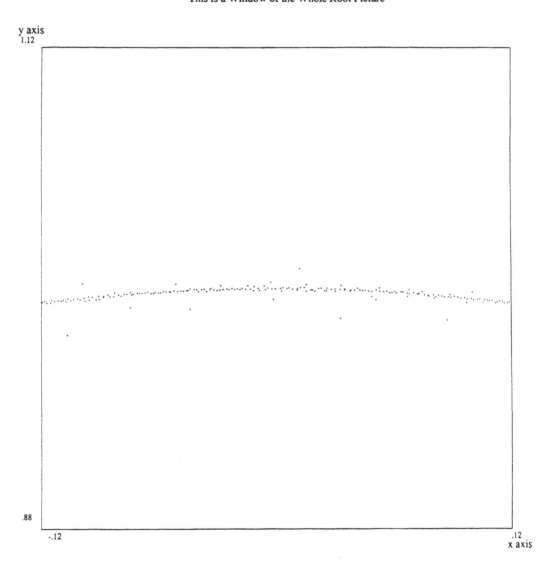

y axis
1.12

.88

-.12   .12
                                                                              x axis

Plot Shows Complex Zeros of the Polynomial

The Degree of the Polynomial Is 6999

This is a random polynomial. Seed is -6

This is a Whole Root Picture

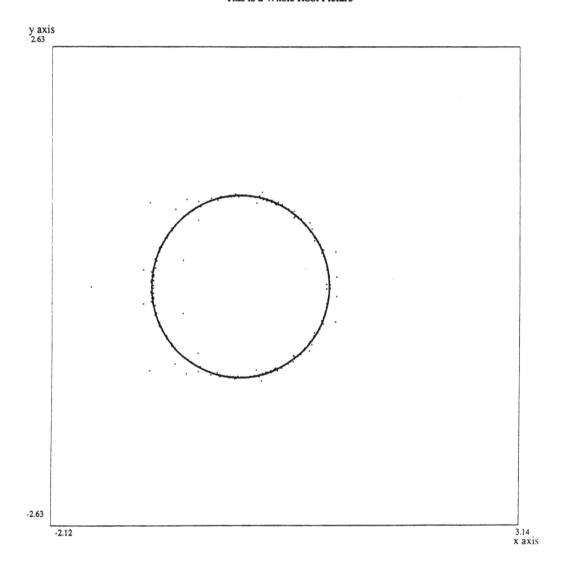

Plot Shows Complex Zeros of the Polynomial

The Degree of the Polynomial Is 6999

This is a random polynomial. Seed is -6

This is a Window of the Whole Root Picture

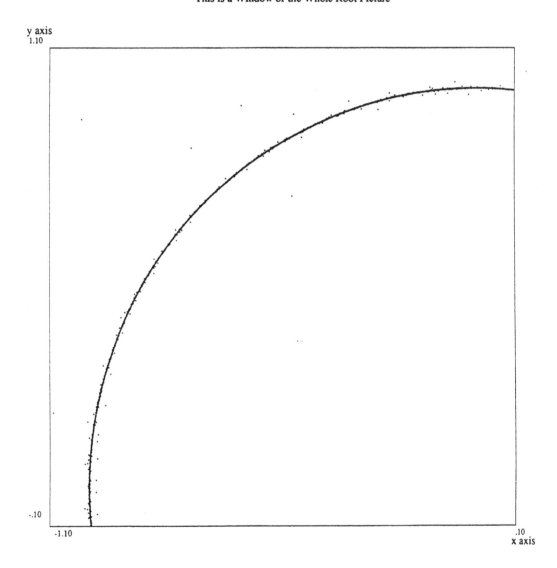

Plot Shows Complex Zeros of the Polynomial

The Degree of the Polynomial Is 6999

This is a random polynomial. Seed is -6

This is a Window of the Whole Root Picture

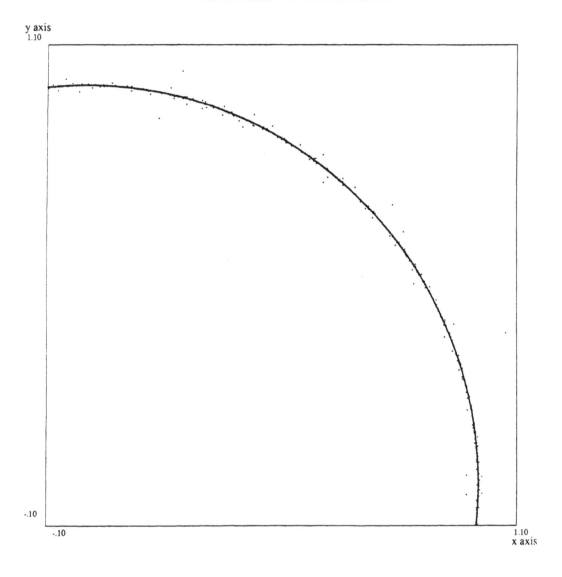

y axis
1.10

-.10

-.10   1.10
                                                                             x axis

Plot Shows Complex Zeros of the Polynomial

The Degree of the Polynomial Is 6999

This is a random polynomial. Seed is -6

This is a Window of the Whole Root Picture

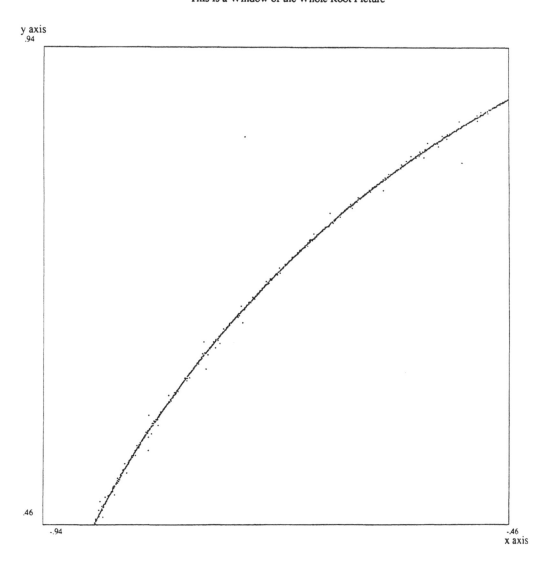

y axis
.94

.46

-.94

-.46
x axis

Plot Shows Complex Zeros of the Polynomial

The Degree of the Polynomial Is 6999

This is a random polynomial. Seed is -6

This is a Window of the Whole Root Picture

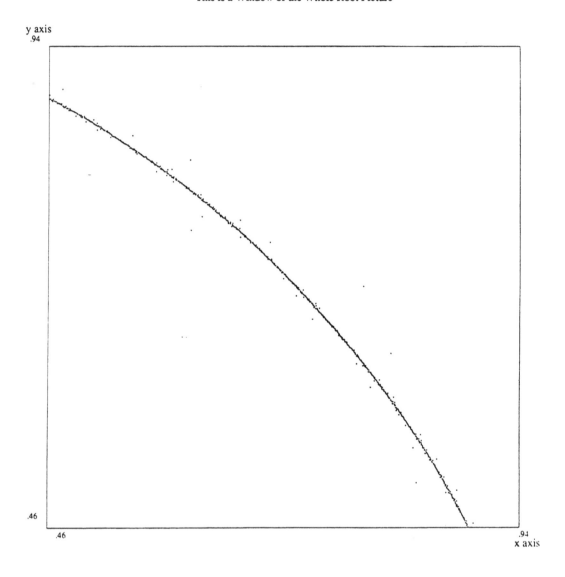

Plot Shows Complex Zeros of the Polynomial

The Degree of the Polynomial Is 6999

This is a random polynomial. Seed is -6

This is a Window of the Whole Root Picture

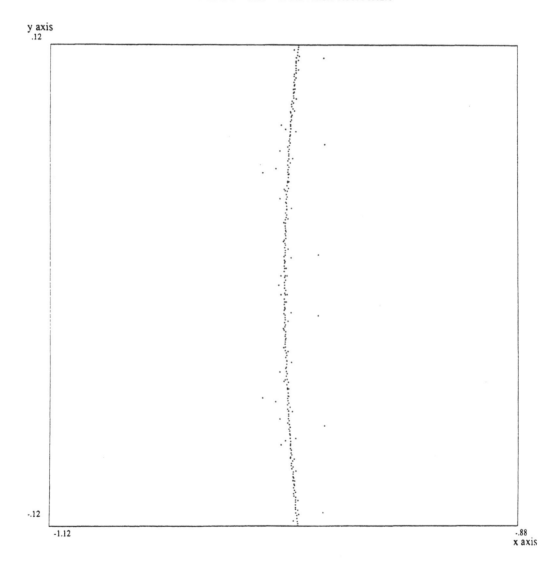

y axis

.12

-.12

-1.12   -.88

x axis

Plot Shows Complex Zeros of the Polynomial

The Degree of the Polynomial Is 6999

This is a random polynomial. Seed is -6

This is a Window of the Whole Root Picture

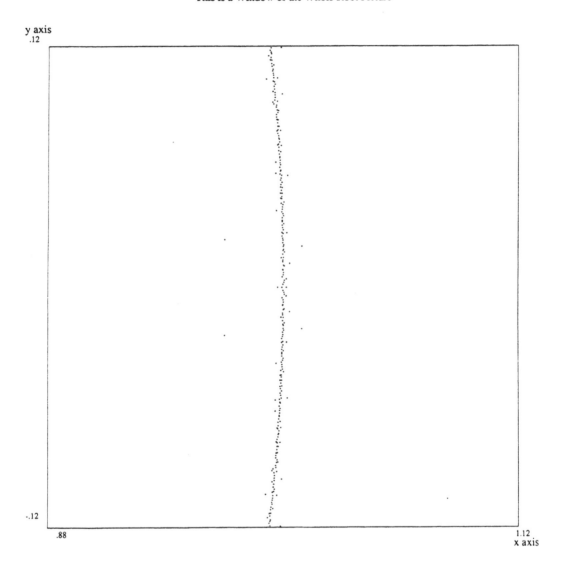

Plot Shows Complex Zeros of the Polynomial

The Degree of the Polynomial Is 7000

This is a random polynomial. Seed is -1

This is a Whole Root Picture

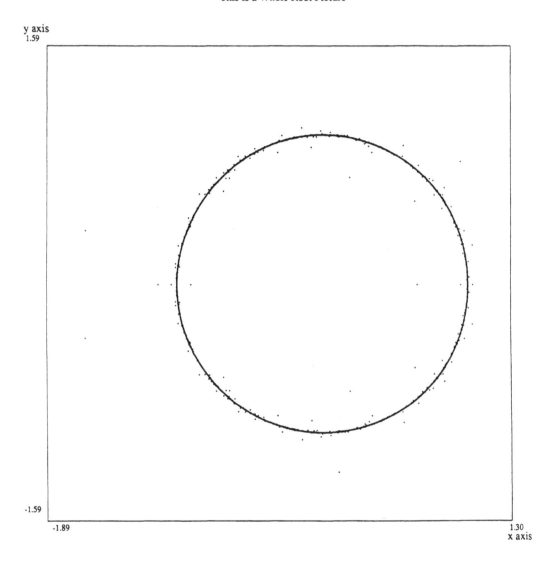

Plot Shows Complex Zeros of the Polynomial

The Degree of the Polynomial Is 7000

This is a random polynomial. Seed is -1

This is a Window of the Whole Root Picture

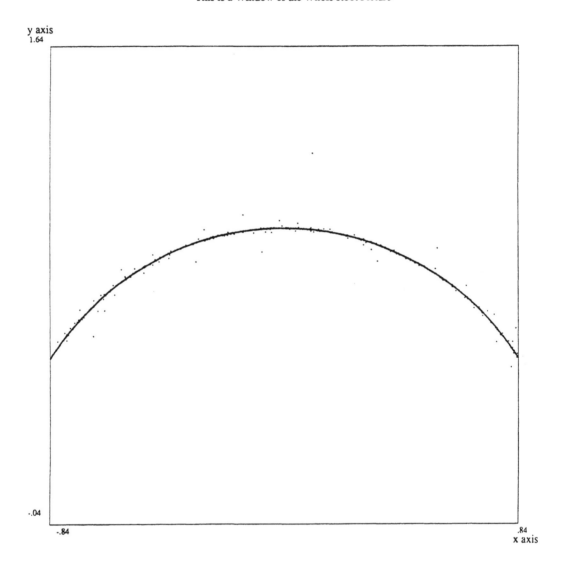

Plot Shows Complex Zeros of the Polynomial

The Degree of the Polynomial Is 7000

This is a random polynomial. Seed is -1

This is a Window of the Whole Root Picture

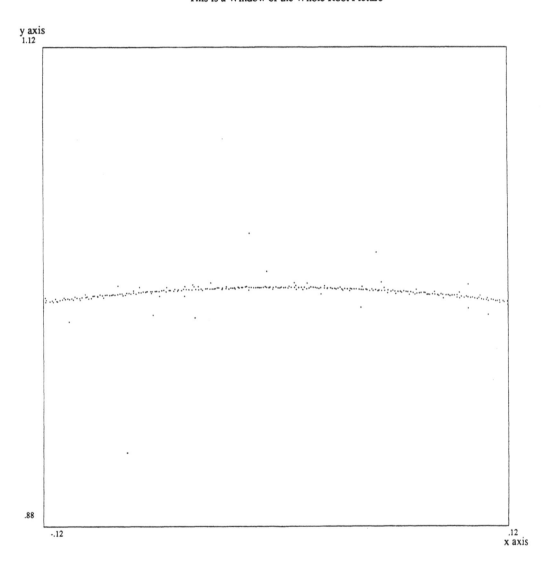

y axis
1.12

.88

-.12

.12
x axis

Plot Shows Complex Zeros of the Polynomial

The Degree of the Polynomial Is 7000

This is a random polynomial. Seed is -1

This is a Window of the Whole Root Picture

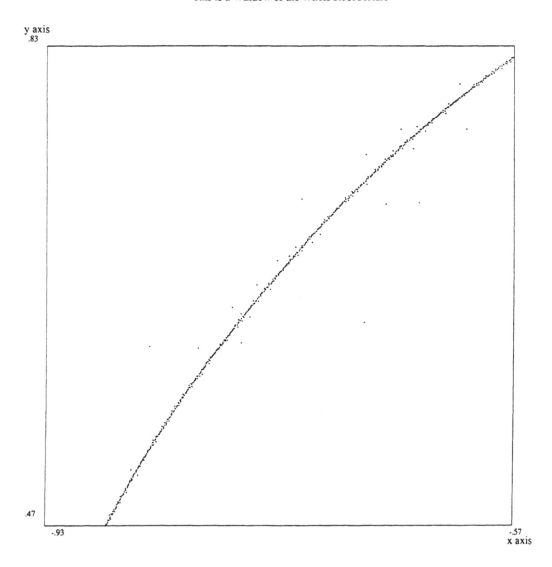

Plot Shows Complex Zeros of the Polynomial

The Degree of the Polynomial Is 7000

This is a random polynomial. Seed is -1

This is a Window of the Whole Root Picture

y axis
.71

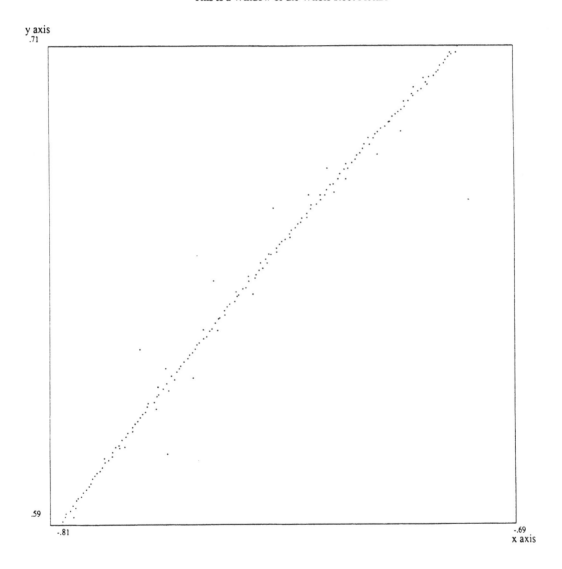

.59

-.81

-.69
x axis

Plot Shows Complex Zeros of the Polynomial

The Degree of the Polynomial Is 7000

This is a random polynomial. Seed is -1

This is a Window of the Whole Root Picture

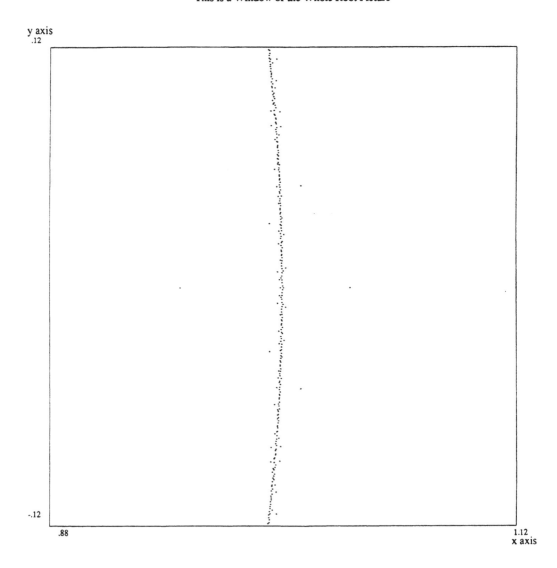

Plot Shows Complex Zeros of the Polynomial

The Degree of the Polynomial Is 7000

This is a random polynomial. Seed is -1

This is a Window of the Whole Root Picture

y axis

.48

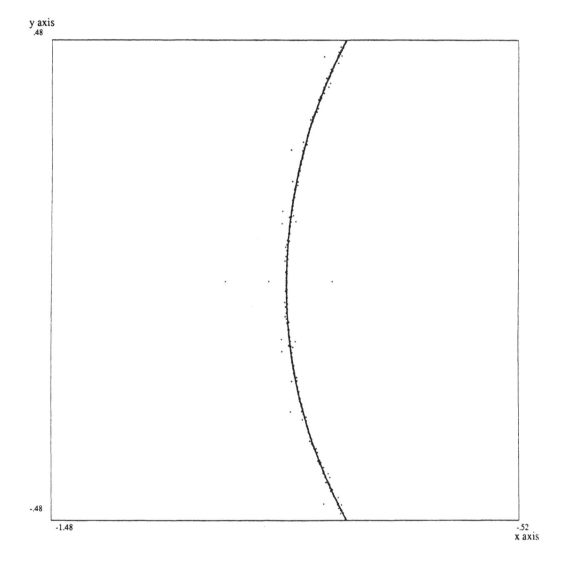

-.48

-1.48

-.52

x axis

Plot Shows Complex Zeros of the Polynomial

The Degree of the Polynomial Is 7999

This is a random polynomial. Seed is -3

This is a Whole Root Picture

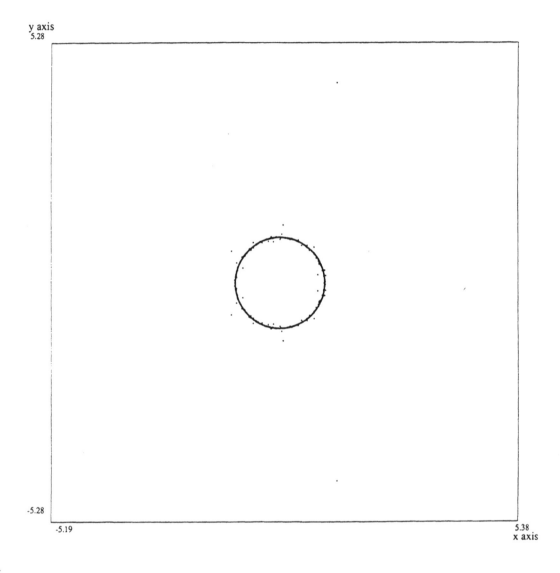

y axis
5.28

-5.28

-5.19   5.38
                                                                         x axis

Plot Shows Complex Zeros of the Polynomial

The Degree of the Polynomial Is 7999

This is a random polynomial. Seed is -3

This is a Window of the Whole Root Picture

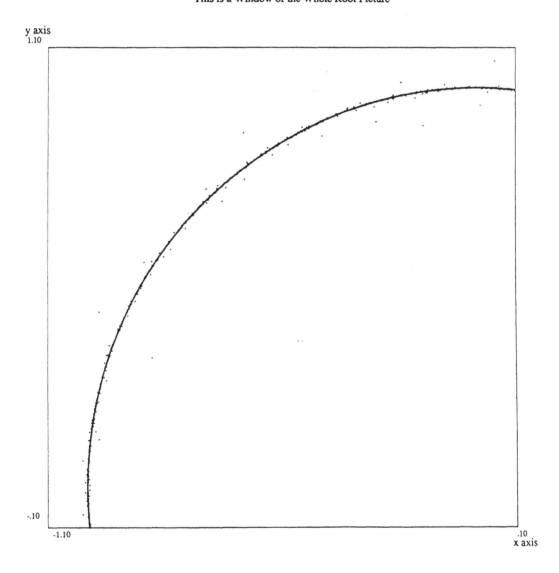

Plot Shows Complex Zeros of the Polynomial

The Degree of the Polynomial Is 7999

This is a random polynomial. Seed is -3

This is a Window of the Whole Root Picture

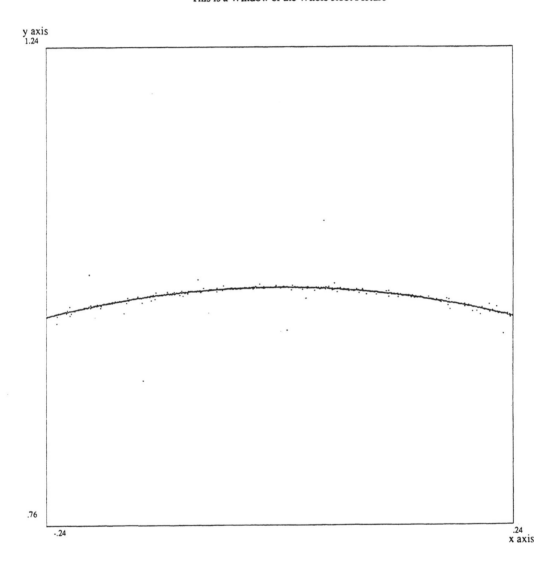

Plot Shows Complex Zeros of the Polynomial

The Degree of the Polynomial Is 7999

This is a random polynomial. Seed is -3

This is a Window of the Whole Root Picture

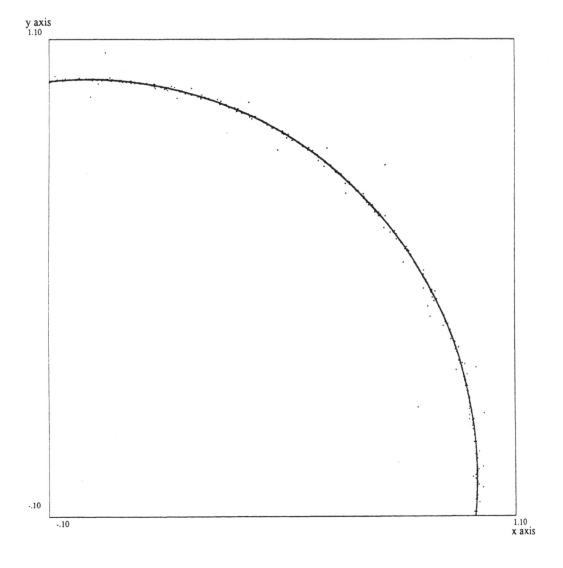

Plot Shows Complex Zeros of the Polynomial

The Degree of the Polynomial Is 7999

This is a random polynomial. Seed is -3

This is a Window of the Whole Root Picture

y axis
.88

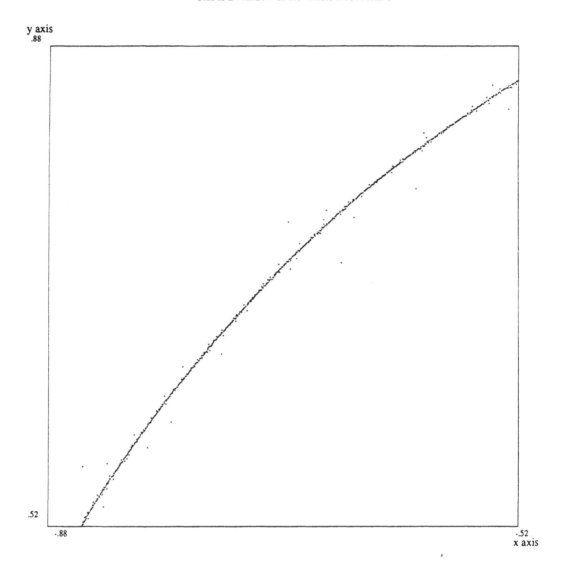

.52

-.88

-.52
x axis

Plot Shows Complex Zeros of the Polynomial

The Degree of the Polynomial Is 7999

This is a random polynomial. Seed is -3

This is a Window of the Whole Root Picture

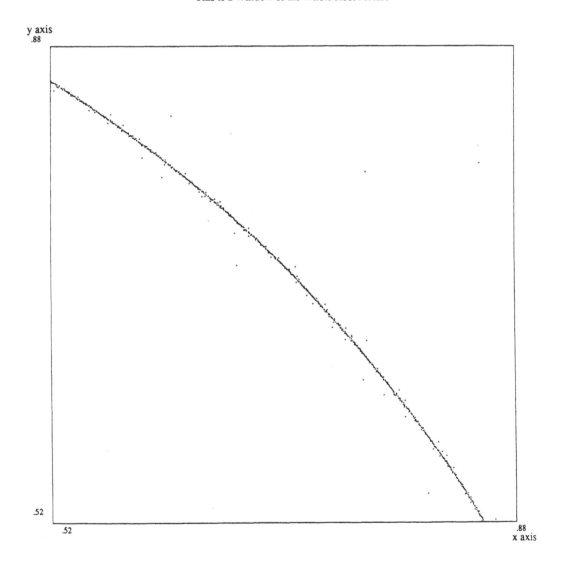

Plot Shows Complex Zeros of the Polynomial

The Degree of the Polynomial Is 7999

This is a random polynomial. Seed is -3

This is a Window of the Whole Root Picture

y axis
.12

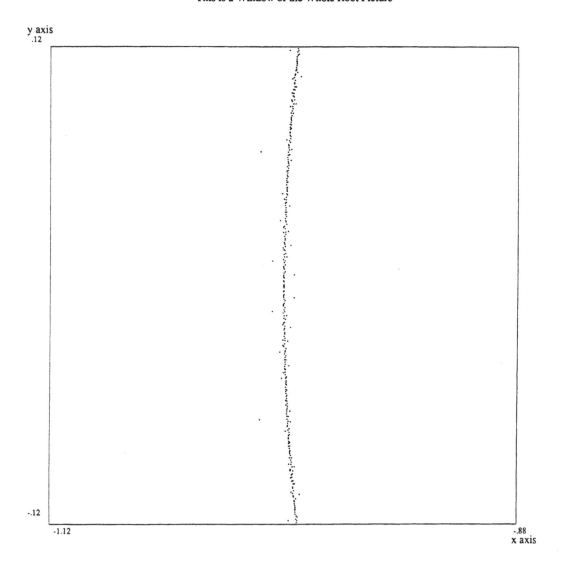

-.12

-1.12

-.88
x axis

Plot Shows Complex Zeros of the Polynomial

The Degree of the Polynomial Is 7999

This is a random polynomial. Seed is -3

This is a Window of the Whole Root Picture

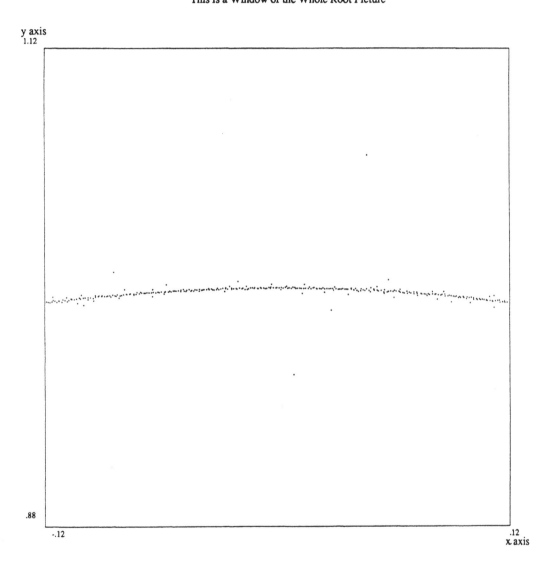

Plot Shows Complex Zeros of the Polynomial

The Degree of the Polynomial Is 7999

This is a random polynomial. Seed is -3

This is a Window of the Whole Root Picture

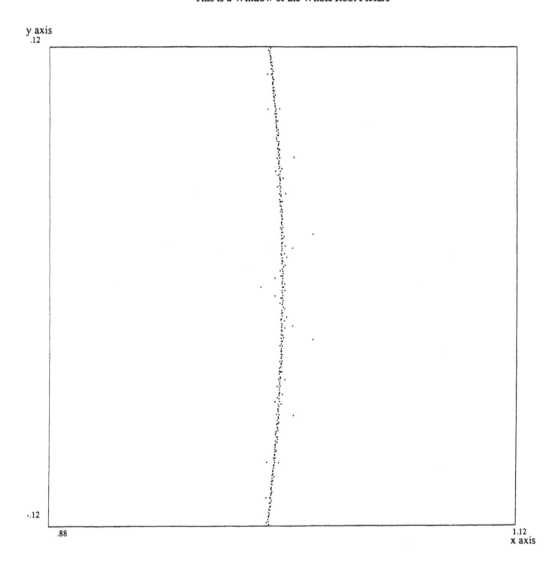

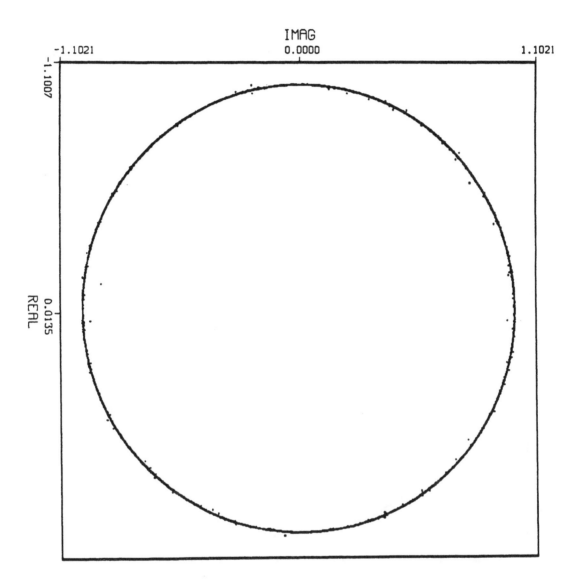

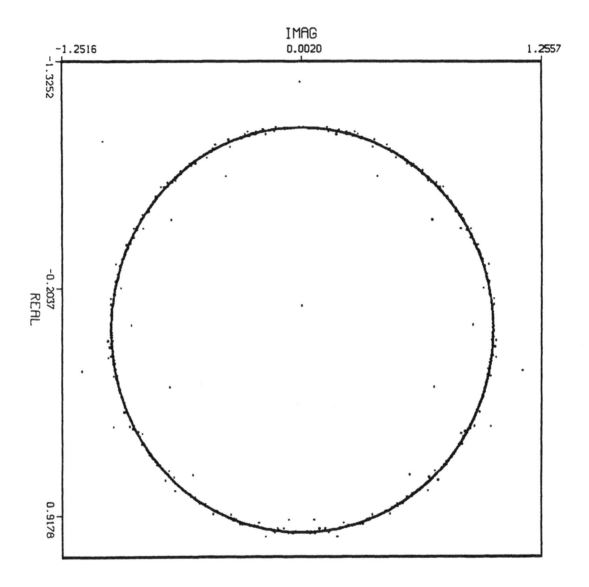

THIS IS A PART OF CRAY-2 COMPUTATION
OF 15,000 ROOTS

REFERENCES

[B1] G. Birkhoff, Current trends in algebra, Amer. Math. Monthly, 79 (1973), 760-780.

[B2] B.C. Brendt, R.J. Evans and B.M. Wilson, Chapter 3 of Ramanujan second notebook, Adv. Math. (to appear).

[B3] R. Birkeland, Résolution de l'équation algébrique trinome par des fonctions hypergéométriques supérieures, C.R. Acad. Sci., 171 (1920), p. 728; p. 1047; and p. 1370.

[B4] G.D. Birkhoff, General theory of linear difference equations, Trans. Amer. Math. Soc., 12 (1911), 243-284.

[B5] R. P,. Brent, Multiplel-precision zero-finding method and the compexity of elementary function evaluation, in Analytic Computational Complexity, J.F. Traub, Ed., Academic Press, 1975, 151-176.

[B6] R. P. Brent, The complexity of multiple-precision arithmetic, in Complexity of Computational Problem Solving, R. S. Anderson, R. P. Brent, Eds., Univ of Queensland Press, Brisbane, 1975, 126-165. [BR] A.T. Bharucha-Reid, M. Sambandham, Academic Press, 1986.

[BOFKT] M. Ben-Or, E. Feig, D. Kozen, P. Tiwari, A fast parallel algorithm for determining all roots of a polynomial with real roots, Proc. XVIII Ann. ACM Symp. Theory Comput., 1986.

[C1] Y.F. Chang, Private communication, Spring, 1986.

[C2] D.V. Chudnovsky, G.V. Chudnovsky, Pole expansions of nonlinear partial differential equations. Nuovo Cimento, 40B (1977), 339-353.

[C3] D.V. Chudnovsky, G.V. Chudnovsky, One-dimension kinetic equation with elliptic potential of iteraction, Lettere Nuovo Cimento, 19 (1977), 300-302.

[C4] D.V. Chudnovsky, G.V. Chudnovsky, On expansion of algebraic functions in power and Puiseux series, Part I and II, J. Complexity, 2(1986) 271-294; and 3(1987), 1-25.

[C5] D.V. Chudnovsky, G.V. Chudnovsky, Computer assisted number theory, Lecture Notes Math., Springer, v. 1240, 1987, 1-68.

[C6] D.V. Chudnovsky, G.V. Chudnovsky, A random walk in higher arithmetic, Adv. Appl. Math., 7(1986), 101-122.

[C7] D.V. Chudnovsky, G.V. Chudnovsky, Applications of Padé approximations to the Grothendieck conjecture on linear differential equations, Lecture Notes Math., Springer, N.Y., 1985, v. 1135, 52-100.

[C8] D.V. Chudnovsky, G.V. Chudnovsky, Sequences of numbers generated by addition in formal groups and new primality and factorization text, IBM Research Report, RC 11262, 7/12/85, 100 pp; Advances in Applied Math., 7 (1986), 187-237.

[C9] D.V. Chudnovsky, G.V. Chudnovsky, Bäcklund transformations for linear differential equations and Padé approximations, I., J. Math. Pures Appl., Paris, 61 (1981), 1- 16.

[C10] D.V. Chudnovsky, G.V. Chudnovsky, Elliptic modular functions and elliptic genera, Topology, 26 (1987) (to appear).

[C11] D.V. Chudnovsky, G.V. Chudnovsky, Padé approximations and diophantine geometry, Proc. Natl. Acad. Sci. USA, 82 (1985), 2212-2216.

[C12] D.V. Chudnovsky, G.V. Chudnovsky, Algebraic complexities and algebraic curves over finite fields, (to appear).

[D1] C. Dandelin, Recherches sur la résolution des équations numériques, Nouv. Mém. Acad. Roy. Sci. et Belles-Lettres de Bruxelles, 3: (1826), 1-71, 153-159.

[D2] D. P. Dalzell, A note on automorphic functions, J. London Math. Soc., 5 (1936), 280-282.

[D3] S.C. Dhar, On the uniformization of a special kind of algebraic curve of any genus, J. London Math. Soc., 10 (1935), 259-263.

[D4] J. Della Dora, E. Fournier, Formal solutions of differential equations in the neighborhood of singular points, in SYMSAC'81 Proceedings (ed P.S. Wang), Assoc. Computer Machinery, 1981, 25- 30.

[F1] F. G. Frobenius, L. Stickelberger, Über die addition und multiplication der elliptischen funktionen, J. Reine Angew. Math., 88 (1880), 146-184.

[F2] G. Faltings, Endlichkeitssätze fur abelsche varietäten über zahlkörpern, Invent. Math., 73 (1983), 349-366.

[G1] C.H. Graeffe, Die Auflösungen der höheren numerischen Gleichungen, als Beantwortung einer von der Königlichen Akademie der Wissenschaften zu Berlin aufgestellten Preisfrage, Friedrich Schulthess, Zürich, 1837.

[H1] A.S. Householder, The Numerical Treatment of a Single Nonlinear Equation, McGraw-Hill, 1970.

[H2] P. Henrici, Applied and Computational Complex Analysis, v. 1-3, John Wiley, 1974-1986.

[H3] R. Harley, On the theory of transcendental solution of algebraic equations, Quart. J. Math., 5 (1862), 337.

[H4] D.A. Hejhal, Monodromy groups and Poincaré series, Bull. Amer. Math. Soc., 84 (1978), 339-376.

[H5] F. Hirzebruch, Topological Methods in Algebraic Geometry, Springer, 1966.

[I1] J.L. Ince, Periodic Lamé functions, Proc. Edinb. Math. Soc., 41 (1923), 94-100 and 60(1939), 47-63.

[JT1] N. A. Jenkins, J.F. Traub, An algorithm for an automatic general polynomial solver, Stanford University Tech. Rep. No. CS71, 1967.

[K1] E. Kaltofen, Polynomial factorization 1982-1986, (these Proceedings).

[K2] D.E. Knuth, The Art of Computer Programming, v. 2, 2nd ed., Addison-Wesley, 1981.

[K3] F. Klein, R. Fricke, Vorlesungen über die Theorie der Automorphen Functionen, bd. 1, Teubner, 1925.

[K4] L. Keen, H.E. Rauch, A.T. Vasques, Moduli of punctured tori and the accessory parameter of Lamé's equations, Trans. Amer. Math. Soc., 255 (1979), 201-229.

[K5] H. vonKoch, Un théorème sur les intégrales irrégulières des équations différentielles linéaires et son application au problème de l'intégration, Ark. för Math., 13 (1918), No. 15, 1-18.

[L1] A.K. Lenstra, H.W. Lenstra Jr., and L. Lovász, Factoring polynomials with rational coefficients, Math. Ann., 261 (1982), 515-534.

[L2] C. Lanczos, Applied Analysis, Prentice-Hall, 1956.

[L3] N. I. Lobachevsky, Algebra ili vyčislenie korney, Collected Scientific Papers, v. 4, Moscow, 1968, p. 356.

[L4] D.H. Lehmer, The complete root-squaring method, J. Soc. Indust. Appl. Math., 11(1963), 705-717.

[L5] J.L. Lagrange, Nouvelle méthode pour résoudre les équations littérales par le moyen de séries, Oeuvres, v. 3, Gauthier-Villars, Paris, 1869, 5-73.

[L6] J. H. Lambert, Observationes variae in mathesin puram, Opera Mathematica, v. 1, Orell Füssli, Zürich, 1946, 16- 51.

[L7] G.L. Lamb, Jr., Elements of Soliton Theory, John Wiley, 1980.

[L8] P. Landweber, R. Stong, Circle action on Spin manifolds and characteristic numbers, Topology 26 (1987), (to appear).

[L9] J. A. Lappo-Danilevsky, Mémoires sur la Théorie des Systèmes des Équations Différentielles Linéaires, Chelsea, 1953.

[M1] Hj. Mellin, Ein Allgemeiner Satz über algebräische Gleichungen, Ann. Acad. Sc. Fennica, 7(1915) no. 7.

[M2] P.J. Myrberg, Über die numerische Ausführung der Uniformisierung, Acta Soc. Sc. Fennicae, 48(1920), No.7.

[M3] M. Mursi, On the uniformization of algebraic curves of genus 3, Proc. Edinburgh Math. Soc., 2(1930), 101-107.

[M4] M. Mursi, A note on automorphic functions, J. London Math. Soc., 6(1931), 166-169.

[M5] M. Magnus, Monodromy groups and Hill's equation, Comm. Pure Appl. Math., 29 (1976), 701-716.

[P1] V. Pan, Fast and efficient algorithms for sequential and parallel evaluation of polynomial zeroes and on matrix polynomials, Proc. 26 IEEE Conf. Comp. Sci. 1985, 522-531.

[P2] A Poincaré, Sur les groupes des équations linéaires, Acta Math., 4 (1884), 201-312.

[R1] S. Ramanujan, Notebooks, 2 volumes, Tate Institute of Fundamental Research, Bombay, 1957.

[R2] R.A. Rankin, The differential equations associated with the uniformization of certain algebraic curves, Proc. Roy. Soc. Edin., 65A(1958), 35-62.

[R3] R.A. Rankin, The schwarzian derivative and uniformization, J. d'Analyse Math., 6 (1958), 149-167.

[S1] S. Smale, The fundamental theorem of algebra and complexity theory, Bull. Amer. Math. Soc., 4 (1981) 1-36.

[S2] A. Schönhage, Equation solving in terms of computational complexity, Proc. Intern. Congress Math., Berkeley, 1986.

[S3] S. Smale, On the efficiency of algorithms of analysis, ibid., 13 (1985), 87-121.

[SS1] M. Shub, S. Smale, Computational complexity on the geometry of polynomials and a theory of cost: II. SIAM J. Comput., 15 (1986), 145-161.

[SS2] A. Schönhage, V. Strassen, Schnelle multiplikation grosser zahlen, Computing, 7(1971), 281-292.

[SW1] H.P.F. Swinnerton-Dyer, Arithmetic groups, in Discrete Groups and Automorphic Functions, Academic Press, 1977, 377-401.

[T1] J.F. Traub, Iterative Methods for the Solution of Equations, Prentice Hall, 1964.

[T2] P. Turan, Power sum method and the approximative solution of algebraic equations, Math. Comp., 29 (1975), 311-316.

[T3] L.A. Takhtadjan, P.G. Zograf, The Liouville equation action–the generating function for accessory parameters, Funct. Anal., 19 (1985), 67-68.

[W1] E.T. Whittaker, On the connection of algebraic functions with automorphic functions, Phil. Trans., 192A (1898), 1-32.

[W2] E.T. Whittaker, On hyperlemniscate functions, a family of auto-morphic functions, J. London Math. Soc., 4(1929), 274-278.

[W3] J. M. Whittaker, The uniformization of algebraic curves, J.London Math. Soc., 5(1920), 150-154.

[W4] E. Witten, Elliptic genera and quantum field theory, Commun. Math. Phys. (to appear).

* This work has been supported in part by the N.S.F., the U.S. Air Force and the Program O.C.R.E.A.E.

Impact of Linear Programming on Computer Development

GEORGE B. DANTZIG   Stanford University, Stanford, California

The history of how L.P. first started in 1947 is a fascinating story. John Von Neumann, Tjalling C. Koopmans, Albert W. Tucker, and many many others played important roles in its early development. It has had a profound effect on economics leading to at least two Nobel Prizes and is ranked by some as the newest yet most potent of all mathematical tools. What few people are aware of is that linear programming also had an impact on the early development of computers.

My job in early 1946 in the Pentagon was to develop an analog device which would accept as input all kinds of equations and inequalities subject to all kinds of side conditions (called ground rules) in order to generate as output a consistent Air Force deployment plan. In the Fall 1946, we began to hear rumors in the Pentagon about a marvelous new invention, the electronic computer. My colleague Marshall Wood and I went to Aberdeen Proving Grounds to see if this device could be used for computation of Air Force plans. Early in 1947, we attended a Symposium on Large-Scale Digital Calculating Machinery at the Harvard Computation Laboratory where Howard Aiken had already built the Mark I Relay Computer.

To Wood and myself this meeting was pure science-fiction fantasy. Were we impressed! Various "Buck Rogers" types spoke about their dreams so vividly, with such conviction, and so realistically, that we instantly became true believers that electronic computers, selectrons, and robots were already reality and concepts that we now called artificial intelligence were just around the corner. After the conference, Marshall and I visited a Boston pub where a jukebox played any tune requested by simply talking into a mouthpiece. "Look," we said to each other, our eyes full of wonder, "it has already happened. Here in this pub there is a voice-recognition device coupled with an electronic gadget which understands speech and selects any record requested and plays it for us." While we were thus

---

This paper is a revision of an address presented to TIMS/ORSA National Meeting in Boston in the Spring 1985 and a talk before the conference at Stanford on Computers & Mathematics, August 1, 1986.

marvelling about this great scientific success, a drunken sailor wandered, more precisely, staggered over to the jukebox, and asked the mouthpiece to play *Swannee River*. He was very very drunk. His tongue was thick, and his words so slurred that they were barely decipherable.

To our amazement, the dulcet tune of Steven Foster's Swannee River issued forth from the box an instant later. No longer were we true believers that the age of miracles had arrived. We knew that no nonhuman could possibly have understood what this drunken sailor was saying. Nevertheless from that point in time on, all our plans assumed that digital electronic computers would some day exist an that we could use them to do our calculations. Meanwhile, we prepared ourselves for that great day by building planning models using punch-card equipment and playing an active role in finding funds to subsidize the development of the computers that would solve Air Force planning problems.

In February 1947 I wrote a letter to my former boss, Tex Thornton, who later headed up Litton Industries. At that time he was still with the Ford Motor Company. I told him about my presentation to General Rawlings, the Airforce Comptroller, on the possibility of a "program integrator" for planning and scheduling. I commented how impressed I was with Rawling's strong backing of our project and that it appeared he would fund the development of computers to the tune of a half-million dollars. This doesn't sound like much today, but it was a *lot* of money in those days.

This letter is interesting. I quote with minor changes some of my early hopes: "I believe the central problem facing us in planning, whether it concerns a large business like Ford, or a corner grocery store, or the plans of a war economy, is that each involves how the relations between the multitude of tiny components affects some desired outcome for the system as a whole. Generally the interrelationship of one component with its neightbor is easy to determine — just as it is easy to determine the position of a tree next to its immediate neighbors in the forest — but it is difficult to integrate this information in order to determine its effect upon the system as a whole."

"Let us suppose, however, that one could design an electronic machine that in a sense simulates the relationships among components; suppose further that this machine can operate so fast and has such a vast memory for detail that it can take information from its memory, perform thousands of calculations per second, and store or print the results; and, finally, suppose that these suppositions are not merely pipe dreams, but such developments as the Harvard Mark II, the ENIAC and the EDVAC, and MIT gadgets are well along this very road, then one can begin to see what this potential contribution will be for our research here in the Pentagon."

If I were asked today who did the most back in 1947 to make all these ideas happen, I would say it was Air Force Comptroller General Ed Rawlings. He had the money and the courage and imagination to allow us to use it. In June 1947 he approved a transfer of $400 thousand to the National Bureau of Standards (NBS) that initiated much of the mathematical research that was done there under John Curtis, and electronic computer research that was done there under Sam Alexander. The Bureau used our funds to support development of UNIVAC and IBM as well as their own computers SEAC and SWAC.

Much of the immediate postwar development of electronic computers can be traced to the direct and indirect sponsorship of the AAF Comptroller's Office.

In July 1948, John Curtis of NBS sponsored an important symposium at UCLA on computers and numerical methods. Speakers included John von Neumann, who spoke on "Electronic Methods of Computation"; D. R. Hartree, Cambridge University on "General Survey of Current British Developments" (re computing machinery) and "Some Unsolved Problems in Numerical Analysis"; D.H. Lehmer, University of California, Berkeley, on "Numerical Methods in Pure Mathematics I"; Hans Rademacher, University of Pennsylvania, on "Numerical Methods in Pure Mathematics II"; Bernard Friedman, New York University, on "Wave Propagation and Hydrodynamics and Electrodynamics"; Solomon Lefschetz, Princeton on "Numerical Calculations in Non-Linear Mechanics"; Herman H. Golstein, Princeton, "Eigenvalues and Eigenvectors for Symmetric Matrices;" I spoke on "Programming in a Linear Structure", the original name for the linear programming field. Others present were Howard Aiken, designer of Harvard's Mark II.

Air Force Project SCOOP (Scientific Computation of Optimal Programs) officially started during this period. In August 1948 Marshall Wood and I briefed the Air Staff on the use of electronic computers in military planning. Here are some quotes from that briefing:

1. "The primary objective of Project SCOOP is the development of an advanced design for an integrated and comprehensive system for the planning and control of all Air Force activities."

2. "The recent development of high-speed digital electronic computers presages an extensive application of mathematics to large-scale management problems of the quantitative type. Project SCOOP is designed to prepare the Air Force to take maximum advantage of these developments."

My task was to explain to the Air Staff in the simplest of terms what this mysterious device called a computer was. See Figure 1.

I then described the fourteen different large-scale digital computers that were under development in 1948 (Figures 2 and 3). Note that most of these were under the sponsorship of the Army, Navy, and Air Force. The three computers being supported by the Air Force were the *NBS Interim* which it was hoped would be ready in 10 to 12 months for calculations of small linear programs; the *UNIVAC* for the Comptroller and the *ERA* for Wright Field. There was also the NBS *SUPERSPEED* that was not yet designed but might be supported in the fiscal year 1950 budget.

Up to that time only one electronic computer had been built, the ENIAC at Aberdeen. There were other computers in operation, but they were electromechanical relay machines. A feature of many of the machines in the design phase or under construction at the time was the planned use of acoustic delay lines made of liquid mercury for internal memory.

The computers listed were being designed to have a wide variety of speeds. The slowest were the relay machines which were not significantly faster than existing punch-card equipment. At the other extreme was the so called "superspeed" computer which was in fact in such a preliminary design stage that it was little more than a gleam in someone's eye. The latter was supposed to perform at the rate of 50 thousand multiplications per second.

In an Air Force memorandum I wrote in June 1949, I noted that *two auxiliary devices* were required before full utilization of a computer could be realized. The *first* was a *magnetic tape* on which data could be stored prior to their introduction into a computer. Tape preparation, however, would not be an easy job in fact it would be a Herculean

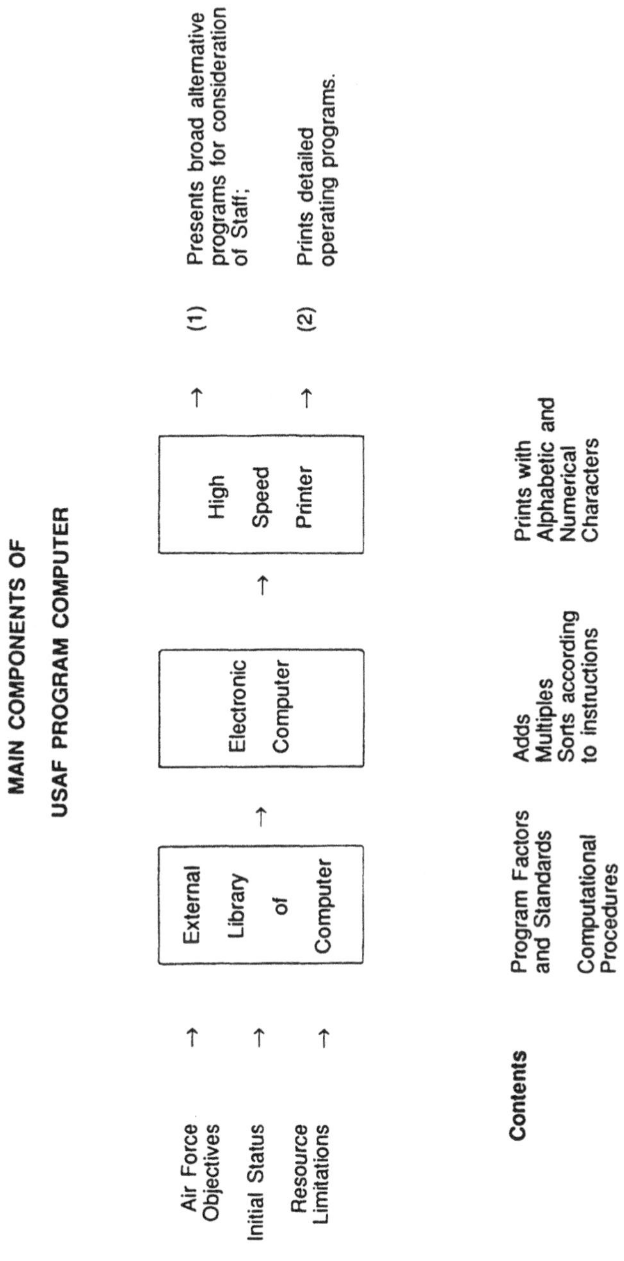

**MAIN COMPONENTS OF
USAF PROGRAM COMPUTER**

External Library of Computer	→	Electronic Computer	→	High Speed Printer

Air Force Objectives →

Initial Status →

Resource Limitations →

(1) Presents broad alternative programs for consideration of Staff;

(2) Prints detailed operating programs.

Contents

Program Factors and Standards

Computational Procedures

Adds
Multiples
Sorts according to instructions

Prints with Alphabetic and Numerical Characters

Figure 1.

LARGE SCALE DIGITAL COMPUTERS
Status as of 1 July 1946

	MACHINE	SPONSOR	BUILDER	LOCATION OR ESTINATION	INPUT AND OUTPUT		STAGE OF COMPLETION
					MEDIUM	WORDS/SEC	
1.	BALISTICS COMPUTER	MDRC MRL	BTL	Ft. Bliss (1) MRL (1)	Punched Paper tape	1.2	In operation
2.	RELAY COMPUTER	BTL MACA BRL	MTL	MACA (1) RBL (1)	Punched paper tape	.33 .55	In operation
3.	KARI I	Navy	IBM Harvard U	Harvard U	Punched cards or paper tape	.40 .40	In operation
4.	MARK II	Navy Bu. Ord.	Harvard U	Dahlgren Bu. Ord.	Paper Tape	.66	In operation
5.	MARK III	Navy Bu. Ord.	Harvard U	Dahlgren Bu. Ord.	Paper or Magnetic tape	13.	Construction
6.	SSEC	AAS Columbia U IBM	IBM	New York IBM	Punched cards or paper tape	25. 20.	In operation
7.	EMIAC	Army Ordance	U of Penn Moore School	Aberdeen BRL	Punched cards	13. 10.	In operation
8.	EDVAC	Army Ordance	U of Penn Moore School	Aberdeen BRL	Paper tape Magetic wire	.30 28.50	Service Test
9.	UNIVAC	Bu. Stand Census USAF	Robert Mauchly Corp.	Washington Pentagon Census	Magnetic tape	833 8.33	Construction
10.	RATHRON	Bu. Stand ONR	Raython Mfg. Co.	Princeton Inst of Numerical Analysis	Magnetic tape	450. 450.	Design
11.	IAS	IAS RCA Army Ord.	Inst. for Advanced Study	Princeton IAS	Magnetic wire	1000.	Design
12.	NBS Interim	MBS and USAF	NBS	Los Angeles	Magnetic tape	350.	Construction
13.	ERA	MBS and ONR	Engineering Research Associates	Washington ONR	Paper Tage	20.	Design Study
14.	MBS Superspeed	USAF	NBS USAF	Washington	Magnetic tape	12000.	Furtue Project

Figure 2.

LARGE SCALE DIGITAL COMPUTERS
Status as of 1 July 1948
(continued)

| MACHINE | INTERNAL MEMORY | | | ARITHMETIC UNIT | | STAGE OF COMPLETION |
	TYPE	CAPACITY	TRANSFER TIME SECONDS	OPERATION	SPEED SECONDS	
1. BALLISTICS COMPUTER	Relay	10 five decimal digits	.4	Addition / Multiplication	2.0	In operation
2. RELAY COMPUTER	Relay	30 seven decimal digits	.7	Addition / Multiplication	0.3 / 1.0	In operation
3. MARK I	Mechanical counters	3.72 twenty three decimal	.3	Addition / Multiplication	0.3 / 3.0	In operation
4. MARK II	Relay	100 ten decimal digits	.033	Addition / Multiplication	.2 / .7	In operation
5. MARK III	Magnetic drum	4000 sixteen decimal digits	.004	Addition / Multiplication	.00066 / .00600	Construction
6. SSBC	Relay / Electronic	150 nineteen and 8 nineteen decimal digits	.07 and .0002	Addition / Multiplication	.001 / .020	In operation
7. EMIAC	Electronic counters	20 ten decimal digits	.0002	Addition / Multiplication	.0002 / .0028	In operation
8. EDVAC	Acoustic	1024 twelve decimal digits	.002	Addition / Multiplication	.00005 / .002	Service Test
9. UNIVAC	Acoustic	1000 eleven decimal digits	.000175	Addition / Multiplication	.00100 / .00224	Construction
10. RATTRON	Acoustic	1000 ten decimal digits	.000197	Addition / Multiplication	.00073 / .00090	Design
11. IAS	Electrostatic	4096 eleven decimal digits	.000025	Addition / Multiplication	.000050 / .000100	Design
12. MBS Interim	Acoustic	500 ten decimal digits	.0002	Addition / Multiplication	.0001 / .0003	Construction
13. ERA	Magnetic drum	16,000 ten decimal digits	.008	Addition / Multiplication	.016	Design Study
14. MBS Superspeed	Acoustic	Undecided	Undecided	Addition / Multiplication	.000020 / .000020	Future Project

Figure 3.

one because the input data of an Air Force planning model could run into hundreds of thousands and perhaps as high as tens of millions of coefficients. The most suitable way to get the data onto tapes was first to punch it into IBM cards and then use special machines to convert the information on cards onto magnetic tapes. The *second* device needed was a *high-speed printer* that would convert the output from the computer into a form where they could be read and distributed to the Air Staff. Early in March 1949 additional funds were transferred to the Bureau of Standards to develop both these devices. This brought the funds transferred to the Bureau of Standards by the end of 1949 to slightly over a half-million dollars.

This period was not without its small financial traumas. There was a sad letter from John Curtis of NBS dated December 1949 begging for $50 thousand to help Harry Huskey's complete his SWAC computer.

June 20, 1950 the NBS SEAC computer was dedicated. In September 1950, I gave a talk before the Association for Computing Machinery on how SEAC could be used to solve linear programs.

Here are some amusing excerpts from a memo dated May 8, 1951, from Air Force General Dau to General Rawlings on how well Air Force Project SCOOP was doing in fulfilling all its promises: "I suppose there is some belief that we have not accomplished all we set out to do.... With respect to SCOOP we are engaged in an educational project which will never end.... I am pleased the unwarranted publicity attached to electronic computers has diminished.... It now appears SEAC will not produce much more than education and experience."

When we speak about the early developments of linear programming and computers, we usually think of the effect that computer's had on linear programming. This effect as we have seen was by no means one way. By January 1952, six months before I left the Air Force and the Pentagon to go to Rand, the Air Force had invested well *over a million dollars* with the National Bureau of Standards. As I noted earlier, this was a lot of money in those days and it had a profound impact on the early development of computers and on computing methodology.

Uses of and Limitations of Computers in Number Theory

P. ERDOS Mathematical Institute, Hungarian Academy of Sciences, Budapest, Hungary

To the memory of my friend Irving Reiner who was one of my child prodigies.

First a disclaimer and apology. I am very far from being a computer expert and it is perhaps even presumptuous for me to write about uses of computers in number theory and even more so to talk about its limitations. Nevertheless since a great deal of my long (too long?) life has been spent thinking about number theory perhaps my remarks will not be entirely useless.

I will discuss first of all the few cases where computers were absolutely essential to prove (or disprove) important results. As far as I know the number of these is small. Then I will mention several cases where computers helped to formulate plausible conjecture or show that there is a great deal of doubt about the truth of certain conjectures and that perhaps one should then try to modify the conjecture or perhaps at least temporarily abandon it. Finally many questions where computers might have helped but so far have not either because the computers are too slow or because there are really too many possibilities so that at least in the foreseeable future there is no hope to find out the truth with computers.

First of all, Halmos once was supposed to have said,

"computers are important but not in mathematics." I do not
think he meant this to be taken entirely seriously. Computers
can clearly be used to make conjectures which could not even
have been suspected without them, e.g. in the theory of iterations
and differential equations   (Ulam   was one of the pioneers
in this), and occasionally computers helped in finding proofs.
The four color theorem of Haken and Appel   seems to be one
of the great exceptions. Another great exception is the
disproof of a conjecture of Mertens by Odlyzko and Riele [12].
Denote by $\mu(n)$ the well known function of Mobius $\mu(n) = 0$
if n has a quadratic factor otherwise $\mu(n) = (-1)^{\omega(n)}$ where
$\omega(n)$ is the number of prime factors of n. It is well known
that $\frac{1}{x} \Sigma_{n=1}^{x} \mu(n) \to 0$ is equivalent to the prime number theorem
and that $\Sigma_{n=1}^{x} \mu(n) = \sigma(x^{1/2+\epsilon})$ for every $\epsilon > 0$ is equivalent
to the Riemann hypothesis. Mertens conjectured more than
100 years ago that for every x

(1) $$\left| \sum_{n=1}^{x} \mu(n) \right| < x^{\frac{1}{2}}.$$

This conjecture was generally disbelieved, but only very
recently using some results of Lovasz, Odlyzko and Riele
disproved (1). In fact they proved that for infinitely many x

$$\sum_{n=1}^{x} \mu(n) > 1.04 \ x^{\frac{1}{2}}.$$

It is generally believed that in fact

(2)
$$\overline{\lim}_{x^{1/2}} \frac{1}{x^{1/2}} \sum_{n=1}^{x} \mu(n) = \infty,$$

but for the present time (2) seems beyond reach. Another

semi-exception is the beautiful result of Hensley and Richards.

Denote by $\pi(x)$ the number of primes not exceeding x.

Hardy and Littlewood conjectured that for every x   and y,

$\pi(x+y) \leq \pi(x) + \pi(y) +1$.This conjecture seemed very plausible

a nd in fact Landau proved it for x = y (in fact Landau proved

it only if x = y and x is sufficiently large, the general

case was settled later with the help of computers by Rosser and

Schoenfeld   ).   Another conjecture of Hardy and Littlewood,

the prime k-tuple conjecture. states that if $a_1, \ldots, a_k$ is a

sequence of integers which does not form a complete set of resi-

dues mod p for any p, then there are infinitely many integers

n   for which all the numbers $n + a_i$, i = 1,2,...,k are primes.

This conjecture is undoubtedly true but is certainly hopeless

since the simplest special case k = 2 would already imply that

every even number is the difference of two primes in an

infinite number of ways. Now Hensley and Richards proved

that these two conjectures are incompatible [9] . In fact they

showed that if the prime k-tuple conjecture holds then for

infinitely many x   and y < x

(3)
$$\pi(x+y) > \pi(x) + \pi(y) + cy/(\log y)^2$$

for a certain positive constant   c.   By the way Richards

remarks that they could have done their work without computers

but probably would not have done so.

Ernst Straus thought that perhaps for $y < x$

(4) $\pi(x+y) \leq \pi(x) + 2\pi(\frac{y}{2})$

holds and in fact (4) is still open as the method of Hensley and

Richards fails to disprove (4).   In a joint paper of Richards

and myself, we disagreed whether (3) can hold for every   c   and

infinitely many values of   x   and   y.   I believed that for

large   c   (3) never holds and if the conjecture of Straus

holds then my conjecture holds.   Richards believes that (3)

holds for all values of c (for suitable   x   and   y).   This

paper with Richards contains several other problems which

I consider very nice.   Let me state here only one of them.

An integer   y   is called good if for all $y \leq x$

$$\pi(x+y) \leq \pi(x) + \pi(y).$$

We conjectured that the density of the good y's is   1.   We

could only prove that the density is positive.

In connection with the prime k‑tuple conjecture I

posed   two problems: first of all for fixed   k   determine

the smallest possible value of a_k presumably as   k   tends

to infinity min $a_k = (1+0(1)) \, k \log k$.   A result of Selberg,

Montgomery and Vaughan implies $a_k < 2\ k\ \log\ k$, but perhaps
here I was more interested to determine min a_k for small
values of k, this question can no doubt be settled by
aid of a good program and a computer. I am not sure for how
large a value of k it will be practical to go.ı Also how
many k-tuples $\{a_1 < a_2 <...< a_k\}$ are there which satisfy the
requirements of the prime k-tuple conjecture? Here of course
some explanation is needed. Two sets of k tuples are
considered the same if they are congruent mod p for all
primes $p \leq k$, the primes $p > k$ can be ignored.

Henceforth I will mainly discuss my own problems. I
am and always have been very interested in extremal problems
in number theory and once, a few years ago, I made the following very plausible conjecture: let $1 \leq a_1 < a_2 <...< a_k \leq x$
be a sequence of integers and assume that $a_i\ a_j/a_i + a_j$ is never
an integer. It is true that max $k = (\frac{1}{2} + 0(1))x^2$. The odd
numbers show that $k \geq \frac{x}{2}$ is certainly possible and I was
convinced that my conjecture is true. Odlyzko found by using
a computer more than 700 numbers not exceeding 1000 .for
which $a_i\ a_j/a_i + a_j$ is never an integer and we now
no longer are convinced that my guess is right, but we have
no other plausible conjecture for max k. By the way, I
suspect that if we require that $2\ a_i\ a_j/a_i + a_j$ should never be
an integer then perhaps k must be $0(x)$, but there is no

numerical or other evidence for my guess. A few days ago

Odlyzko   carried out the computation up to 10000 and found

6805 integers $\leq$ 10000 but he used the greedy algorithm and

this does not necessarily give the best results, thus the slow

drop towards $\frac{1}{2}$ may not have any significance.

One of my first theorems which dates back to 1932

states as follows: let $a_1 < a_2 < \ldots$ be a sequence of integers

no one of which divides the other. Then there is an absolute

constant C for which

$$\sum_{k} \frac{1}{a_k \log a_k} \leq C.$$

I suspect that $C = \sum \frac{1}{p \log p}$ where p runs through the sequence

of primes. I am farily sure that this question can be decided

with the help of a good computer and a good program and I

hope that the answer will be affirmative. For many related

problems and results see my joint paper with Sárközy and

Szemerédi [8] and for the older literature the book of

Halberstam and Roth "Sequences."

Silverman and I posed a few years ago the

following question: let $a_1 < a_2 < \ldots$ be an infinite sequence

of integers, assume that $a_1 + a_j$ is never a square. What is

the largest possible value of the density of our sequence?

The problem of course also has a finite version: let

$a_1 < \ldots < a_k \leq x$ assume that $a_i + a_j$ is never a square.
Determine or estimate max k. Clearly the integers $\equiv$ 1(mod 3)
have this property and at first we though that this is best
possible, but soon Massias found by some computation that
the following 11 residues mod 32, 1(mod 4), 14, 26, 30 mod 3
have the same property, i.e. no two of these residues add up
to a square. Lagarias,Odlyzko and Shearer proved that if our
sequence is periodic then its density is indeed at most $\frac{11}{32}$.
Their proof is quite complicated. Very likely this is true
for a non-periodic sequence too i.e. $\frac{11}{32}$ is the largest possible
value of the density of our sequence and perhaps even in the
finite version max k $= \frac{11}{32}x + 0(1)$. Lagarias, Odlyzko and
Shearer proved that the density of our sequence is $\leq 0.475x$,
the proof again is very difficult. Their proof also applies
for k-th powers instead of squares but the following problem
is left open. Let $n_1 < n_2 < \ldots$ be a sequence of integers which
is uniformly distributed mod d for every d and which satisfies $n_{i+1}/n_i \to 1$. Is it then true that if A is a sequence
of integers so that $a_u + a_v \neq n_i$ for every $a_u, a_v \in A$ and every
i then the density of A is less than $\frac{1}{2}$?

An old theorem of mine [3] states that every large integer
$\not\equiv$ 1(mod 4) is the sum of a square of an odd prime and a
squarefree number. (I am afraid that in my paper "I forgot"

the condition $\not\equiv 1 \pmod 4$.) I have sharpened two earlier results of Estermann who proved that every large integer is
the sum of a square and a squarefree number and a prime and a
squarefree number. In all three cases it might be worth
while to determine all the exceptional numbers. This would
only be a perhaps tiresome exercise. Perhaps the following
question is of some interest: let $A = \{a_1 < a_2 < \ldots\}$ be a
sequence of integers and $b_1 < b_2 < \ldots$ is the sequence of integers no one of which divides any of the a's. For which A
is it true that every large n is of the form $a_i + b_j$?

Let me state a question where I hoped a computer can help
but I was probably too optimistic. Wieferich proved that every
integer is the sum of 9 or fewer cubes (here and later we mean
positive cubes) and Landau
proved that every large integer is the sum of ≤ 8 cubes.
Dickson proved (this was long before the modern computers) that
23 and 239 are the only integers which need 9 cubes. Linnik,
Gordon Pall and later in a much simpler form, Watson proved that
every large integer is the sum of ≤ 7 cubes and computation
showed that it is virtually certain that every integer > 454
is the sum of ≤ 7 cubes, further computation convinced us (here
"us" stands for experienced computer experts, I am unfortunately
not one of them) that every number > 8042 is the sum of ≤ 6
cubes and every number > 1290740 is the sum of ≤ 5 cubes. It is
generally believed that every large number is the sum of four

cubes. I hoped to check by computer up to about 10^{12} or 10^{13}

and hope that the integers needing 5 cubes will disappear by

then. While present computers are not fast enough to do this,

a computation to 10^9 or even 10^8 might be sufficient to find

the last one needing five cubes. It is sad that we all seem to be too

stupid to prove that every large integer is the sum of four

positive cubes. In fact it is not even known that every

large integer is the sum of 4 positive or negative cubes. It

might be worthwhile to show this for all integers $\leq 10^6$ or

10^5.

Another great mystery is: is the density of integers

which are the sum of 3 (positive or negative) cubes, or more generally

is it true that the density of integers which are the sum of k

k-th powers is positive?

Ivić and I once conjectured that every large integer is

the sum of three squarefull numbers, i.e. numbers of the form

$x^2 y^3$. The proof is nowhere in sight but computer experiments

make this conjecture a virtual certainty.[*] I once conjectured

that there are infinitely many integers which are not the sum

of k or fewer integers all prime factors of which have

exponent $\geq$ k and I stated that I can prove this since the number

of solutions of $\sum_{i=1}^{k} a_i \leq x$ is less than $(1-\varepsilon)x$ where all

prime factors of the a's occur with an exponent $\geq$ k. I

---

[*]Recently Heath-Brown proved this conjecture.

stated that this follows by a simple computation but Schinzel

checked it and found that I am wrong.   Now the problem remains.

Is my statement correct?   Here   numerical experiments might

be worthwhile at least for k = 3.

Selfridge and I proved that the product of consecutive

integers is never a power.   We needed only the help of a

calculator not an electronic computer.

In a forthcoming joint paper with Lacampagne and Selfridge [5]

we studied prime factors of binomial coefficients.   An old

and hopeless conjecture of Schinzel states that for every   k

there are infinitely many values of   n   for which $\binom{n}{k}$ is the

product of   k   primes > k.   We define the deficiency of

$\binom{n}{k}$ as follows: $\binom{n}{k}$ must first of all have all its prime

factors > k and its deficiency is   r   if it is the product

of k-r   primes > k. E.g. $\binom{47}{11}$ =   47 43 41 37 29 23 19 13 i.e. its

deficiency is   4.   The great surprise was when we found that

$\binom{284}{28}$ has deficiency 9, probably this is the largest

possible deficiency, certainly this is true for n < 1000 and

we hope to push the search further.   It seems quite likely

that there are only a finite number of integers   n   and   k

whose deficiency is ≥ 1, unfortunately we do not see how to

attack this problem.   On the other hand it is an easy exercise

to prove that for fixed   k   and n > k! + k the deficiency of $\binom{n}{k}$

must be 0, and for fixed   k   it might be of some interest

to find the largest n with positive deficiency. I expect that

it will be much smaller than lcm(1,...,k): but we did not yet

investigate this question. Another interesting question is:

let n_h be the smallest integer for which

$p(\binom{n}{h}) > h$ where $p(m)$ is the smallest prime factor of m.

All we know that $n_h > h^{1+c}$ for some positive c. It seems

clear that n_h increases much faster. Preliminary computer

trials seem to confirm this--no doubt $n_h > \exp h^c$ for some

positive c and probably $c > 1 - \epsilon$ for every $\epsilon > 0$ for all

$h > h_0$ and perhaps $n_h > \exp c \dfrac{h}{\log h}$.

There is another perhaps not uninteresting way of defin-

ing the deficiency of $\binom{n}{h}$. Assume that $p(\binom{n}{h}) > h$ and that

there are r of the integers n, n-1 ... n - h + 1 all whose

prime factors are $\leq h$ -in this case we say that the deficiency

of $\binom{n}{h}$ is r. It is now easy to see that for infinitely many

pairs n and h the deficiency is ≥ 1 (here too of course we

must have n < h!). Probably there are again

only finitely many pairs n and h for which the deficiency

is ≥ 1.

Let me state now a few rather special problems. Let

$n = \prod p_i^{\alpha_i}$ put

$$u(n) = \prod_{\alpha_i > 1} p_i^{\alpha_i} \quad \text{and} \quad v(n) = \prod_{\alpha_i = 1} p_i$$

u(n) is the squareful part and v(n) the squarefree part of n.

Put $A(n;h) = \prod_{i=0}^{h} (n+i)$. I thought that n = 24 is the only

integer for which for every k   u(A(n;k)) > v(A(n;k)), but

Selfridge   pointed   out to me that n = 48 has the same pro-

perty and Massias found no further exceptions $\leq$ 1000 and hopes

to push the conputations as long as his computer and patience

will hold out. Unfortunately I do not see how to prove any-

thing here. Perhaps I can explain how I came to believe in

this conjecture. It is easy to see that for large k

u(A(n;k))>v(A(n;k))in fact this happens already for k = $_{\epsilon}n^{1/4}$[*].

On the other hand it is not hard to see that for almost all

n there is a k > $n^{1/2-\epsilon}$, for which v(A(n;k))>u(A(n;k)).By the way

when I first thought about this problem I noticed that n = 8

is a "near miss" u(8 9 10 11 12 13 14) < v(8 9 10 11 12 13 14)

since $2^7 3^3$ < 5, 7 11 13. I think by the way that for every

n there are integers k_1 and k_2 so that

$$u\left(\prod_{i=-k_2}^{+k_2} (n-i)\right) < v\left(\prod_{i=-k_1}^{+k_1} (n-i)\right)$$

but of course I can not prove this. An old and no doubt

hopeless conjecture of mine states: for every k   an

ϵ > 0 if n > $n_0(\epsilon,k)$ we have

$$u(A(n;k)) < n^{2+\epsilon}.$$

If true this is certainly best possible since there are

_____

[*]And infinitely many n.

infinitely many consecutive integers both of which are squareful . When I first met Mahler at the international conference in Oslo I (foolishly) asked are there infinitely many consecutive squareful integers? Mahler immediately answered of course the equation $x^2 - 8y^2 = 1$ has infinitely many solutions. Denoting $A(x)$ the number of integers n for which

$$u(n) = n, \quad u(n+1) = n + 1.$$

Perhaps $A(x) < (\log x)^{\alpha}$ for some α? ($\alpha = 1$??.) Tenenbaum has a surprisingly complicated proof that $A(x) < cx^{1/3}$. It might be worthwhile to calculate $A(x)$ for as large values of x as possible. I am not sure if this has been done already, but it would be perhaps worthwhile.

Denote by $u_r(n)$, $(n = \prod p_i^{\alpha_i})$, $u_r(n) = \prod_{\alpha_i \geq r} p_i^{\alpha_i}$.
A nice and no doubt very difficult questions is: determine the smallest ϵ_r for which for every k and $n > n_0(k,r)$

$$u_r((A(n;k))) < n^{1+\epsilon_r}.$$

I expect that for all $r > 2$, $\epsilon_r < 2$ but perhaps for all $r, \epsilon_r > 1$, this latter statement if true is perhaps not hopeless, but I certainly do not see how to attack it.

Finally, I would like to state a random collection of

some problems which are all very difficult and some more
computations might throw some light on them. Let W(n) be the
smallest integer for which if one divides the integers $\leq$ W(n)
into two classes at least one of them contains an arithmetic
progression of n terms. (W(n) is named after van der Waerden.)
Is W(n) primitive recursive?* $W(n)/2^n \to \infty$, W(5) = 178;
it might be nice to have some good estimate for W(6); the exact
value may be beyond our reach.

Is it true that for every k there are k primes in an
arithmetic progression. The current record is k = 19 due to
Pritchard and Weintraub. Here is my 3000 dollar conjecture
which if true would give an affirmative answer. Let
$a_1 < a_2 < \ldots$ be a sequence of integers for which $\sum \frac{1}{a_n} = \infty$
then the a's contain arbitrarily long arithmetic progressions.

Is it true that there is an r for which every integer
is the sum of a prime and r powers of 2? A theorem of
Gallagher implies that for every ϵ there is an r for which
the lower density of the numbers representable as the sum
of a prime and r powers of 2 is $> 1 - \epsilon$. Probably for
large r $(r \geq 2?)$ every arithmetic progression contains
infinitely many such numbers false for r = 1 (Van-der Corput
and Erdos).

An old conjecture of mine states that every sufficiently

---
*Very recently, Shelah proved this.

large integer n $\not\equiv$ 0 (mod 4) is the sum of a squarefree number

and a power of 2. I could not even prove that this holds for

almost all integers. Here perhaps I should mention one of my

favorite problems: a system of congruences

$$a_i \pmod{n_i}, \ n_1 < n_2 < \ldots < n_k$$

is called a covering congruence if every integer t satisfies

at least one of these congruences. Is it true that for every

integer m there is such a system for which n_1 = m ? I

offer 1000 dollars for a proof or disproof. If the answer

affirmative let f(m) be a smallest k   for which such a

system exists. Estimate or determine f(m). Many further

problems which can be attacked (at least partially) by computa-

tional methods can be found in my book with Graham and in a

more recent book by Guy.

I would like to mention three more problems. I

conjectured that 105 is the largest integer n   for which all the num-

bers n-2^k are prime, $1 \leq k \leq \frac{\log n}{\log 2}$. This conjecture has been

checked up to at least 10^6, but seems to be quite hopeless.

In our book Graham and I posed the following problem.

Let t be an odd number such that $2^\alpha t + 1$ is never a prime

(α = 1,2,...). Must there be a set of covering congruences

responsible for this? I.e. is there an N so that for every

t $(2^\alpha t + 1, N) > 1$? The answer is almost certainly negative.

This type of problem can be posed in many forms but it always seems hopeless.

In my work with J.L. Nicolas on extremal values of various number theoretic functions computers were often useful to make (or break) plausible conjectures, which we quite often later could settle. Here is a nice problem of ours of a slightly different nature. Is it true that there is an absolute constant c so that for any choice of t primes $p_1 < p_2 < \ldots < p_t$ we have

(A)
$$\sum_{1 \le i < j \le t} \frac{1}{p_j - p_i} < c_t.$$

I at first thought that (A) holds if we only assume that the only condition we have to assume about the p's is that every interval of length x contains fewer than $c\, x/\log x$ p's. This is well known to be satisfied for the primes. Ruzsa found the following ingenious counterexample: let the p's be the integers of the form $\sum_{i=1}^{L} \epsilon_i 2^i$, $\epsilon_i = 0$ or 1, but $\epsilon_1 = 0$ if $i = 2^r$. Perhaps if we further assume that the p's satisfy the conditions of the prime k-tuple conjecture i.e. for no prime q does our set $p_1, \ldots, p_t$ contain all the residues, then (A) holds: we could not decide this question.

Finally many challenging and interesting problems are mentioned in our paper with Penney and Pomerance but perhaps this paper is already too long and the interested reader can look up our paper.

To end the paper (every good or bad thing except Mathematics has to end sometime) let me make two remarks. Gauss allegedly once was asked how he guessed his many wonderful theorems and was supposed to have answered "by systematic trial and error". This was of course long before the age of computers. On the other hand, a word of caution if you are not Gauss you should be careful not to jump to conclusions from numerical evidence. The well known "law of small numbers" may trick you. As far as I know Guy was the first to use this terminology. Many of us (even Riemann and perhaps Gauss himself) occasionally formulated wrong conjectures. Let me illustrate the possibility of making a wrong conjecture in one case where I happened to be right in my guess. Pomerance asked the following question: Consider the longest arithmetic progression

$$a + b, \ a + 2b, \ldots, \ a + nb, \ (a,b) = 1, \ 0 < b < a$$

which are all composite but $a + (n+1)b$ is a prime.
Is it true that the longest progressions always occur
if a is a prime? Several colleagues believed this but I suspected the law of small numbers and Odlyzko found several counterexamples. $a = 8207$, $b = 3251$, $n = 135$ is the smallest counterexample. Odlyzko suggests that the values of a which establish new records will be very thin the values of the records

n have probably also density 0 but are probably not so thin.

An old conjecture of Mirsky and myself stated that for

infinitely many integers n we have $d(n) = d(n+1)$ where $d(n)$

is the number of divisors of n. This problem seemed unattain-

able at first but Claudia Spiro first proved that

$d(n) = d(n+5040)$ has infinitely many solutions and using

and further developing   her method Heath-Brown proved our

conjecture. One of the key lemmas of Heath-Brown was that

for every k there are k integers $n_1 < n_2 < \ldots < n_k$ for

which for every $1 \leq i < j \leq k$

$$\frac{n_i}{n_j - n_i}$$

is an integer. It would be of some interest to get upper and

lower bounds for n_k. Heath-Brown proved

$$\exp(ck) \log k < n_k < \exp k^{3+\varepsilon}.$$

It would be of interest to determine n_k for small values

of k (say for all $k \leq 10$). $n_6 = 240$ was shown by Balog with a

very small computer. Also it would be of interest to get

upper and lower bounds for $n_k - n_1$.

One key lemma in the proof of Claudia Spiro was to find

8 primes $p_1, p_2, \ldots, p_8$ for which the least common multiple

of $(p_j - p_i)$ is 5040.   Narkievicz and I tried to get upper and lower bounds for

$$\min_{1 \leq i < j \leq k} (p_j - p_i) = H(k)$$

in terms of k.   In particular for $k = 8, H(8) = 5040$ is almost certain.   Trivially $H(8) \geq 2520$ but to prove that it is 5040 would need some computation.   If it is 2520 then the 8 primes must be incongruent mod 16.[*]

I discuss only problem which I worked on myself.   Thus e.g. I omit the important new results on primality testing and factoring not because I do not think they are important but because I do not know so much about them.

---

[*]Recently $H(8) = 50,400$ has been proved.

References

[1] J. Bohman and C. E. Froberg, Numerical investigation of Waring's problem for cubes, *BIT* **21** (1981), 118-122.

[2] W. J. Ellison, Waring's problem, *Amer. Math. Monthly* **78** (1971), 10-36.

[3] P. Erdős, The representation of an integer as the sum of the square of a prime and of a squarefree integer, *J. London Math. Soc* **10** (1935), 243-245.

[4] P. Erdős, On integers of the form $2^k + p$ and some related problems, *Summa Brasiliensis Math.* **II** (1950), 113-123.

[5] P. Erdős, C. B. Lacampagne and J. L. Selfridge, Prime factors of binomial coefficients and related problems, *Acta Arith.* **49** (1988), 507-523.

[6] P. Erdős, D. E. Penney and Carl Pomerance, On a class of relatively prime sequences, *J. Number Theory* **10** (1978), 451-474.

[7] P. Erdős and I. Richards, Density functions for prime and relatively prime sequences, *Monatch. Math.* **83** (1977), 99-112.

[8] P. Erdős, A. Sárközy and E. Szemerédi, On divisibility properties of sequences of integers, in *Number Theory (Colloq. J. Bolyai Math. Soc., Debrecen, 1968)*, North Holland, Amersterdam (1970), 35-49.

[9] D. Hensley and I. Richards, Primes in intervals, *Acta Arith.* **25** (1979), 345-391.

[10] J. C. Lagarias, A. M. Odlyzko and J. B. Shearer, On the density of sequences of integers the sum of no two of which is a square. I. Arithmetic Progressions, *J. Combin. Theory Ser. A* **33** (1982), 167-185.

[11] J. C. Lagarias, A. M. Odlyzko and J. B. Shearer, On the density of sequences of integers the sum of no two of which is a square. II. General Sequences, *J. Combin. Theory Ser. A* **34** (1983), 123-139.

[12] A. M. Odlyzko and H. J. J. te Riele, Disproof of the Mertens conjecture, *J. Reine Angew. Math.* **357** (1985), 138-160.

Strip Mining in the Abandoned Orefields of Nineteenth Century Mathematics

WILLIAM GOSPER   Symbolics, Inc., Mountain View, California

0. Apology

This was my 1984 (NYU) talk. My 1986 (Stanford) conference talk began

"The last time Dick [Jenks] asked me to do this, my talk was so weird, and my slides so complicated that I left much of my audience exhausted and confused. By way of remedy, we have amassed here today an array of video equipment so expensive that it can actually reduce mathematics to the intellectual level of daytime TV. So kick back, shift your brain into neutral, and we'll watch a few pictures whose only purpose is to leave you *relaxed* and confused."

I then had the great fortune to project onto a huge, bright screen, a live, relatively disaster free, forty minute color animation. (Let me assure those who attended that the backup videotape *would* have been a disaster by comparison. You just can't tape most of that stuff, or even convert it to current broadcast standard.)

Although the talk was fun, it contained little suitable for print publication. "The medium was the message." That is, the novelty was not so much in the mathematical concepts presented, but rather in the way that motion (particularly zooming) and color could vivify those concepts.

We look forward to an era of electronic publication in which widely available, muscular, high definition graphics engines will let us vivify for each other genuinely interesting mathematics.

Meanwhile, all I can offer is atonement for yet another misdeed at the aforementioned (NYU) conference—due to a two-year separation from TEX, I never contributed to the proceedings. Here then is a slightly less incoherent rendition of my 1984 talk.

1. Teaser

By strip mining, I meant applying the power of a modern symbolic processor, particularly via undetermined coefficients, to old fashioned investigations that were computationally inaccessible to the old fashioned investigators.

I feel sometimes like a kid at the controls of huge, bucket wheel excavator, chewing indiscriminately into apparently mined out formations, yet finding paydirt through sheer, childish brutality.

Random excavations, such as with Berlekamp's factoring algorithm, produce random nuggets:

$$\frac{(x-a)^3}{(z-x)^3(x-y)^3} + \frac{(y-a)^3}{(x-y)^3(y-z)^3} + \frac{(z-a)^3}{(y-z)^3(z-x)^3} = 3\frac{(x-a)(y-a)(z-a)}{(x-y)^2(y-z)^2(z-x)^2}$$

$$\frac{1}{(w-x)^3(w-y)^3(w-z)^3} + \frac{1}{(x-y)^3(x-z)^3(x-w)^3}$$

$$+ \frac{1}{(y-z)^3(y-w)^3(y-x)^3} + \frac{1}{(z-w)^3(z-x)^3(z-y)^3}$$

$$= 3\frac{(w-x-y+z)(w-x+y-z)(w+x-y-z)}{(w-x)^2(w-y)^2(w-z)^2(x-y)^2(x-z)^2(y-z)^2},$$

as does fooling with the trigonometric simplifiers:

$$\sin(x-w)\sin(x-y)\sin(y-w)\sin(y-z)\sin(z-w)\sin(z-x)$$

$$+ \cos(x-w)\cos(x-y)\sin(y-w)\cos(y-z)\cos(z-w)\sin(z-x)$$

$$+ \cos(x-w)\sin(x-y)\cos(y-w)\cos(y-z)\sin(z-w)\cos(z-x)$$

$$+ \sin(x-w)\cos(x-y)\cos(y-w)\sin(y-z)\cos(z-w)\cos(z-x) = 0.$$

These results are just like the factorization of $x^2 - y^2$, or the sine addition formula, only millions of times less useful. They lead one to speculate whether algebraic engines in the Renaissance might have made this sort of thing into an art form.

More systematic excavations churn up bigger clumps of pay dirt:

$$\sum_{n\geq 0}(1 - q^{6n+\frac{1}{2}})\frac{(q^{-1/2}, q^{3/2}; q)_{2n}}{(q^2; q)_{4n}}\frac{(q^{5/2}/a, a; q^5)_n}{(q^{3/2}/a, a/q; q)_n}q^n$$

$$= \frac{(\sqrt{q}; q)_\infty (a; q^5)_\infty}{(q^2, a/q; q)_\infty}\left(\frac{(a/\sqrt{q}; q)_\infty (q^5; q^5)_\infty}{(q^{5/2}, aq^{5/2}; q^5)_\infty} - \frac{a}{q}\frac{(\sqrt{q}; q)_\infty (q^{5/2}/a; q^5)_\infty}{(q^{3/2}/a; q)_\infty}\sum_{n\geq 0}(-a^2)^n q^{5n^2/2}\right)$$

is but one of dozens of identities that can be extracted from a single mathematical structure uncovered by a brute force attack on certain simultaneous polynomial equations. I call such a structure a "path invariant matrix system." Points in some space are linked by edges that are labeled with matrices in such a way that the matrix product taken along any path depends only on the endpoints. The path invariance condition on minimal loops induces polynomial equations among the matrix elements. A successful solution then provides a limitless space of closed contours, each yielding a correct identity, several of which may be concise enough to be interesting.

2. Matrices

Before getting into path invariance, let me extol some notational, analytic, and computational virtues of matrix products. First, notice how cumbersome are the conventional notations for an identity such as

$$\prod_{k \geq 1} \begin{pmatrix} \dfrac{k^2}{(4k+2)^2} & \dfrac{30k-11}{16k(2k-1)} \\ 0 & 1 \end{pmatrix} = \begin{pmatrix} 0 & \zeta(3) \\ 0 & 1 \end{pmatrix}.$$

Hypergeometrically, you get a ridiculous $\frac{19}{16}{}_6F_5[\frac{1}{16}]$. As a straight sum, you can do as well as

$$\sum_{k \geq 1} \frac{30k-11}{4(2k-1)k^3 \binom{2k}{k}^2} = \zeta(3),$$

but, taken literally, this says something rather silly: add up a sequence of terms whose kth contains products of k factors, in this case $\binom{2k}{k}^2$. Now most of us know the trick of incrementally computing each term from the previous, but this is nowhere hinted in the sum. Furthermore, you'll need a less-than-obvious auxiliary variable to avoid the wasted effort of cancelling out factors introduced in the term just preceding, or, more seriously, to avoid division by 0 if the numerator of the preceding term happens to vanish (*e.g*, if the 11 above were instead a 90).

Now suppose that you want a very precise value of $\zeta(3)$, and thus wish to sum n terms of this series. You will find it dramatically cheaper to write out instead the first n matrices, and then pairwise multiply to form $\lceil n/2 \rceil$ products, and repeat until only one matrix remains. (In 1985, I used this technique, essentially due to R. Schroeppel, to temporarily steal the π computation record from Japan.)

This representation of simple sums as 2×2 matrix products is merely a special case (a vanishing off-diagonal) of the well known representation of nested homographic functions. When a diagonal element vanishes, you have a continued fraction. It is strange that so fruitful a representation is applied so rarely to sums. Perhaps discouraging is the apparent nonlinearity of the matrix product form, (partially) concealing such familiar operations as termwise differentiation or combination with other series. We shall see that a small bag of tricks, again based on path invariance, more than remedies these problems.

Even greater simplifications are possible with larger matrices:

$$\sum_{n=0}^{m} a_n x^{n-1} \sum_{k=0}^{n-1} \frac{b_k}{k!} = \prod_{n=0}^{m} \begin{pmatrix} \dfrac{x}{n+1} & b_n & 0 \\ 0 & x & a_n \\ 0 & 0 & 1 \end{pmatrix}_{1,3},$$

where the subscript$_{1,3}$ means the upper right element. Here, the matrix product not only vaporizes a nested loop, but also dispels an artificial asymmetry between a_i and b_i. They can, in fact, be switched via summation by parts, but then they are asymmetrical the other way!

And here is a way to get factorable multiple sums:

$$\sum_{i=0}^{m}\sum_{j=0}^{m}a_ib_j = \prod_{n=0}^{m}\begin{pmatrix} 1 & a_n & b_n & a_nb_n \\ 0 & 1 & 0 & b_n \\ 0 & 0 & 1 & a_n \\ 0 & 0 & 0 & 1 \end{pmatrix}_{1,4}.$$

Perhaps most important of all are those sadly neglected recurrences that can be expressed by products of denser, (typically) nontriangular matrices. Who knows what we might find, once emancipated from our feebly notated sums, (scalar) products, and continued fractions? One enticement is a smooth generalization of the notion of "closed form": a recurrence is in simpler (and therefore more canonical) form when it is the product of smaller or sparser matrices. Fully closed form is a 1×1, *i.e.* scalar product. The importance of such products, by the way, is almost purely their uniqueness and comparability, since they usually converge more slowly than most of the higher order recurrences that they canonicalize.

3. Path Invariance

This idea was crystalized by Kevin Karplus while a student in the only course I ever taught. Label the edges of some sort of directed, multiply connected graph with matrices so that their product, taken along any connected sequence of edges, depends only on the endpoints. Usually, the graph is piecewise grid-like, and it is sufficient to demonstrate path invariance around each of the grid cells, and around the (usually) triangular cells where the grids patch together. For example, a purely two dimensional, path invariant grid would look like

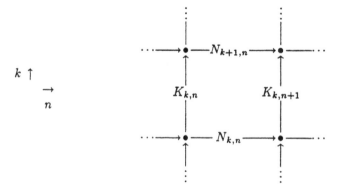

where the matrices $N_{k,n}$ and $K_{k,n}$ satisfy the invariance

$$N_{k,n} K_{k,n+1} = K_{k,n} N_{k+1,n},$$

a sort of discrete Cauchy-Riemann condition. For example, we might have

$$N_{k,n} \overset{\text{def}}{=} \begin{pmatrix} \dfrac{q^k}{1 - q^{n+1}} & 1 - q^k \\ 0 & 1 \end{pmatrix}, \qquad K_{k,n} \overset{\text{def}}{=} \begin{pmatrix} \dfrac{q^n}{1 - q^{k+1}} & 1 - q^n \\ 0 & 1' \end{pmatrix}.$$

Then

$$N_{k,n} K_{k,n+1} = K_{k,n} N_{k+1,n} = \begin{pmatrix} \dfrac{q^{n+k+1}}{(1 - q^{n+1})(1 - q^{k+1})} & 1 \\ 0 & 1 \end{pmatrix}.$$

If we now equate the two path products along a rectangle starting at $n = 0, k = a$ and closing at $n = n_{\max}, k = k_{\max}$, and then let and $(k_{\max}, n_{\max}) \to (\infty, \infty)$ (however they wish, in this case), then we have, after a limit interchange justifiable when $|q| < 1$,

$$\prod_{n \geq 0} \begin{pmatrix} \dfrac{q^a}{1 - q^{n+1}} & 1 - q^a \\ 0 & 1 \end{pmatrix} \prod_{k \geq a} \begin{pmatrix} 0 & 1 \\ 0 & 1 \end{pmatrix} = \prod_{k \geq a} \begin{pmatrix} \dfrac{1}{1 - q^{k+1}} & 0 \\ 0 & 1 \end{pmatrix} \prod_{n \geq 0} \begin{pmatrix} 0 & 1 \\ 0 & 1 \end{pmatrix}.$$

Expanding the infinite products, (and, for convergence, assuming $|q^a| < 1$, *i.e.* $\Re a > 0$),

$$\begin{pmatrix} 0 & (1 - q^a)\left(1 + \frac{q^a}{1-q}\left(1 + \frac{q^a}{1-q^2}\left(1 + \cdots\right)\right)\right) \\ 0 & 1 \end{pmatrix} \begin{pmatrix} 0 & 1 \\ 0 & 1 \end{pmatrix}$$

$$= \begin{pmatrix} \dfrac{1}{1 - q^{a+1}} \dfrac{1}{1 - q^{a+2}} \dfrac{1}{1 - q^{a+3}} \cdots & 0 \\ 0 & 1 \end{pmatrix} \begin{pmatrix} 0 & 1 \\ 0 & 1 \end{pmatrix},$$

or, equating the upper right elements and dividing by $1 - q^a$,

$$1 + \frac{q^a}{1 - q} + \frac{q^{2a}}{(1 - q)(1 - q^2)} + \frac{q^{3a}}{(1 - q)(1 - q^2)(1 - q^3)} + \cdots = \frac{1}{1 - q^a} \frac{1}{1 - q^{a+1}} \frac{1}{1 - q^{a+2}} \cdots .$$

Given path invariance, we can adjoin the reverses of those edges with invertible labels. And we can freely shortcut any connected sequence of edges by a single edge, since its associated matrix is uniquely determined. In particular, we can draw the diagonals J_j connecting those nodes where $k - n = a$:

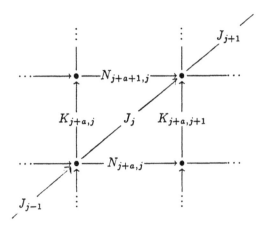

Then

$$J_j \stackrel{\text{def}}{=} N_{j+a,j} K_{j+a,j+1} = K_{j+a,j} N_{j+a+1,j} = \begin{pmatrix} \dfrac{q^{2j+a+1}}{(1-q^{j+1})(1-q^{j+a+1})} & 1 \\ 0 & 1 \end{pmatrix},$$

and the shortcut across the whole rectangle is equivalent to the previous edge traversals:

$$\prod_{j \geq 0} J_j = \prod_{n \geq 0} N_{a,n} \prod_{k \geq a} K_{k,\infty} = \prod_{k \geq a} K_{k,0} \prod_{n \geq 0} N_{\infty,n},$$

that is

$$1 + \frac{q^{a+1}}{(1-q)(1-q^{a+1})} + \frac{q^{2a+4}}{(1-q)(1-q^2)(1-q^{a+1})(1-q^{a+2})}$$

$$+ \frac{q^{3a+9}}{(1-q)(1-q^2)(1-q^3)(1-q^{a+1})(1-q^{a+2})(1-q^{a+3})} + \cdots$$

$$= (1-q^a)\left(1 + \frac{q^a}{1-q} + \frac{q^{2a}}{(1-q)(1-q^2)} + \frac{q^{3a}}{(1-q)(1-q^2)(1-q^3)} + \cdots\right)$$

$$= \frac{1}{1-q^{a+1}} \frac{1}{1-q^{a+2}} \frac{1}{1-q^{a+3}} \cdots.$$

The quadratically progressing exponent in the diagonal sum enhances convergence both numerically, and in the symbolic power series expansion at $q = 0$.

4. One step transformations

The grids for most of the standard series transformations (such as Kummer's or Euler's) are just "railroad tracks". One rail (call it the $j = 0$ rail) is labeled with the original matrix sequence $N_{0,n}$, and the "ties", labeled with matrices J_n, mechanize the transformation to the $j = 1$ rail. This same geometry mechanizes the termwise transformation of a composition of arbitrary homographic functions (possibly a sum!) into a (typically, non-regular) continued fraction. That is, let

$$N_{0,n} := \begin{pmatrix} a_n & b_n \\ c_n & d_n \end{pmatrix}$$

be the matrix for the homographic function

$$f_n(x) := \frac{a_n x + b_n}{c_n x + d_n},$$

and define

$$g_n := a_n + \frac{c_n}{c_{n-1}} d_{n-1}, \quad h_n := -\frac{c_{n+1}}{c_n}(a_n d_n - b_n c_n).$$

Here, we must have $c_n \neq 0$, so any sum must be via $b_n = 0$. Then the "tie" matrices are

$$J_n := \begin{pmatrix} 1 & a_n - g_n \\ 0 & c_n \end{pmatrix}$$

and the continued fraction ($j = 1$) rail has matrices

$$N_{1,n} := \begin{pmatrix} g_n & h_n \\ 1 & 0 \end{pmatrix}.$$

Path invariance follows from mechanically verifying $N_{0,n} J_{n+1} = J_n N_{1,n}$. Then the equivalence of the paths $(0,1) \to (0,m) \to (1,m)$ and $(0,1) \to (1,1) \to (1,m)$, after multiplying on the right by the equation

$$\begin{pmatrix} \left(y + \frac{d_m}{d_{m+1}}\right) c_{m+1} \\ 1 \end{pmatrix} =: \begin{pmatrix} x \\ 1 \end{pmatrix}$$

and equating the ratios of the upper and lower elements, gives the desired identity:

$$f_1(f_2(\ldots(f_m(y))\ldots)) = -\frac{d_0}{c_0} + \frac{1}{c_1}\left(g_1 + \cfrac{h_1}{g_2 + \cfrac{h_2}{\ddots \atop g_m + \cfrac{h_m}{x}}}\right).$$

To convert a sum from the standard form ($c_n = 0$), adjoin it as a $N_{-1,n}$ "third rail" via ties $J'_n := \left(\begin{smallmatrix} 0 & 1 \\ 1 & 0 \end{smallmatrix}\right)$, or any other matrix which unzeros the lower left element of $N_{0,n} = J'^{-1}_n N_{-1,n} J'_{n+1}$.

Even the artifice of multiplication by $\left(\begin{smallmatrix} z \\ 1 \end{smallmatrix}\right)$ fits into the path invariance scheme, merely by connecting a new, "black hole" node to all existing nodes via edges labeled with this (uninvertible) matrix.

Another transformation, summation by parts, is applicable either before or after conversion to ratio form, via a "double decker" railroad track in the form of a square tube. Let p_n and q_n be two arbitrary functions of the summation index n. Traditional summation by parts transpires on the bottom ($j = 0$) track, via the matrices

$$N_{0,0,n} := \begin{pmatrix} 1 & q_n(p_{n+1} - p_n) \\ 0 & 1 \end{pmatrix}, \qquad N_{0,1,n} = \begin{pmatrix} 1 & p_{n+1}(q_{n+1} - q_n) \\ 0 & 1 \end{pmatrix}$$

$$K_{0,0,n} := \begin{pmatrix} -1 & -p_n q_n \\ 0 & 1 \end{pmatrix},$$

where the subscript order is j, k, n. Now suppose that

$$\rho_n := \frac{p_{n+1}}{p_n}, \quad \text{and} \quad \sigma_n := \frac{q_{n+1}}{q_n}$$

are "nicer" functions of n than the corresponding p_n and q_n, perhaps by virtue of factorials or nth powers in the latter. Then one can jump to the upper track, where summation by parts looks like

$$N_{1,0,n} = \begin{pmatrix} \rho_n \sigma_n & \rho_n - 1 \\ 0 & 1 \end{pmatrix}, \qquad N_{1,1,n} = \begin{pmatrix} \rho_n \sigma_n & \rho_n(\sigma_n - 1) \\ 0 & 1 \end{pmatrix}$$

$$K_{1,0,n} = \begin{pmatrix} -1 & -1 \\ 0 & 1 \end{pmatrix}$$

if one labels the "vertical ties" with

$$J_{0,k,n} := \begin{pmatrix} p_n q_n & 0 \\ 0 & 1 \end{pmatrix}.$$

For example, with

$$\rho_n := \frac{n+c}{n+e} \frac{n+d}{n+c+d-e}, \qquad \sigma_n := \frac{n+a+1}{n+1} \frac{n+b+1}{n+a+b+1},$$

the path equivalence $(1,0,0) \to (1,1,0) \to (1,1,\infty) = (1,0,0) \to (1,0,\infty) \to (1,1,\infty)$, *i.e.*

$$K_{1,0,0} \prod_{n\geq 0} N_{1,1,n} = \left(\prod_{n\geq 0} N_{1,0,n} \right) K_{1,0,\infty},$$

gives, after negating the upper right elements,

$$
_4F_3 \left[\begin{matrix} a, & b, & c, & d \\ & a+b+1, & e, & c+d-e \end{matrix} \right] = \frac{(a+b)!\,(c+d-e-1)!\,(e-1)!}{a!\,b!\,(c-1)!\,(d-1)!}
$$
$$
+ \frac{e-c}{e} \frac{e-d}{e-c-d} {}_4F_3 \left[\begin{matrix} a+1, & b+1, & c, & d \\ & a+b+1, & e+1, & c+d-e+1 \end{matrix} \right].
$$

When only one of σ_n and ρ_n is nice, it is straightforward to construct diagonal ties linking an upper rail with a lower one, getting a hybrid, partially rationalized summation by parts.

5. Telescopy

Telescopy takes place between a single rail and a black hole. The problem is to find columnar, "diving in" matrices which path-invariantly connect the black hole to all the nodes along the rail. Then the telescoping identity is the equivalence between diving straight in, and performing the sum prior to diving in.

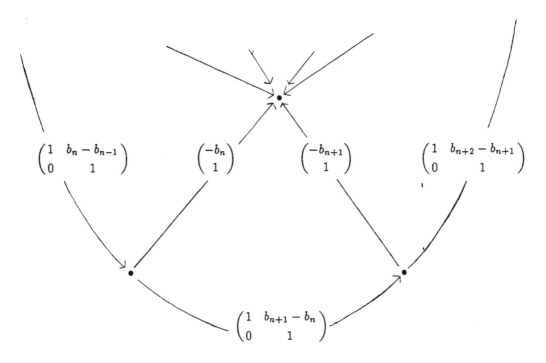

A nontrivial choice of b_n might be

$$b_n := \frac{2^{-n}}{1 - x^{2^{-n}}} \quad \Rightarrow \quad \sum_{n \geq 1} \frac{2^{-n}}{1 + x^{2^{-n}}} = \frac{1}{\log x} + \frac{1}{1 - x},$$

when the alternate paths dive in from $n = 1$ and $n = \infty$.

Rationalized telescopy differs from regular telescopy only in the contents of the matrices.

$$\begin{pmatrix} \frac{1 + f_n}{f_{n+1}} & 1 \\ 0 & 1 \end{pmatrix} \begin{pmatrix} -f_{n+1} \\ 1 \end{pmatrix} = \begin{pmatrix} -f_n \\ 1 \end{pmatrix} \quad \Rightarrow \quad \prod_{n=0}^{m} \begin{pmatrix} \frac{1 + f_n}{f_{n+1}} & 1 \\ 0 & 1 \end{pmatrix} \begin{pmatrix} -f_{m+1} \\ 1 \end{pmatrix} = \begin{pmatrix} -f_0 \\ 1 \end{pmatrix}$$

An interesting choice of f_n is

$$f_n := -\frac{x^{2^{-n}} + c}{c^2 + c} \quad \Rightarrow \quad \sum_{n \geq 0} (\sqrt{x} - c)(\sqrt{\sqrt{x}} - c) \ldots (x^{2^{-n}} - c) = \frac{\frac{x}{c} + 1}{c + 1}, \quad |1 - c| < 1.$$

Any hypergeometric sum, that is, one whose term ratio is a rational function of the summation index n, can be expressed in numerous ways as a product of matrices of polynomials in

n. We now have infallible algorithms (decision procedures) for determining, where possible, arbitrary partial sums of such series. At least one of these algorithms hinges on minimizing the degrees of the on-diagonal polynomials at the expense of the non-zero off-diagonal one. This canonicalization proceeds in a finite number of steps, and can be visualized as path invariant hopping between parallel rails to the minimal one. The minimality then reduces the search for the telescoper ("diving in") matrix to a single system of simultaneous linear equations.

6. "Residues", Euler-Maclaurin summation

Let f be sufficiently tame, and define

$$N_{k,n} := \begin{pmatrix} 1 & \int_0^1 f^{(k)}(t+n) B_k(t)\, dt \\ 0 & 1 \end{pmatrix}, \quad K_{k,n} := \begin{pmatrix} -\dfrac{1}{k+1} & -\dfrac{B_{k+1}}{k+1} f^{(k)}(n) \\ 0 & 1 \end{pmatrix}.$$

Then path invariance,

$$N_{k,n} K_{k,n+1} = K_{k,n} N_{k+1,n}, \qquad\qquad k > 0$$

follows from integration by parts, except when $k = 0$, where

$$N_{0,n} K_{0,n+1} = E_n K_{0,n} N_{1,n},$$

$$E_n := \begin{pmatrix} 1 & f_{n+1} \\ 0 & 1 \end{pmatrix} = N_{0,n} K_{0,n+1} N_{1,n}^{-1} K_{0,n}^{-1},$$

because 1 is the only non-negative integer k for which $B_k(0) \neq B_k(1)$. Thus, path invariance breaks down between $k = 0$ and $k = 1$, and the residues are tallied by the E_n, which commute to the left. In the final formula, these residues comprise the $\sum f(n)$ to which the $\int f(t)$ (from the $k = 0$ path) is the first approximation.

$$\prod_{n=a}^{b} N_{0,n} \prod_{k=0}^{c} K_{k,b+1} = \prod_{j=a}^{b} E_j \prod_{k=0}^{c} K_{k,a} \prod_{n=a}^{b} N_{c+1,n}$$

$$\Leftrightarrow \quad \int_a^{b+1} f(t)\, dt + \sum_{k=0}^{c} \frac{(-)^{k+1}}{(k+1)!} B_{k+1} f^{(k)}(b+1)$$

$$= \sum_{n=a}^{b} f(n+1) + \sum_{k=0}^{c} \frac{(-)^{k+1}}{(k+1)!} B_{k+1} f^{(k)}(a) + \frac{(-)^{c+1}}{(c+1)!} \sum_{n=a}^{b} \int_0^1 f^{(c+1)}(t+n) B_{c+1}(t)\, dt$$

7. $_3F_2[1]$ Rosetta stone

$$N_{g,h,i,j,k,n} := \begin{pmatrix} \dfrac{n+h}{n+1}\dfrac{n+i}{n+g}\dfrac{n+j}{n+k} & 0 & n \\[2ex] 0 & \dfrac{n+h}{n+1}\dfrac{n+i}{n+g}\dfrac{n+j}{n+k} & 1 \\[2ex] 0 & 0 & 1 \end{pmatrix}$$

$K_{g,h,i,j,k,n} :=$

$$\begin{pmatrix} \dfrac{(k-h)(k-i)(k-j)+hij}{k(k+g-h-i-j-1)(k+n)} & \dfrac{hij}{(k+g-h-i-j-1)(k+n)} & \dfrac{n(n+g-1)}{k+g-h-i-j-1} \\[2ex] \dfrac{1}{k+n} & \dfrac{k}{k+n} & 0 \\[2ex] 0 & 0 & 1 \end{pmatrix}$$

$J_{g,h,i,j,k,n} :=$

$$\begin{pmatrix} \dfrac{(j-k-g+h+i+2)(j+n)}{(j-g+1)(j-k+1)} & \dfrac{hi(j+n)}{(j-g+1)(j-k+1)} & n\dfrac{n+g-1}{j-g+1}\dfrac{n+k-1}{j-k+1} \\[2ex] -\dfrac{(j-k-g+h+i+2)(j+n)}{j(j-g+1)(j-k+1)} & \dfrac{j+n}{j}\left(1-\dfrac{h}{j-g+1}\dfrac{i}{j-k+1}\right) & -\dfrac{n}{j}\dfrac{n+g-1}{j-g+1}\dfrac{n+k-1}{j-k+1} \\[2ex] 0 & 0 & 1 \end{pmatrix}$$

This six-dimensional path invariant grid codifies all of the $_3F_2[1]$ contiguity relations (and, as a limiting case, all the $_2F_1[z]$ and $_1F_1[z]$). The three undisplayed matrices follow from symmetry: $G_{g,h,i,j,k,n} = K_{k,h,i,j,g,n}$, etc. The $_3F_2[1]$ arises in element$_{2,3}$ of $\prod_{n\geq 0} N_{...,n}$. Any contiguous sum can be reached in a finite number of multiplications by the parameter gunching matrices. The contiguity identity follows by closing the contour at $n = \infty$.

Contours that range indefinitely in indices other than n yield identities involving the general $_3F_2[1]$ which are unsurprisingly resistant to conventional notation, due to the non-triangularity of the non-N matrices. But even if we never become facile with them, at least some such identities should be valuable for their rapid numerical convergence.

Perhaps more interestingly, by conceding linear constraints among the parameter indices, we can triangularize some of their matrices, and get all of the familiar $_3F_2[1]$ identities (Dixon's, Saalschütz's, Whipple's, Watson's, etc.). When you change coordinates, Whipple's becomes Watson's, and Saalschütz's becomes the very well poised $_5F_4$.

The imposition of a linear constraint entails the replacement of several indices by a new one. This corresponds to recoordinatization and then dimension reduction in the grid. An example is given in section 9 ("An application").

As often happens when a system like the above is symmetric in several indices, one can find a corresponding system with those symmetric indices collapsed into one, and running at fractional speed, e.g. with (g, k) replaced by $(\frac{k}{2}, \frac{k+1}{2})$, or (h, i, j) replaced by $(\frac{i}{3}, \frac{i+1}{3}, \frac{i+2}{3})$. Another way to put it is that, in the (g, k) case, the matrix $G_{k,h,i,j,k,n} K_{k+1,h,i,j,k,n}$ can be factored into the form $M(2k)M(2k+1)$. Such parameter specializations often simplify the quest for triangularizations, and lead to interesting identities upon subsequent coordinate changes.

Letting parameters, say h and g, blow up in a fixed relative ratio z yields a four-dimensional system for $_2F_1[z]$, which is easier to strip mine for triangularizations, particularly when z is negotiable. Further specializing via the abovementioned "fractional parameter collapse" led me to

$$_2F_1\left[\begin{matrix} -\frac{a}{2}, & \frac{1-a}{2} \\ & 2a+\frac{3}{2} \end{matrix}\,\middle|\,\frac{1}{5}\right] = \sqrt{\pi\frac{5+\sqrt{5}}{5}}\,\frac{2^{5a+\frac{3}{2}}}{5^{3a+1}}\,\frac{(2a+\frac{1}{2})!}{(a-\frac{1}{5})!(a+\frac{1}{5})!}$$

$$_2F_1\left[\begin{matrix} -\frac{a}{2}, & \frac{1-a}{2} \\ & 2a+\frac{5}{2} \end{matrix}\,\middle|\,\frac{1}{5}\right] = \sqrt{\pi\frac{5-\sqrt{5}}{5}}\,\frac{2^{5a+\frac{7}{2}}}{5^{3a+2}}\,\frac{(2a+\frac{3}{2})!}{(a+\frac{2}{5})!(a+\frac{3}{5})!}.$$

while strip mining the remaining degrees of freedom. (Professor P. Karlsson has lately informed me that these were dug up by W. Heymann in 1898 ([1],[2]), *i.e.*, with his bare hands!)

I also have the q-versions of the $_3F_2[1]$ system. These have the annoying property of sometimes leaving factors of $1-q$ instead of 0 when subjected to the transformations which triangularize their $q = 1$ cousins, which is one way to explain why we yet lack q-generalizations for some of our more exotic $_2F_1$ identities.

8. Continued fractions

This is three N-K planes, $j = -1, 0, 1$, joined by two J matrices. In the top $(j = 1)$ plane, the k direction computes a continued fraction, and on the bottom $(j = -1)$, it's the n direction.

$$N_{1,k,n} = \begin{pmatrix} (1-n-k)z & a((k-1)^2 + b)z^2 \\ 1 & a(n-k+1)z \end{pmatrix},$$

$$K_{1,k,n} = \begin{pmatrix} ((a-1)k - (a+1)n + 1)z & a(k^2 + b)z^2 \\ 1 & 0 \end{pmatrix}$$

$$J_{0,k,n} := \begin{pmatrix} z^{n+k} & ((a+1)n + k - 1)z^{n+k+1} \\ 0 & z^{n+k+1} \end{pmatrix}$$

$$N_{0,k,n} := \begin{pmatrix} an & a\big(b - (a+1)kn\big) \\ 1 & -n - (a+1)k \end{pmatrix}, \quad K_{0,k,n} := \begin{pmatrix} ak & a\big(b - (a+1)nk\big) \\ 1 & -k - (a+1)n \end{pmatrix}$$

$$J_{-1,k,n} := \begin{pmatrix} z^{-n-k} & \big(1 - (a+1)k - n\big)z^{-n-k} \\ 0 & z^{-n-k-1} \end{pmatrix}$$

$$N_{-1,k,n} = \begin{pmatrix} \big((a-1)n - (a+1)k + 1\big)z & a(n^2 + b)z^2 \\ 1 & 0 \end{pmatrix},$$

$$K_{-1,k,n} = \begin{pmatrix} (1 - k - n)z & a\big((n-1)^2 + b\big)z^2 \\ 1 & a(k - n + 1)z \end{pmatrix}$$

Thus, to get a relation between two continued fractions, we must j-hop between planes before switching directions. If k runs from c to ∞ and n runs from d to ∞, and the corresponding continued fractions converge, then they are insensitive to whatever subsequent matrices accrue on the right by way of path closure. $I.e.$, the two paths can be $(j, k, n) = (0, c, d) \to (1, c, d) \to (1, \infty, d)$ and $(0, c, d) \to (-1, c, d) \to (-1, c, \infty)$. This yields the identity

$$f(c, d) = f(d, c)$$

where

$$f(x, y) := \big((a+1)y + x - 1\big)z$$

$$+ \big((a-1)x - (a+1)y + 1\big)z + \cfrac{a(x^2 + b)z^2}{\big((a-1)(x+1) - (a+1)y + 1\big)z + \cfrac{a\big((x+1)^2 + b\big)z^2}{}}.$$

Note that z can be simply canceled out of this identity, but it is handy to leave it in. Since we never travel in the $j = 0$ plane, you might consider discarding it and coalescing the two J matrices. My only objections to this are that it would destroy some symmetry, and more importantly, it would conceal the simple derivation, which just uses the technique in section 4 to force alternately the K and N matrices into continued fraction form.

Another application of this particular path invariant system is a direct interderivation of the series and continued fraction forms for arctan, in the form

$$1 + \cfrac{x^2}{3 + \cfrac{4x^2}{5 + \cfrac{9x^2}{\ddots}}} = \frac{x}{\arctan x} = \cfrac{\dfrac{2y}{1 - y^2}}{2 \arctan y}$$

$$= \frac{1}{(1 - y^2)\left(1 - \dfrac{y^2}{3} + \dfrac{y^4}{5} - \overset{!}{\dots}\right)},$$

where

$$x := \frac{2y}{1 - y^2}.$$

We merely remain in the $j = -1$ plane and get our series by choosing the parameters to annihilate the upper right element of K:

$$a := y, \qquad b := 0, \qquad z := -\frac{2}{1 - y^2}.$$

This gives

$$F_n := N_{-1, \frac{1}{2}, n} = \begin{pmatrix} 2n - 1 & n^2 \left(\dfrac{2y}{1 - y^2}\right)^2 \\ 1 & 0 \end{pmatrix}$$

and

$$G_k := K_{-1, k, 1} = \begin{pmatrix} \dfrac{2k}{1 - y^2} & 0 \\ 1 & -\dfrac{2ky^2}{1 - y^2} \end{pmatrix}.$$

Then the ratio of the lefthand elements of $F_1 F_2 F_3 \dots$ gives the continued fraction, while the corresponding ratio from $G_{\frac{1}{2}} G_{\frac{3}{2}} G_{\frac{5}{2}} \dots$ gives the reciprocated series.

9. An application

$$J_{j,k,n} := \begin{pmatrix} \dfrac{1-q^{j-k+n+\frac{1}{2}}}{1-q^{2j+2}}\dfrac{1-q^{j+k-n+\frac{1}{2}}}{1-q^{j-k-n+\frac{1}{2}}} & \dfrac{1-q^{2k}}{1-q}\dfrac{1-q^{2n}}{1-q^{j-k-n+\frac{1}{2}}}q^{j-k-n+\frac{1}{2}} \\ 0 & 1 \end{pmatrix}$$

$$K_{j,k,n} := \begin{pmatrix} \dfrac{1-q^{k-n+j+\frac{1}{2}}}{1-q^{2k+2}}\dfrac{1-q^{k+n-j+\frac{1}{2}}}{1-q^{k-n-j+\frac{1}{2}}} & \dfrac{1-q^{2n}}{1-q}\dfrac{1-q^{2j}}{1-q^{k-n-j+\frac{1}{2}}}q^{k-n-j+\frac{1}{2}} \\ 0 & 1 \end{pmatrix}$$

$$N_{j,k,n} := \begin{pmatrix} \dfrac{1-q^{n-j+k+\frac{1}{2}}}{1-q^{2n+2}}\dfrac{1-q^{n+j-k+\frac{1}{2}}}{1-q^{n-j-k+\frac{1}{2}}} & \dfrac{1-q^{2j}}{1-q}\dfrac{1-q^{2k}}{1-q^{n-j-k+\frac{1}{2}}}q^{n-j-k+\frac{1}{2}} \\ 0 & 1 \end{pmatrix}$$

This pretty, three-dimensional system, when path multiplied around an infinite rectangle based at $j = a, k = b, n = 0$ in either the k-n or n-j plane, gives Andrews's q-gneralization of Bailey's $_2F_1[\frac{1}{2}]$ theorem ([3]). (And, by symmetry, there are four other ways to do it.)

By making the coordinate transformation $n \leftarrow n + j + k + \frac{1}{2}$, you get Andrews's q-gneralization of Gauss's $_2F_1[\frac{1}{2}]$ (also [3]). To preserve path invariance, this coordinate change requires that the $J_{j,k,n}$ and $K_{j,k,n}$ matrices be replaced by $J_{j,k,n}N_{j+1,k,n}$ (or $N_{j,k,n}J_{j,k,n+1}$) and $K_{j,k,n}N_{j,k+1,n}$ (or $N_{j,k,n}K_{j,k,n+1}$) respectively, followed by the actual substitution $n \leftarrow n + j + k + \frac{1}{2}$. The multiplication of J and K by N reflects that every incrementation of j or k must also increment what was formerly n.

A more interesting coordinate change, in greater detail, is

$$\begin{Bmatrix} J_{j,k,n} \\ K_{j,k,n} \\ N_{j,k,n} \end{Bmatrix} \leftarrow \begin{Bmatrix} J_{j,k+j,n+k+j}\,K_{j+1,k+j,n+k+j}\,N_{j+1,k+j+1,n+k+j} \\ K_{j,k+j,n+k+j}\,N_{j,k+j+1,n+k+j} \\ N_{j,k+j,n+k+j} \end{Bmatrix}$$

which gives

$$J_{j,k,n} = \begin{pmatrix} -\dfrac{1-q^{j-n+\frac{1}{2}}}{1-q^{2(j+1)}}\dfrac{1-q^{j+n+\frac{1}{2}}}{1-q^{2(j+k+1)}}\dfrac{1-q^{j+2k+n+\frac{1}{2}}}{1-q^{2(j+k+n+1)}}q^{3j+2k+n+\frac{3}{2}} & -\dfrac{1-q^{3j+2k+n+\frac{1}{2}}}{1-q} \\ 0 & 1 \end{pmatrix}$$

$$K_{j,k,n} = \begin{pmatrix} \dfrac{1-q^{2k+j+n+\frac{1}{2}}}{1-q^{2(k+j+1)}}\dfrac{1-q^{2k+j+n+\frac{3}{2}}}{1-q^{2(k+j+n+1)}}q^{2j} & -\dfrac{1-q^{2j}}{1-q} \\ 0 & 1 \end{pmatrix}$$

$$N_{j,k,n} = \begin{pmatrix} \dfrac{1-q^{n+j+\frac{1}{2}}}{1-q^{n-j+\frac{1}{2}}}\dfrac{1-q^{n+j+2k+\frac{1}{2}}}{1-q^{2(n+j+k+1)}} & \dfrac{1-q^{2j}}{1-q}\dfrac{1-q^{2(j+k)}}{1-q^{n-j+\frac{1}{2}}}q^{n-j+\frac{1}{2}} \\ 0 & 1 \end{pmatrix}.$$

Then a rectangle in the j-k plane, based at $j = 0, k = (b-a)/2, n = a - \frac{1}{2}$, gives

$$\prod_{j \geq 0} \begin{pmatrix} -\dfrac{1-q^{j+a}}{1-q^{2j+2}}\dfrac{1-q^{j+1-a}}{1-q^{2j+b-a+2}}\dfrac{1-q^{j+b}}{1-q^{2j+b+a+1}}q^{3j+b+1} & \dfrac{1-q^{3j+b}}{1-q} \\ 0 & 1 \end{pmatrix}_{1,2}$$

$$= \frac{\frac{b-a}{2}!_{q^2}}{(-\frac{1}{2})!_{q^2}}\frac{\frac{b+a-1}{2}!_{q^2}}{(b-1)!_q}(1+q)^b,$$

where

$$z!_q := (1-q)^{-z}\prod_{n \geq 1}\frac{1-q^n}{1-q^{n+z}} \quad \text{`` = ''} \quad \frac{1-q}{1-q}\frac{1-q^2}{1-q}\cdots\frac{1-q^z}{1-q}.$$

Alternatively, writing $q^a =: A, q^b =: B$, and multiplying through by $1-q$,

$$\sum_{j\geq 0}(1 - Bq^{3j})\frac{(A, q/A, B; q)_j}{(q^2, ABq, Bq^2/A; q^2)_j}(-Bq^{(3j-1)/2})^j = \frac{(B; q)_\infty}{(ABq, Bq^2/A; q^2)_\infty}.$$

Letting $q \to 1$ in the penultimate equation, and dividing through by b,

$${}_4F_3\left[\begin{matrix} a, & 1-a, & b, & \frac{b}{3}+1 \\ & \frac{b-a}{2}+1, & \frac{b+a+1}{2}, & \frac{b}{3} \end{matrix}\middle| -\frac{1}{8}\right] = \frac{\frac{b-a}{2}!\; \frac{b+a-1}{2}!}{(-\frac{1}{2})!\; b!}2^b.$$

This gives a rapidly convergent (three bits per term) series for the useful but very slowly (if at all) convergent ${}_2F_1[a, b; c + 1|1]$:

$$\frac{c!\,(c-a-b)!}{(c-a)!\,(c-b)!} = {}_2F_1\left[\begin{matrix} a, & b \\ & c+1 \end{matrix}\middle| 1\right]$$

$$= \frac{{}_4F_3\left[\begin{matrix} \frac{1}{2}+a+b, & \frac{1}{2}-a-b, & 2c-a-b+\frac{1}{2}, & \frac{2c-a-b}{3}+\frac{7}{6} \\ & c+1, & c-a-b+1, & \frac{2c-a-b}{3}+\frac{1}{6} \end{matrix}\middle| -\frac{1}{8}\right]}{{}_4F_3\left[\begin{matrix} a-b+\frac{1}{2}, & b-a+\frac{1}{2}, & 2c-a-b+\frac{1}{2}, & \frac{2c-a-b}{3}+\frac{7}{6} \\ & c-a+1, & c-b+1, & \frac{2c-a-b}{3}+\frac{1}{6} \end{matrix}\middle| -\frac{1}{8}\right]}$$

$$= \frac{{}_4F_3\left[\begin{matrix} c-2b-a+1, & -c+2b+a, & c-a, & \frac{c-a}{3}+1 \\ & c-b-a+1, & b+\frac{1}{2}, & \frac{c-a}{3} \end{matrix}\middle| -\frac{1}{8}\right]}{{}_4F_3\left[\begin{matrix} c-2b+1, & -c+2b, & c, & \frac{c}{3}+1 \\ & c-b+1, & b+\frac{1}{2}, & \frac{c}{3} \end{matrix}\middle| -\frac{1}{8}\right]}2^a$$

$$= \frac{{}_4F_3\left[\begin{matrix} c+a+1, & -c-a, & c-a, & \frac{c-a}{3}+1 \\ & c+1, & \frac{1}{2}-a, & \frac{c-a}{3} \end{matrix}\middle| -\frac{1}{8}\right]}{{}_4F_3\left[\begin{matrix} c-b+a+1, & -c+b-a, & c-b-a, & \frac{c-b-a}{3}+1 \\ & c-b+1, & \frac{1}{2}-a, & \frac{c-b-a}{3} \end{matrix}\middle| -\frac{1}{8}\right]}2^{-b}$$

Even if you neglect to use matrix products to evaluate the ${}_4F_3$s, these are notably cheaper than four invocations of the Γ function, especially when a, b, and c might be complex. Also, the (complex) Beta function is the special case

$$\mathrm{B}(a, b) = \frac{a+b}{ab\,{}_2F_1[-a, -b; 1|1]}.$$

10. A commercial for q-Trigonometry

In the preceding section, we saw both the factorial and q^2-factorial of $-1/2$. The former is $\sqrt{\pi}$; why not call the latter $\sqrt{\pi}_q$? More generally, why don't we have a q-generalization of the factorial reflection formula $z!\,(-z)! = \pi z/\sin \pi z$? I propose

$$\Pi_q := \frac{\pi_q}{1-q^2} = q^{\frac{1}{4}}\frac{(-\frac{1}{2})!^2_{q^2}}{1-q^2} = q^{\frac{1}{4}} \prod_{n\geq 1} \frac{(1-q^{2n})^2}{(1-q^{2n-1})^2} = \frac{q\text{-Wallis product}}{1-q}$$

$$= q^{\frac{1}{4}}(1+q+q^3+q^6+q^{10}+\cdots)^2 = \frac{\vartheta_2 \vartheta_3}{2} = \frac{\vartheta_2^2(0,q^{\frac{1}{2}})}{4}$$

$$= \lim_{z\to 0} \frac{\sin_q \pi z}{q^{-z}-q^z} = -\frac{\pi}{2}\frac{\sin'_q 0}{\ln q} = \frac{\eta(q^2)^4}{\eta(q)^2}$$

where

$$\sin_q \pi z := \frac{q^{z(z-1)}\pi_q}{(z-1)!_{q^2}\,(-z)!_{q^2}} = q^{(z-1/2)^2} \prod_{n\geq 1} \frac{(1-q^{2n-2z})(1-q^{2n+2z-2})}{(1-q^{2n-1})^2}$$

$$= iq^{z^2}\frac{\vartheta_1(iz\ln q)}{\vartheta_4} = -\sin_q \pi(z+1)$$

$$= \frac{\displaystyle\sum_{-\infty<n<\infty} (-1)^n q^{(n-z+\frac{1}{2})^2}}{\displaystyle\sum_{-\infty<n<\infty} (-q)^{n^2}}$$

and

$$\cos_q \pi z := \sin_q \pi(\tfrac{1}{2}-z) = q^{z^2} \prod_{n\geq 1} \frac{(1-q^{2n-2z-1})(1-q^{2n+2z-1})}{(1-q^{2n-1})^2}$$

$$= q^{z^2}\frac{\vartheta_4(iz\ln q)}{\vartheta_4} = \frac{\displaystyle\sum_{-\infty<n<\infty} (-1)^n q^{(n-z)^2}}{\displaystyle\sum_{-\infty<n<\infty} (-q)^{n^2}}$$

While Π_q is expressible with two Dedekind η functions, Jacobi's *æquatio identica satis abstrusa* ([4]) returns the favor:

$$\eta(q) := q^{1/24} \prod_{n\geq 1} 1-q^n = \left(\frac{\Pi_q^5}{\Pi_{q^2}^2} - 16\,\Pi_q\,\Pi_{q^2}^2\right)^{1/6}.$$

So we see that $\sin_q \pi z$ and $\cos_q \pi z$ are period 2, unit amplitude functions with many of the properties of their $q \to 1$ ancestors:

$$\sin_q(x+y) = \frac{\sin_q(x-y)}{\sin_{q^2}(x-y)}(\sin_{q^2} x \cos_{q^2} y + \cos_{q^2} x \sin_{q^2} y)$$

$$\cos_q(x+y) = \frac{\cos_q(x-y)}{\cos_{q^2}(x-y)}(\cos_{q^2} x \cos_{q^2} y - \sin_{q^2} x \sin_{q^2} y)$$

$$\cos_q 2z = (\cos_{q^2} z)^2 - (\sin_{q^2} z)^2 = \cos 2z \prod_{n \geq 0}(\sin^2_{q^{2-n}} z + \cos^2_{q^{2-n}} z)$$

$$= (\cos_q z)^4 - (\sin_q z)^4$$

$$\sin_q 2z = \frac{\Pi_q}{\Pi_{q^2}} \sin_{q^2} z \cos_{q^2} z$$

$$= \frac{1}{2} \frac{\Pi_q}{\Pi_{q^4}} \sqrt{(\sin_{q^4} z)^2 - (\sin_{q^2} z)^4} = \frac{1}{2} \frac{\Pi_q}{\Pi_{q^4}} \sqrt{(\cos_{q^4} z)^2 - (\cos_{q^2} z)^4}$$

$$\sin_q 3z = \frac{\Pi_q}{\Pi_{q^3}}(\cos_{q^3} z)^2 \sin_{q^3} z - (\sin_{q^3} z)^3$$

$$= \frac{1}{3} \frac{\Pi_q}{\Pi_{q^9}} \sin_{q^9} z - \left(1 + \frac{1}{3} \frac{\Pi_q}{\Pi_{q^9}}\right)(\sin_{q^3} z)^3$$

$$\sin_q 5z = \frac{\Pi_q}{\Pi_{q^5}}(\cos_{q^5} z)^4 \sin_{q^5} z - \sqrt{\frac{\Pi_q^3}{\Pi_{q^5}^3} - 2\frac{\Pi_q^2}{\Pi_{q^5}^2} + 5\frac{\Pi_q}{\Pi_{q^5}}}\,(\cos_{q^5} z)^2 (\sin_{q^5} z)^3$$

$$+ (\sin_{q^5} z)^5$$

$$\sin_q(x-w)\sin_q(x-y)\sin_q(y-w)\sin_q(y-z)\sin_q(z-w)\sin_q(z-x)$$

$$+ \cos_q(x-w)\cos_q(x-y)\sin_q(y-w)\cos_q(y-z)\cos_q(z-w)\sin_q(z-x)$$

$$+ \cos_q(x-w)\sin_q(x-y)\cos_q(y-w)\cos_q(y-z)\sin_q(z-w)\cos_q(z-x)$$

$$+ \sin_q(x-w)\cos_q(x-y)\cos_q(y-w)\sin_q(y-z)\cos_q(z-w)\cos_q(z-x) = 0$$

I actually discovered this last formula in q-land, via a sequence of symmetrizing generalizations guided by the additional structure imposed by the qs.

Strip mining over undetermined linear combinations of extensive Taylor expansions turned up the empirical relations

$$\sum_{n\geq1}\frac{q^n}{(1-q^n)^2}-2\sum_{n\geq1}\frac{q^{2n}}{(1-q^{2n})^2}=\frac{1}{24}\Big(\frac{\Pi_q^4}{\Pi_{q^2}^2}-1\Big)+\frac{2}{3}\Pi_{q^2}^2$$

$$\sum_{n\geq1}\frac{q^n}{(1-q^n)^2}-3\sum_{n\geq1}\frac{q^{3n}}{(1-q^{3n})^2}=\frac{(\Pi_q^2+3\Pi_{q^3}^2)^2}{12\,\Pi_q\,\Pi_{q^3}}-\frac{1}{12}$$

$$\sum_{n\geq1}\frac{q^n}{(1-q^n)^2}-4\sum_{n\geq1}\frac{q^{4n}}{(1-q^{4n})^2}=\frac{1}{8}\Big(\frac{\Pi_q^4}{\Pi_{q^2}^2}-1\Big)$$

$$\sum_{n\geq1}\frac{q^{2n}}{(1-q^{2n})^2}-9\sum_{n\geq1}\frac{q^{18n}}{(1-q^{18n})^2}=\frac{\Pi_{q^3}^3}{\Pi_q}+\frac{1}{3}\Big(\frac{\Pi_{q^3}^3}{\Pi_{q^9}}-1\Big)$$

$$\sum_{n\geq1}\frac{q^{2n-1}}{(1-q^{2n-1})^2}-2\sum_{n\geq1}\frac{q^{4n-2}}{(1-q^{4n-2})^2}=\Pi_{q^2}^2=\sum_{n\geq1}\frac{(2n-1)q^{2n-1}}{1-q^{4n-2}}$$

$$\sum_{n\geq1}\frac{q^{2n-1}}{(1-q^{2n-1})^2}-5\sum_{n\geq1}\frac{q^{10n-5}}{(1-q^{10n-5})^2}=\Pi_{q^5}^2\,\frac{\frac{\Pi_{q^5}^2}{\Pi_{q^{10}}^2}+16\frac{\Pi_{q^{10}}^2}{\Pi_{q^5}^2}}{\frac{\Pi_q}{\Pi_{q^5}}-4-\frac{\Pi_{q^5}}{\Pi_q}}.$$

$$\sum_{n\geq0}\frac{1-q^{12n+1}}{1-q^2}\frac{(2n-\frac{3}{2})!_{q^2}(2n+\frac{1}{2})!_{q^2}}{(4n+1)!_{q^2}}q^{4n^2}=-q^{1/4}\sqrt{\pi_q\pi_{q^5}\frac{1-q^2}{1-q^{10}}}$$

This last is the $a\to0$ case of the summation identity of Section **1**. Notice how much easier it is to take the $q\to1$ limit in this form than in that.

A limiting case of a certain two parameter path invariance result (omitted) can be written

$$\sum_{k\geq0}\frac{(a+k-1)!_q(k-a)!_q}{(2k)!_q}q^{k^2}=\cdot\frac{\pi_{q^{1/2}}}{\sin_{q^{1/2}}\pi a}\frac{\sin_{q^{3/2}}\pi\frac{a+1}{3}}{\sin_{q^{3/2}}\frac{\pi}{3}}q^{(a-1)a/3}.$$

Although Professor Andrews has uncovered an equivalent result in Ramanujan's "lost" "notebook" ([5]), it is worth noting how the q-trigonometric version clearly reveals this $f(a)$ as period 2 poles times period 6 zeros times a quadratic power of q.

Professors Andrews and Berndt also inform me that J. W. L. Glaisher had at least the first half of the identity

$$\Pi_q^4=\sum_{n\geq1}\frac{n^3q^n}{1-q^{2n}}=\sum_{n\geq1}\mathrm{Li}_{-3}(q^{2n-1})$$

by 1905 ([6]). Combinatorially, it says that the number of ways to express $n - 1$ as the sum of 8 triangular numbers equals n^3 times the sum of the cubes of the reciprocals of the odd divisors of n. But my real reason for restating it is to lead into a

11. Commercial for negapolylogs!

(If Sesame Street can run commercials for letters of the alphabet, I demand equal time for my favorite special functions.) Running the polylog recurrence in reverse, we will just differentiate our way back through a sequence of rational functions:

$$\mathrm{Li}_0(x) = x\frac{1}{1-x}$$

$$\mathrm{Li}_{-1}(x) = x\frac{1}{(1-x)^2}$$

$$\mathrm{Li}_{-2}(x) = x\frac{x+1}{(1-x)^3}$$

$$\mathrm{Li}_{-3}(x) = x\frac{x^2+4x+1}{(1-x)^4}$$

$$\mathrm{Li}_{-4}(x) = x\frac{x^3+11x^2+11x+1}{(1-x)^5}$$

But they are *useful* rational functions. While yet retaining certain loggish tendencies:

$$\mathrm{Li}_{-k}(e^z) = \frac{d^k}{dz^k}Li_0(e^z) \qquad\qquad\qquad k \geq 0$$

$$= \frac{k!}{(-z)^{k+1}} - \sum_{n\geq 0}\frac{B_{n+k+1}(1)}{n+k+1}\frac{z^n}{n!}.$$

(Note the gap of k powers of z.) The generating function is

$$\sum_{k\geq 0}\frac{\mathrm{Li}_{-k}(e^{-z})}{k!}t^k = \frac{1}{e^{z-t}-1}$$

A sample application is the expansion of tan (or sec or cot or ...) about an arbitrary point,

$$\frac{d^k}{dx^k}\tan x = (2i)^{k+1}\operatorname{Li}_{-k}(-e^{2ix}),$$

which would otherwise invoke misleadingly transcendental polygammas. And here is the nub of the "q-Stirling" formula

$$\log\prod_{n\geq z+y}1-q^n \ \asymp_z \ \sum_{k\geq 0}\frac{B_k(y)}{k!}(\ln q)^{k-1}\operatorname{Li}_{2-k}(q^z).$$

And we can use negapolylogs to compute Bernoulli numbers and special values of the Euler polynomials.

$$\operatorname{Li}_{-k}(-1) = (-1)^k(2^{k+1}-1)\frac{B_{k+1}}{k+1} \ = \ \frac{(-1)^{k+1}}{2}E_k(0) = -\frac{E_k(1)}{2}$$

$$\operatorname{Li}_{-k}(-1) = -2^{k-1}E_k(1) + \frac{i}{2}E_k$$

$$\operatorname{Li}_{-k}(e^{\frac{\pi i}{3}}) = \frac{1}{4}\Big((1-3^{k+1})E_k(1) + i\,3^{k+\frac{1}{2}}(1+(-1)^k)E_k(\tfrac{1}{3})\Big)$$

$$\operatorname{Li}_{-k}(e^{\frac{2\pi i}{3}}) = \frac{1}{4}\Big(\frac{1-3^{k+1}}{2^{k+1}-1}E_k(1) + i\,3^{k+\frac{1}{2}}\frac{1+(-1)^k}{2^{k-1}+1}E_k(\tfrac{1}{3})\Big)$$

Finally, do not confuse these with negapolygammas, ψ_{-k}, which become "more transcendental" (deeply nested integrals) with diminishing subscript, and are effectively the logarithms of the "higher factorials" $1^{1^{k-1}}2^{2^{k-1}}\ldots$ .... But that's another story.

References

[1] W. Heymann, *Ueber hypergeometrische Functionen, deren letztes Element speciell ist, nebst einer Anwendung auf Algebra*, Jahresbericht der Technischen Staatslehrenstalten in Chemnitz 1897-1898, (1898), 3-44

[2] W. Heymann, *Über hypergeometrische Funktionen, deren letztes Element speziell ist*, Z. Math. Phys. 44, (1899), 280-288.

[3] G. Andrews, *On the q-Analog of Kummer's Theorem and Applications*, Duke Mathematical Journal 40, No. 3 (September 1973), 525-528.

[4] E. T. Whittaker and G. N. Watson, (A Course of) Modern Analysis, (1902-1984), 470.

[5] G. Andrews, *Ramanujan's "Lost" Notebook II*, Adv. in Math. 41, (1981), 173-185 (Eq. (1.10.)$_R$ of Thm. 1).

[6] J. W. L. Glaisher, *Representations of a number as a sum of four squares*, Quarterly J. of Pure and Applied Math. 36 (1905), 355.

Polynomial Factorization 1982-1986

ERICH KALTOFEN   Rensselaer Polytechnic Institute, Troy, New York

1. Introduction

In late 1981 I wrote the survey article [41] on the problem of factoring polynomials for the Computer Algebra book edited by B. Buchberger, G. Collins, and R. Loos. In the conclusion of that survey, I raised six opens problems, all of which were related to finding polynomial-time solutions to problems in polynomial factorization. During the tremendous research effort of the last five years, all six problems have been satisfactorily solved. This paper complements my previous survey and summarizes the major advances that now make polynomial factoring a fine example of a classical computational question for which efficient and polynomial time algorithms have been designed.

Before we describe the new results, however, I want to make a few additions to my discussion on the earlier developments [41]. Von Schubert's finite step algorithm of 1793 for factoring univariate integral polynomial is merely a generalization of Newton's method published in the chapter "invention of divisors" in his Arithmetica Universalis in 1707. The idea is simple. In order to find a factor of degree d, one evaluates the polynomial at $d+1$ integers and interpolates all possible combinations of integer factors of the resulting values. Van der Waerden [88] points out that this method is valid over any unique factorization domain with finitely many units, and that then the decidability of factoring in the domain implies the decidability of polynomial factoring. He suggests, however, that over infinite fields the factoring problem may be undecidable given only effective procedures for the field arithmetic (including zero testing). This is indeed so, and Fröhlich and Shepherdson [26] show the existence of such "explicit" fields over which irreducibility testing is undecidable. Of course this undecidability result stands in the shadow of the polynomial-time solution for most natural fields such as the rational numbers.

From the work by Zassenhaus [93] and Lenstra [62] in 1981 it became clear that the polynomial-time resolution of univariate integer polynomial factorization hinged on progress for certain diophantine approximation questions, whereas the author's early results on Hilbert

* This material is based upon work supported by the National Science Foundation under Grant No. DCR-85-04391 and by an IBM Faculty Development Award. This article was written for a tutorial on polynomial factorization given at the conference "Computers and Mathematics," at Stanford University on August 2, 1986.

Irreducibility (cf. [42] and [47], §7) indicated that the question of multivariate factoring may be polynomial-time reducible to univariate factoring. The missing link in the univariate polynomial factoring algorithm was finally supplied by Lovász in late 1981, who constructed a remarkable algorithm for finding relatively short vectors in multidimensional integer lattices. The resulting polynomial factoring algorithm by A. K. Lenstra, H. Lenstra, and Lovász [67] could decompose an integer polynomial of degree n with coefficients absolutely bounded by B into irreducible non-constant integer polynomials in $O(n^{12} + n^9(\log B)^3)$ binary steps, this when using classical long integer arithmetic procedures. Polynomial-time reductions from multivariate to univariate integer polynomial factorization were discovered at about the same time. The algorithm by the author [47], is based on Zassenhaus's root approximation scheme, which in those circumstances happens in the power series domain and then leads to a linear system rather than a diophantine approximation problem. This algorithm only requires field arithmetic and a univariate factoring oracle, provided the field has sufficiently many elements. Another algorithm by Christov and Grigoryev [18] uses the Hilbert irreducibility results mentioned earlier and a polynomial-time bivariate polynomial factoring algorithm over finite coefficient fields, which is derived from the lattice approach. In the multivariate case, the number of possible monomial coefficients can grow exponentially with the number of variables. A densely represented, or shortly a dense multivariate polynomial is one where all coefficients are counted towards the input size, even if a large proportion of them is zero. The algorithms just described have polynomial running time in the dense size of the input polynomial.

Other dense multivariate polynomial factoring algorithms over finite fields were found by von zur Gathen and the author [31] and by A. K. Lenstra [66]. For large coefficient fields, the algorithms are randomized in the Las Vegas sense, "always correct and probably fast." Only recently has it been established that multivariate irreducibility testing over large finite fields can be accomplished in polynomial time without probabilistic choices [52]. There are also several generalizations of the integer lattice algorithm to dense multivariate factoring over the rational numbers [64] and [39].

Factoring polynomials over algebraic extensions is polynomial-time reducible to factoring over the ground field by virtue of the classical Kronecker reduction, and the timing analysis with Trager's improvement can be found in [59]. This approach is also described in [18]. An alternate way to factor over algebraic number fields in polynomial-time, again based on lattice reduction, is given by A. K. Lenstra [63], [65]. It turns out that the reduction from multivariate to univariate polynomial factorization also gives an absolute irreducibility test [44].

The high exponents in the running time bounds may indicate that the polynomial-time solutions are quite impractical at present. This is probably true for the original algorithms, but the short lattice vector algorithm, aside form many intriguing diophantine approximation applications, also has its function in polynomial factorization. A. K. Lenstra's lattice procedure to recover algebraic numbers from their modular images [62] seems an efficient way to eliminate the exponentially complex Chinese remaindering step in the Weinberger-Rothschild [91] factoring algorithm over algebraic number fields [1]. There is also the possibility now to factor univariate integer polynomials via complex root approximation and lattice reduction [82], [55]. In particular, Schönhage's algorithm has asymptotic running time $O(n^8 + n^5(\log B)^3)$, n = input degree, B = coefficient bound. This approach may be superior in practice on hard-to-factor inputs [53]. M. Monagan's experiments [74] also indicate that the Adleman-Odlyzko irreducibility test [6] combined with a probabilistic integer primality test is quite practical. The new absolute irreducibility criteria [44], [86], Chapter 3, and [23] are also very useful in practice.

If the number of variables is allowed to grow with the input size, then the dense representation often consumes too much space. The first less expensive representation studied is the sparse one, in which only monomials with non-zero coefficients are stored. Zippel's sparse interpolation [94] and lifting algorithms [95] were the first methods based on randomization rather than heuristics. However, a rigorous probability analysis of the sparse lifting process required an effective probabilistic version of the Hilbert irreducibility theorem. Already in 1981, Heintz and Sieveking [37] provided such a theorem for algebraically closed fields, and in 1983 von zur Gathen proved a suitable version for arbitrary coefficient fields. In retrospect, the author's effective Hilbert irreducibility theorem also lends itself to an even simpler probabilistic version [45]. In [32] sparse polynomials are described that possess irreducible factors with super-polynomially more terms. These examples imply that any sparse Hensel Lifting scheme can have more than polynomial running time on certain inputs. In order to deal with this phenomenon it became clear that the sparse representation had to be replaced by a more powerful one.

The usage of "straight-line programs" as a means to compute certain polynomials has been developed in the framework of complexity theory in the past decade, refer for example to [83] and [84]. In 1983 von zur Gathen [28] combined his probabilistic Hilbert irreducibility theorem with the probabilistic method of straight-line program evaluation [80] and [40] to find the factor degree pattern of polynomials defined by straight-line computations. A previously known operation on polynomials in straight-line representation is that of taking first order partial derivatives [9]. Although there is evidence that other operations such as higher partial derivatives are inherently complex [87] (cf. also §5), the greatest common divisor problem of polynomials is straight-line program representation could shown by the author in late 1984 to be in probabilistic polynomial-time [49]. In 1985 the author could also show that straight-line programs for the irreducible factors of a polynomial given by a straight-line program can be found in probabilistic polynomial-time [50]. With Zippel's 1979 sparse polynomial interpolation algorithm [94] our factorization result resolves all problems left open in [95], [32], and [46]. We note that, unlike the randomized solutions for factorization of univariate polynomials over large finite fields, the probabilistic solutions are of the Monte-Carlo kind, "probably correct and always fast." The failure probability can, of course, be made arbitrarily small.

2. The Lattice Reduction Algorithm

In this section we present a key algorithm in diophantine optimization, which enabled A. K. Lenstra, H. Lenstra, and L. Lovász [67] to prove that polynomial factoring is in polynomial time. Given a set of n linearly independent m-dimensional integer vectors, the algorithm finds a vector in the Z-span of these vectors, the Euclidean length of which is within a factor of $2^{(n-1)/2}$ of the length of the shortest vector in this lattice. Aside from polynomial factorization, where the algorithm has at least two applications, one for factoring primitive integer polynomials and one for recovering algebraic numbers from their modular images, this short vector algorithm has found many other applications. The algorithm can be used to find simultaneous rational approximations to a set of reals [67], Proposition 1.39, and [57], to break some variants of the Merkle-Hellman knapsack based encryption scheme [3], [75], to solve certain subset sum problems [58], to determine integer relations among reals [36] (see also §3), to find a lattice point close to a given point [8], to name but a few.

We now introduce the notion of reduced basis. Let $b_1,...,b_n \in Z^m$, $m \geq n$, be linearly independent over Q. By $b_1^*,...,b_n^*$ we denote the orthogonalization of this basis, namely

$$b_i^* = b_i - \sum_{j=1}^{i-1} \mu_{i,j} b_j^*, \ 1 \leq i \leq n, \tag{2.1}$$

$$\mu_{i,j} = \frac{(b_i, b_j^*)}{\|b_j^*\|^2}, \ 1 \leq j < i \leq n, \tag{2.2}$$

where $(,)$ denotes the scalar product and $\| \ \|$ the square norm. The basis $b_1,...,b_n$ is called *reduced* if

$$\|b_k^*\|^2 \geq \frac{1}{2}\|b_{k-1}^*\|^2 \text{ for } 1 < k \leq n. \tag{2.3}$$

Lemma ([67], Proposition 1.11): Let $b_1,...,b_n \in Z^m$ form a reduced basis. Then for any non-zero vector $x \in Zb_1 + \cdots + Zb_n$

$$2^{n-1}\|x\|^2 \geq \|b_1\|^2.$$

Hence $\|b_1\|$ is within a factor of $2^{(n-1)/2}$ of the norm of the shortest vector in the lattice spanned by $b_1,...,b_n$.

Proof: Let $x = \sum_{i=1}^{l} r_i b_i = \sum_{i=1}^{l} r_i^* b_i^*$ with $r_i \in Z$, $r_i^* \in Q$ and $r_l \neq 0$, $1 \leq l \leq n$. Since b_i^* is orthogonal to b_i for $i = 1,...,l-1$, $(x, b_l^*) = r_l(b_l, b_l^*) = r_l^*(b_l^*, b_l^*)$. But $(b_l, b_l^*) = (b_l^*, b_l^*) \neq 0$. Therefore, $r_l^* = r_l$, which is an integer $\neq 0$. It follows that $\|x\|^2 = \sum_{i=1}^{l} (r_i^*)^2 \|b_i^*\|^2 \geq (r_l^*)^2 \|b_l^*\|^2 \geq \|b_l^*\|^2 \geq \|b_1^*\|^2/2^{l-1}$, the last inequality by using (2.3) inductively. Since $b_1^* = b_1$ the lemma is proven. $\square$

For reference we shall give the reduction algorithm, incorporating a slight improvement of [43].

Algorithm *Lattice Basis Reduction*

<u>Input:</u> n linearly independent m dimensional integer vectors $b_1,\ldots,b_n$.

<u>Output:</u> $b_1,\ldots,b_n$ forming a reduced basis for the integer lattice spanned by the input vectors. By the above lemma, b_1 must therefore be a short vector.

Step I (Gram-Schmidt Initialization): The arrays μ and β are initialized such that $\mu_{i,j}$, $1 \le j < i \le n$, and $\beta_l = \|b_l^*\|^2$, $1 \le l \le n$, satisfy (2.1) and (2.2):

FOR $i \leftarrow 1$ TO n DO

    (The following loop is skipped for $i = 1$.)
    FOR $j \leftarrow 1$ TO $i-1$ DO

$$\mu_{i,j} \leftarrow \frac{1}{\beta_j}\left[(b_i,b_j) - \sum_{l=1}^{j-1}\mu_{j,l}\mu_{i,l}\beta_l\right].$$

$$\beta_i \leftarrow \|b_i\|^2 - \sum_{l=1}^{i-1}\mu_{i,l}^2\beta_l.$$

$k \leftarrow 2$. (Used as counter in next step.)

Step R (Reduction Loop): At this point μ and β correspond to the orthogonalization of $b_1,\ldots,b_n$. Moreover $b_1,\ldots,b_{k-1}$ is reduced, i.e.

$$\|b_l^*\|^2 \ge \tfrac{1}{2}\|b_{l-1}^*\|^2,\ 1 < l \le k-1,\ \text{and}\ |\mu_{i,j}| \le \tfrac{1}{2},\ 1 \le j < i \le k-1. \tag{2.4}$$

IF $k=n+1$ THEN RETURN $(b_1,\ldots,b_n)$.
IF $k=1$ THEN $k \leftarrow 2$; *incremented* $\leftarrow$ *true*; GOTO step R.
IF *incremented* THEN perform step A given below.

Now (2.4) is also valid for $i = k$, i.e.

$$|\mu_{k,j}| \le \tfrac{1}{2} \text{ for } 1 \le j < k. \tag{2.5}$$

IF $\beta_k \ge \tfrac{1}{2}\beta_{k-1}$ THEN $k \leftarrow k+1$; *incremented* $\leftarrow$ *true*; GOTO step R.

At this point

$$\|b_k^*\|^2 < \tfrac{1}{2}\|b_{k-1}^*\|^2 \le \left[\tfrac{3}{4} - \mu_{k,k-1}^2\right]\|b_{k-1}^*\|^2, \tag{2.6}$$

the last inequality because of (2.4) for $j = k-1$.

Interchange b_{k-1} and b_k and update the arrays μ and β such that (2.1) and (2.2) is satisfied for the new order of basis vectors. The only entries which change are β_{k-1}, β_k and $\mu_{i,j}$, $i = k-1, k$, $1 \le j < i$, as well as $\mu_{i,k-1}$, $\mu_{i,k}$, $k < i \le n$. Let γ and ν denote the updated contents of β and μ, then, according to [67], 1.22 and Figure 1,

$$\gamma_{k-1} = \beta_k + \mu_{k,k-1}^2\beta_{k-1},\quad \nu_{k,k-1} = \frac{\mu_{k,k-1}\beta_{k-1}}{\gamma_{k-1}},\quad \gamma_k = \frac{\beta_{k-1}\beta_k}{\gamma_{k-1}},$$

$$
\left.
\begin{aligned}
v_{i,k-1} &= \mu_{i,k-1} v_{k,k-1} + \mu_{i,k} \beta_k / \gamma_{k-1} \\
v_{i,k} &= \mu_{i,k-1} - \mu_{i,k} \mu_{k,k-1}
\end{aligned}
\right\} \text{ for } k < i \le n,
$$

$$
v_{k-1,j} = \mu_{k,j}, \ v_{k,j} = \mu_{k-1,j} \text{ for } 1 \le j < k-1.
$$

$k \leftarrow k-1$; *incremented* $\leftarrow$ *false*; GOTO step R.

Step A (Adjust $\mu_{k,j}$): This step replaces b_k by $b_k - \sum_{l=1}^{k-1} \lambda_l b_l$, $\lambda_l \in \mathbf{Z}$, such that the new $\mu_{k,j}$ satisfy (2.4). The replacements of the entries in b_k are carried out modulo M, where M is chosen so large that the final b_k agrees with its modular image.

Substep AM (Initialize the modulus):

$M \leftarrow \left\lceil \sqrt{(k+3) \max\{\beta_1, \ldots, \beta_k\}} \right\rceil$. (For efficiency in modular operations,' M can also be chosen the least power of the radix larger than that.)

FOR $l \leftarrow k-1$ DOWNTO 1 DO substep AL.

Substep AL (Make $\mu_{k,l}$ absolutely smaller than 1/2):

IF $|\mu_{k,l}| \ge \frac{1}{2}$ THEN

> $r \leftarrow \text{ROUNDED}(\mu_{k,l})$. $\text{ROUNDED}(x)$ denotes the largest integer z s.t. $|z-x| \le 1/2$.
> Replace b_k by $(b_k - r \, b_l)$ modulo M.
> (Notice that in this case $\lambda_l = r$, otherwise $\lambda_l = 0$.)
>
> FOR $j \leftarrow 1$ TO $l-1$ DO $\mu_{k,j} \leftarrow \mu_{k,j} - r\mu_{l,j}$.
>
> $\mu_{k,l} \leftarrow \mu_{k,l} - r$.

Substep AB: Balance the residual entries of b_k mod M such that each modulus $> \frac{1}{2}M$ is replaced by its negative equivalent $\le \frac{1}{2}M$. $\square$

We will not prove the basis reduction algorithm correct or analyze its complexity, but refer to [67]. However, for later reference we shall state its binary running time in a more refined way. Let us assume that we carry out the arithmetic on the rationals in β_k and $\mu_{i,j}$ by representing their integer numerators and denominators. Let d_l denote the Gramian of the vectors $\{b_1, \ldots, b_l\}$, that is

$$
d_l = \det\left[(b_i, b_j) \right]_{1 \le i,j \le l} = \prod_{i=1}^{l} \|b_i{}^*\|^2 \quad \text{for } 1 \le l \le n.
$$

Then $\|b_l{}^*\|^2 = d_l/d_{l-1}$, $2 \le l \le n$. Moreover,

$$
d_j \mu_{i,j} \in \mathbf{Z} \quad \text{and} \quad \mu_{i,j}^2 \le d_{j-1} \|b_i\|^2 \quad \text{for } 1 \le j < i \le n.
$$

These relationships is always true at the top of step R. Throughout the algorithm $\|b_i\|^2 < nB$, where B is the the square of the length of the longest vector in the original basis.

Furthermore, the Gramians only change when b_k is exchanged with b_{k-1}, in which case the new value of d_{k-1} is at least 3/4 times smaller than the old value. Therefore we have the following theorem.

Theorem: Let B be as above, and let d_l be the Gramian of the first l input vectors. Then the Lattice Basis Reduction Algorithm requires

$$O(n^2m + nm \log(\prod_{l=1}^{n}d_l))$$

integer arithmetic operations. The integers on which these operations are performed have binary length

$$O(\log(\max\{n, B, d_1, ..., d_n\})).$$

In general, the number of arithmetic operations and the integer lengths can be bounded by $O(n^3m \log(B))$, $O(n \log(B))$, respectively.

In practice, the β's and μ's should be computed in big floating point arithmetic. Roundoff errors can be corrected by checking whether the final lattice is reduced, but at the moment we have no well-tested recommendations what mantissa length should be used (perhaps $2n + \log_2(B)$ bits). Schnorr [77] has theoretically justified the use of floating point arithmetic and cut the integer lengths in the algorithm to $O(n + \log(B))$. For particular lattices, Schönhage's improvements [82] speed up the short vector construction further.

Although computational solutions for finding short lattice vectors have been carried out earlier, e.g. by Dieter [22] and by Ferguson and Forcade [24], the Basis Reduction algorithm is the first guaranteed polynomial-time solution. Aside from improving the running time, one can ask to improve the ratio (length of computed short vector)/(length of shortest vector) while retaining polynomial running time. The Basis Reduction algorithm can be improved to produce a ratio $\leq \gamma^{(n-1)/2}$, where $\gamma > 4/3$. All one has to change is the comparison in step R to $\beta_k \geq \beta_{k-1}/\gamma$. Schnorr [78] constructs for every $\varepsilon > 0$ a polynomial-time basis reduction algorithm such that the ratio is $(1+\varepsilon)^n$. The running time of these algorithm are in fact only a constant times slower the than Basis Reduction algorithm, where the constant depends on ε.

3. Root Approximation Algorithms

In this section, we will describe two algorithms with which we can establish that dense multivariate rational polynomials can be factored in polynomial time. The first algorithm splits univariate integer polynomials, and the second reduces the problem of multivariate factoring to univariate factoring. Both algorithms make use the following idea. Let $f(x) \in F[x]$ be the polynomial to be factored. First, we find an approximation ζ to a root $f(\zeta)=0$. Then we try to find another polynomial $\hat{g}(x) \in F[x]$, $\deg(\hat{g}) < \deg(f)$, whose coefficients are small and for which $\hat{g}(\zeta)$ is approximately zero. Clearly, the minimal polynomial g of ζ is a candidate for $\hat{g}$. The key argument will show that for sufficiently good approximation ζ the only polynomial $\hat{g}$ with small enough coefficients for which $\hat{g}(\zeta)$ remains bounded is $\hat{g} = g$.

The notions of "small" will be different for $F = \mathbf{Q}$ and $F = \mathbf{Q}(y_1,\ldots, y_r)$. For $F = \mathbf{Q}$, we use the distance in the complex plane as our valuation and for $F = \mathbf{Q}(y_1,\ldots, y_r)$ we use the order of the multivariate Taylor series approximation. We now present the details for the algorithm in $\mathbf{Q}[x]$.

Let $f(x) \in \mathbf{Z}[x]$ be a primitive polynomial, $n = \deg(f)$. Let $\zeta \in \mathbf{C}$ with $f(\zeta) = 0$ and let $g(x) = g_0 + g_1 x + \cdots + g_m x^m \in \mathbf{Z}[x]$ be the minimal polynomial of ζ, $m \leq n$. Furthermore, for $k > 0$, let $\hat{\alpha}_i$, $\hat{\beta}_i \in \mathbf{Z}$, $0 \leq i \leq m$, satisfy

$$| \, 2^k \operatorname{Re}(\zeta^i) - \hat{\alpha}_i \, | \leq \frac{1}{2}, \quad | \, 2^k \operatorname{Im}(\zeta^i) - \hat{\beta}_i \, | \leq \frac{1}{2}.$$

Now consider the $m+3$ dimensionals lattice spanned by the $m+1$ columns of

$$L_m = \begin{bmatrix} 1 & 0 & 0 & \cdots & 0 \\ 0 & 1 & 0 & & \vdots \\ 0 & 0 & 1 & & \\ \vdots & \vdots & \vdots & & 0 \\ 0 & 0 & 0 & \cdots & 1 \\ 2^k (=\hat{\alpha}_0) & \hat{\alpha}_1 & \hat{\alpha}_2 & \cdots & \hat{\alpha}_m \\ 0 (=\hat{\beta}_0) & \hat{\beta}_1 & \hat{\beta}_2 & \cdots & \hat{\beta}_m \end{bmatrix}$$

The vector

$$L_m \times \begin{bmatrix} g_0 \\ \vdots \\ g_m \end{bmatrix} = \begin{bmatrix} g_0 \\ \vdots \\ g_m \\ \sum_{i=0}^m g_i \hat{\alpha}_i \\ \sum_{i=0}^m g_i \hat{\beta}_i \end{bmatrix} = \begin{bmatrix} g_0 \\ \vdots \\ g_m \\ \hat{\alpha} \\ \hat{\beta} \end{bmatrix}$$

is an element in this lattice. We first observe that

$$\|g\| \leq \binom{2m}{m}^{\frac{1}{2}} \|f\| =: B_1(m, \|f\|),$$

§4.6.2, Exercise 20. Now with $i = \sqrt{-1}$

$$\hat{\alpha}^2 + \hat{\beta}^2 = \left| \sum_{i=0}^m g_i (\hat{\alpha}_i + i\hat{\beta}_i) \right|^2 = \left| \sum_{i=0}^m g_i (\hat{\alpha}_i + i\hat{\beta}_i - 2^k \zeta^i) \right|^2$$

$$\leq \left[\sum_{i=0}^m \left| g_i(\hat{\alpha}_i - 2^k \operatorname{Re}(\zeta^i)) + i g_i (\hat{\beta}_i - 2^k \operatorname{Im}(\zeta^i)) \right| \right]^2$$

$$\leq (\sum_{i=0}^m |g_i|)^2 \leq (m+1) \|g\|^2.$$

The last inequality follows from

$$(\sum_{i=1}^m w_i)^2 \leq m \sum_{i=1}^m w_i^2 \quad \text{for } w_i \geq 0. \tag{3.1}$$

Therefore,

$$\sqrt{\alpha^2 + \beta^2 + g_0^2 + \cdots + g_m^2} \leq \sqrt{m+2} \, \|g\| \leq \sqrt{m+2} \, B_1 =: B_2(m, \|f\|),$$

which means that no matter how large k is chosen, the lattice L_m contains a vector of length $\leq B_2$. Assume now that $[\bar{\alpha}, \bar{\beta}, \bar{g}_0, \ldots, \bar{g}_m]^T \in \mathbf{Z}^{m+3}$ is a short vector in L_m, that is

$$\sqrt{\bar{\alpha}^2 + \bar{\beta}^2 + \bar{g}_0^2 + \cdots + \bar{g}_m^2} \leq 2^{m/2} B_2 =: B_3(m, \|f\|). \tag{3.2}$$

We show that for sufficient large k, $\bar{g}(x) = \sum_{i=0}^{m} \bar{g}_i x^i$ is a scalar multiple of g. Let $s(x)$, $t(x) \in \mathbf{Z}[x]$ such that

$$\rho = \mathrm{resultant}(g(x), \bar{g}(x)) = s(x)g(x) + t(x)\bar{g}(x).$$

Since the coefficients of s and t are minors of the Sylvester matrix [12], we get from Hadamard's determinant inequality that

$$|t| \leq \|g\|^{\deg(\bar{g})} \|\bar{g}\|^{\deg(g)} \leq (B_1 B_3)^m =: B_4(m, \|f\|),$$

where $|t|$ is the height (infinity norm) of the polynomial t. Therefore,

$$|t(\zeta)| \leq B_4 \frac{|\zeta|^m - 1}{|\zeta| - 1} =: B_5(m, \|f\|, |\zeta|).$$

Moreover by (3.1) and (3.2)

$$2^k |\bar{g}(\zeta)| \leq \left| \sum_{i=0}^{m} \bar{g}_i (2^k \zeta^i - \alpha_i - i\beta_i) \right| + \left| \sum_{i=0}^{m} \bar{g}_i \alpha_i \right| + \left| \sum_{i=0}^{m} \bar{g}_i \beta_i \right|$$

$$\leq \sum_{i=0}^{m} |\bar{g}_i| + |\bar{\alpha}| + |\bar{\beta}| \leq \sqrt{m+3} \, B_3.$$

Hence,

$$|\rho| = |t(\zeta)\bar{g}(\zeta)| \leq \frac{\sqrt{m+3} \, B_3 B_5}{2^k} =: \frac{B_6(m, \|f\|, |\zeta|)}{2^k}. \tag{3.3}$$

If we choose k so large that the RHS of (3.3) is less than 1, then by the integrality of ρ we get $\rho = 0$. Hence $\bar{g}$ must divide g, and since g is irreducible, our claim is established.

From the lattice reduction algorithm and complex root approximation procedures, we can now conclude that factoring $f(x)$ can be done in polynomial-time. The powers ζ^i of the root ζ are needed to precision $2^{-k_{\max}}$, where

$$k_{\max} = \log_2(B_6(n-1, \|f\|, |\zeta|)) = O(n^2 + n \log\|f\|).$$

There are many simple ways of showing that this complex root approximation problem is in polynomial-time. We refer to Schönhage's monograph [81] for an exhaustive study on the computational complexity of the fundamental theorem of algebra. Once the $\hat{\alpha}_i$ and the $\hat{\beta}_i$ are known with respect to $k_{\max}$, we compute a short vector in $L_2, L_3, \ldots, L_{n-1}$ until a factor g is found. Notice that if we use the basic reduction algorithm, we will always get $\bar{g} = \pm g$. Also the reduction of L_{i+1} implicitly performs the reduction of L_i. Since g is irreducible, we can continue the process with another root of f/g.

The running time of this method is dominated by the lattice reductions. Let g_r be a factor of f with degree $m_r = \deg(g_r)$. It turns out that the Gramians of L_{m_r} can be explicitly computed. As in §2, we denote by d_i the Gramian spanned by the first i columns of L_{m_r}, $1 \le i \le m_r$. It can be shown [82], Lemma 6.1, that

$$d_i = A_i B_i - C_i^2 \quad \text{where} \quad A_i = 1 + \sum_{l=0}^{i-1} \hat{\alpha}_l^2, \ B_i = 1 + \sum_{l=0}^{i-1} \hat{\beta}_l^2, \ C_i = \sum_{l=0}^{i-1} \hat{\alpha}_l \hat{\beta}_l.$$

Not only does this formula show that $\log(d_i) = O(k_{\max})$, but it also provides a quick way to compute the initial $\beta_i = d_i/d_{i-1}$, $1 \le i \le m+1$. Incidentally, a similar formula yields the initial $\mu_{i,j}$, which we give for the convenience to implementors:

$$\mu_{i,j} = \frac{1}{d_j}(\hat{\alpha}_{i-1}\hat{\alpha}_{j-1}B_{j-1} + \hat{\beta}_{i-1}\hat{\beta}_{j-1}A_{j-1} - (\hat{\alpha}_{i-1}\hat{\beta}_{j-1} + \hat{\alpha}_{j-1}\hat{\beta}_{i-1})C_{j-1}),$$

$1 \le j < i \le m+1$. By the theorem in §2, the reduction of L_{m_r} now costs $O(m_r^3 k_{\max})$ arithmetic steps (including divisions with remainder) on integers of size $O(k_{\max})$. Since $\sum_r m_r = n$, the total requirement is $O(n^5 + n^4 \log|f|)$ steps on integers of size $O(n^2 + n\log|f|)$. This root approximation algorithm is due to Schönhage [82], who also presents a modification of the reduction algorithm itself to obtain the following theorem. An asymptotically slower version is described in [55].

Theorem (Schönhage): For any $\varepsilon > 0$, the irreducible factors of a primitive polynomial $f \in \mathbf{Z}[x]$, $n = \deg(f)$, can be found in

$$O(n^{6+\varepsilon} + n^4 (\log|f|)^{2+\varepsilon})$$

binary steps.

Once the $\hat{\alpha}_i$ and $\hat{\beta}_i$ are known, the dependency on f is only in the bound B_1. Assume $2^k > B_6$ and the length of the short vector found in L_m is $> B_3$. Then no polynomial of degree $\le m$ and Euclidean norm $\le B_1$ can have α as its root. The method therefore can be used to provide evidence that certain numbers are transcendental in the spirit of Ferguson and Forcade [24]. One can employ the algorithm also to find the defining equations for certain fields. Yui and the author, for instance, use it to find equations for Hilbert class fields [54]. Those polynomials possess exactly one real root, which is easily computed. Finally, Kannan, A. K. Lenstra, and Lovász [55] show that the bits of certain transcendental numbers are not computationally unpredictable as defined by Micali and Blum [11].

Our algorithm essentially finds an integer relation among $\alpha, \ldots, \alpha^m$ or proves that none below a given norm bound exists. It lies at hand to generalize this problem to a set of complex numbers $\gamma_1, \ldots, \gamma_m$. The difficulty is to establish the non-existence of a norm-bounded relationship, which is resolved in [36].

We now come to the multivariate case. We shall restrict ourselves to bivariate polynomials. The full multivariate case is dealt with in [47]. Let $\bar{f}(y, x) \in F[y, x]$, F a field. Then by transforming and squarefree decomposing the polynomial we can reduce the problem to factoring $f(y, x) = x^n + f_{n-1}(y)x^{n-1} + \cdots + f_0(y) \in F[y, x]$ with $f(0, x)$ squarefree. We shall be more specific about that. Let $\tilde{f}(y, x) = \sum_{i=0}^{n} \sum_{j=0}^{d} f_{i,j} y^j x^i$ be a squarefree factor of $\bar{f}(y, x)$, which can be found by multivariate GCD computations or Hensel lifting as described in [41]. We now choose $a, b \in F$ such that $f(y, x) = \tilde{f}(y+bx+a, x)$ has $\deg_x(f) = \deg(\tilde{f})$ and $f(0, x)$ is squarefree. The first condition implies that the leading coefficient of f in x is a field element, by which we can divide the polynomial f. Notice that b has to satisfy

$$\sum_{i+j=\deg(\tilde{f})} f_{i,j} b^j \neq 0,$$

and a must not be the zero of a certain discriminant (cf. [47], Lemma 1). If $g(y, x)$ is now an irreducible factor of f, then $\tilde{g}(y, x) = g(y-bx-a, x)$ is one for $\tilde{f}$. If the field F has at least $2(n+d)d$ elements, suitable a and b can always be found. The only problem occurs if $F = \mathbf{F}_q$, the field with q elements, with q being too small. Then we perform our transformations in $\mathbf{F}_{q^p}$, p a prime $> \max(n, d)$. The irreducible factors of $\tilde{f}(y, x)$ in $\mathbf{F}_{q^p}[y, x]$ actually lie in $\mathbf{F}_q[y, x]$, due to the following useful lemma.

Lemma (von zur Gathen [28], Theorem 7.1): Let $f(x_1, \ldots, x_r) \in \mathbf{F}_q[x_1, \ldots, x_r]$ be irreducible, p a prime $> \deg_{x_i}(f)$ for all $1 \leq i \leq r$. Then f is irreducible in $\mathbf{F}_{q^p}[x_1, \ldots, x_r]$.

We now describe our factorization algorithm [47], which works for arbitrary fields.

Algorithm *Bivariate Factoring*

<u>Input:</u> $f(y, x) \in F[y, x]$ monic in x such that $f(0, x)$ is squarefree.

<u>Output:</u> $g(y, x) \in F[y, x]$ irreducible that divides $f(y, x)$.

Step N (Newton iteration): We compute the approximation to a root of f in $\bar{F}[[y]]$.

Find an irreducible factor $h(x)$ of $f(0, x)$. Now there exists a root $\alpha = \sum_{j=0}^{\infty} a_j y^j \in G[[y]]$, where $G = F[z]/(h(z))$ and $a_0 = z \bmod h(z)$. Let $n = \deg_x(f)$, $d = \deg_y(f)$, $l = \deg(h)$. The maximum needed precision for the root is $k = \lceil (2n-1)d/l \rceil$. By Newton iteration (cf. Lipson [68], §IX.3.3) we find $a_0 = z, a_1, \ldots, a_k$ such that

$$f(y, a_0 + a_1 y + \cdots + a_k y^k) \equiv 0 \bmod y^{k+1}.$$

Step M (Minimal polynomial determination): Compute powers $\alpha_k^{(i)}(y) \in G[y]$ of the root approximation. $\alpha_k^{(0)}(y) \leftarrow 1$.

FOR $i \leftarrow 1, \ldots, l-1$ DO $\alpha_k^{(i)}(y) \leftarrow (a_0 + \cdots + a_k y^k) \alpha_k^{(i-1)}(y) \bmod y^{k+1}$.
FOR $m \leftarrow l, \ldots, n-1$ DO Step L.
At this point, the polynomial f is known to be irreducible.

Step L: First compute the next power of the root approximation.

$$\alpha_k^{(m)}(y) \leftarrow (a_0 + \cdots + a_k y^k)\alpha_k^{(m-1)}(y) \bmod y^{k+1}.$$

The precision needed to find a minimal polynomial of degree m is $\kappa \leftarrow \lceil (n+m)d/l \rceil$. Try to solve the equation

$$\alpha_\kappa^{(m)}(y) + \sum_{i=0}^{m-1} g_i(y)\alpha_\kappa^{(i)}(y) \equiv 0 \bmod y^{\kappa+1} \tag{3.4}$$

for polynomials $g_i(y) \in F[y]$, $\deg(g_i) \le d$. Let $g_i(y) = \sum_{0 \le j \le d} g_{i,j} y^j$ and let

$$\alpha_\kappa^{(i)}(y) =: \sum_{j=0}^{\kappa} \left[\sum_{\lambda=0}^{l-1} a_{j,\lambda} z^\lambda \right] y^j \in F[z,y], \quad a_{j,\lambda} = 0 \text{ for } j < 0.$$

Then (3.4) leads to the linear system over F

$$a_{j,\lambda}^{(m)} + \sum_{i=0}^{m-1} \sum_{\mu=0}^{d} a_{j-\mu,\lambda} g_{i,\mu} = 0 \tag{3.5}$$

for $j = 0, \ldots, \kappa$, $\lambda = 0, \ldots, l-1$ in the variables $g_{i,j}$, $i = 0, \ldots, m-1$, $j = 0, \ldots, d$. If this linear system has a solution, which then must be unique, we return the irreducible polynomial

$$g(x) \leftarrow x^m + \sum_{i=0}^{m-1} g_i(y)x^i. \quad \square$$

The proof that a solution to the system (3.4) corresponds to the minimal polynomial of α is similar to the univariate case and can be found in [47], Theorem 1. In order to prove that the algorithm works in polynomial-time for $F = \mathbf{Q}$, the size of the numerators and denominators needs to be bounded. We will not present the fairly intricate analysis here but refer to [47], §6. For $f(x, y) \in \mathbf{Z}[y, x]$ the intermediate integers can be shown to be no more than $O(n^4 d^3 \log|f|)$ bits in length. Since we know that $|g| \le \sqrt{6}^{n+d}|f| =: B_7(n, d, |f|)$ [34], Chapter III, §4, Lemma II, a randomized approach becomes asymptotically significantly faster. For we can choose a random prime p with $2B_7 < p \le 4B_7$. Then for sufficiently large n and d with high probability Steps N, M, and L, when carried out in the homomorphic image $\mathbf{F}_p$ of $\mathbf{Z}$, result in the polynomial $g \bmod p$, from which g is readily recovered. Although the justification of the following theorem needs to be pieced together from the above, the transformation to monicity [56], §4.6.2, Exercise 18, and probability estimates derived according to [44], Theorem 4, we nonetheless have:

Theorem: Let $T(\deg(f), \log|f|)$ be a function dominating the binary running time for factoring $f \in \mathbf{Z}[x]$, and let ω be the exponent for matrix multiplication (classically $\omega = 3$, at the moment the best is $\omega = 2.376 + o(1)$ [21]). Then for any $\varepsilon > 0$ the irreducible factors of a coefficient primitive polynomial $f \in \mathbf{Z}[y, x]$, $\delta = \deg(f)$, can be found by randomization in

$$T(\delta, \delta^{1+\varepsilon} + \log|f|) + O(\delta^{2\omega+2+\varepsilon} + \delta^{2\omega+1}(\log|f|)^{1+\varepsilon})$$

expected binary steps.

Algorithm Bivariate Factoring generalizes to an arbitrary number of variables [47]. The main advantage of this algorithm over other polynomial-time solutions [18], [64], and [39] is that it is field independent. Nonetheless, it is of little practical significance since combinational explosion in the multivariate Hensel algorithm is unlikely to occur because of theory of Hilbert irreducibility as described in the survey [41] (see also §5). In conclusion to this section, we can state that dense multivariate polynomials over the prime fields F_p and Q can be factored in polynomial-time.

4. Factoring over Finite Fields

Historically, it should be added to our survey [41] and Knuth's book [56], §4.6.2, that the "Q-matrix" construction, which is the basis of Berlekamp's algorithm, was first discovered by Butler in 1954 [14]. A nice generalization of that construction can be found in [35]. Using asymptotically fast polynomial arithmetic procedures, linear algebra algorithms, and multipoint polynomial evaluation with Zassenhaus's improvement [56], §4.6.2, Exercise 14, the running time of Berlekamp's algorithm in $F_p[x]$ is

$$O(n^\omega + \log(p)n^{1+\epsilon} + \max(p,n)(\log n)^{2+\epsilon})$$

deterministic arithmetic steps in F_p. The first two of the three terms correspond to the complexity of Butler's irreducibility test, but for that particular problem the distinct degree factorization is asymptotically faster. The usage of randomization in order to get the expected running time polynomial in $\log(p)$ is already considered a classic in the theory of randomized algorithms. A beautifully simple approach is due to Cantor and Zassenhaus [17], §3, in conjunction with Rabin's [76] analysis, or Ben-Or's refinement [10]. Its expected arithmetic running time is $O(\log(p)n^{2+\epsilon})$, but more importantly it requires only to store $O(n)$ field elements at a time.

For large p, the probabilistic algorithms are of course the only practical choice, although it might not be clear which of several randomization schemes [76], [17], [61], [16], or [92], is preferable.

As a consequence of our Bivariate Factoring algorithm, dense multivariate factoring over finite fields also becomes a polynomial-time problem. The analysis can be found in [31], different algorithms are discussed in [18] and [66]. For large p, the algorithms are probabilistic in the Las Vegas sense, but again irreducibility testing can be done polynomially in $\log(p)$ without random choices [52].

New progress towards the removal of random choices in the univariate case can be reported. If one allows preprocessing depending on the field F_p only and with unlimited computational resources, a so-called splitting set of cardinality $\leq 2 \log_2 p$ can be found, such that the random choices in certain probabilistic algorithms can be restricted to this set [2], [16]. A special problem is that of taking squareroots. Then the Tonelli-Shanks method [56], §4.6.2, Exercise 15, requires as its splitting set a single quadratic non-residue. This algorithm has been generalized to k-th roots, the splitting set being d-th non-residues, d all primes dividing both $p-1$ and k [5]. Assuming an extended Riemann hypothesis such splitting sets are deterministically constructible, in fact under such an assumption any polynomial $f \in \mathbf{Z}[x]$ with Abelian Galois group can be factored mod p in $(\deg(f)\log(p))^{O(1)}$ deterministic steps [38]. If $p-1$ is a "smooth" integer, i.e. $p-1$ only has prime factors of order $(\log p)^{O(1)}$, Moenck's algorithm [73] requires a primitive root and this requirement can even be shown to be necessary to polynomial-time factoring [29].

Finally, the theory of elliptic curves also has made its entry into factoring in $\mathbf{F}_p[x]$. Schoof [79] uses this theory to show that squareroots of $a \mod p$ can be taken deterministically in $O(\sqrt{|a|}(\log p)^8)$ steps.

A somewhat different question is the generation of irreducible polynomials of degree n in $\mathbf{F}_p[x]$. The probability that a randomly picked monic n-degree polynomial is irreducible in $\mathbf{F}_p[x]$ is asymptotically $1/n$ and Rabin's probabilistic generation uses such an estimate [76], Lemma 2. An improvement to this probability is reported in [15]. Recently, deterministic algorithms have been invented under hypothesis by von zur Gathen [30] and by Adleman and H. Lenstra [4].

5. Multivariate Factoring

As we have discussed already in §3, factoring of dense multivariate polynomials over the usual coefficient fields can be accomplished in polynomial-time. However, if the number of indeterminates is high, the dense representation causes exponential expression swell compared to more compact representations such as sparse ones. Note that there are $\binom{n+d}{d}$ monomials of total degree $\leq d$ in n indeterminates, although in the sparse representation only a few may be non-zero. Sparse lifting procedures strive to preserve the sparseness of input and output. Additional insight has been gained towards the well-known complication [41] arising during this process.

a) *The leading coefficients problem*: There are two new techniques for dealing with it. One is to use Viry's translation

$$\tilde{f}(x_1, x_2, \ldots, x_n) = f(x_1, x_2 + b_2 x_1, \ldots, x_n + b_n x_1),$$

where the b_i are random elements such that $\tilde{f}$ becomes monic in x_1 [71]. The drawback of this method is that $\deg_{x_1}(\tilde{f})$ might be substantially larger than $\deg_{x_1}(f)$ making the univariate factoring step costly. Another method by the author [48] finds the leading coefficients by lifting from a single univariate factorization and appears to be the algorithm of choice within a sparse lifting procedure. In the univariate case Wang's [89] idea appears quite useful for predicting integer leading coefficients.

b) *The extraneous factors problem*: Controlling this problem by randomization and Hilbert irreducibility has been theoretically justified. An effective theorem reads like that.

Theorem (cf. [45] and [51]): Let $f \in F[x_1, \ldots, x_n]$, F a perfect field, $d = \deg(f)$, $R \subset F$. The factor degree pattern of f is a lexicographically ordered vector $((d_i, e_i))_{i=1,\ldots,r}$ such that for $f = \prod_{i=1}^{r} h_i^{e_i}$, $h_i \in F[x_1, \ldots, x_n]$,

$$h_i \text{ irreducible}, \quad d_i = \deg(h_i) \geq 1, \quad h_i/h_j \notin F, \quad \text{for } 1 \leq i \neq j \leq r.$$

Let $a_1, a_3, \ldots, a_n, b_3, \ldots, b_n \in R$ be randomly selected elements,

$$f_2 = f(x_1 + a_1, x_2, b_3 x_1 + a_3, \ldots, b_n x_1 + a_n).$$

Then

$$\text{Prob}(f \text{ and } f_2 \text{ have the same factor degree pattern}) \geq 1 - \frac{4d \, 2^d + d^3}{\text{card}(R)}.$$

Although this and all other known effective theorems [37], [28] only reduce to bivariate factoring, in practice one maps directly to the univariate case by letting $x_2 = b_2 x_1 + a_2$. The similarity of the used evaluations with those of Viry's can be taken advantage of for controlling the leading coefficients problem as well. Again, the classical mapping

$$f(x_1, x_2, \ldots, x_n) \rightarrow f(x_1, a_2, \ldots, a_n), \quad a_i \in F,$$

leads to a more efficient lifting procedure [48], although effective bounds for the probabilities for getting extraneous factors are not known.

c) *The bad zeros problem*: During the lifting process, the coefficients of monomials $x_1^{e_1} \cdots x_i^{e_i}$ need to be computed. In order to combat inefficiency, one should lift variable by variable, i.e. compute the factorization of $f(x_1, \ldots, x_i, a_{i+1}, \ldots, a_n)$ explicitly. Even then, in the presence of many factors, the problem of collecting like terms has exponential complexity [32]. We refer to Lugiez's lifting scheme [71] and to Luck's heuristics [70] for suggestions to control this problem in practice. We note that the reference as "bad zeros problem" is somewhat a misnomer chosen here for historial reasons.

Another newly discovered issue is that sparse polynomials can have dense factors [32]. For instance, the first factor in

$$(d + \prod_{i=1}^{n}(1+x_i+\cdots+x_i^{d-1})) \prod_{i=1}^{n}(x_i-1), \quad d \text{ prime} \tag{5.1}$$

is irreducible over $\mathbf{Q}$ and contains d^n non-zero monomials, whereas the product has $t < 4^n$ non-zero monomials. Since $d^n > t^{1/2 \log_2 d}$, the number of monomials in the first factor grows by more than $t^{O(1)}$. Therefore, any sparse lifting procedure will need more than polynomial running time with respect to the input size.

Furthermore, consider the Vandermonde determinant

$$\det\left(\begin{bmatrix} 1 & x_1 & x_1^2 & \cdots & x_1^{n-1} \\ 1 & x_2 & x_2^2 & \cdots & x_2^{n-1} \\ \vdots & \vdots & \vdots & & \vdots \\ 1 & x_n & x_n^2 & \cdots & x_n^{n-1} \end{bmatrix}\right) = \prod_{i>j}(x_i - x_j).$$

If the corresponding Vandermonde matrix were multiplied with a unimodular matrix, the determinant of the resulting matrix would be inaccessible to a sparse factorization procedure by virtue of its $n!$ non-zero monomials. In general, symbolic objects represented by formulas (cf. (5.1)) or determinants cannot be dealt with by sparse techniques alone.

A computational model for evaluating polynomials and rational functions is that by a straight-line program. In order to deal with the issue of denseness we have adopted straight-line programs as a representation for polynomials. Let us give a small example for this representation, as it would appear in our system [25].

```
(c1) matrix([1,x12,x13],[x21,2,x23],[0,x32,3]);
```

```
                    [   1     x12    x13  ]
                    [                     ]
(d1)                [  x21     2     x23  ]
                    [                     ]
                    [   0     x32     3   ]
```

The next instruction converts the determinant of d1 to straight-line format and then optimizes the resulting program

```
(c2)  straightopt3(polytostraight('determinant(%)));
7(26%) instructions saved.
```

```
              v1  := 0
              v2  := 1
              v3  := x12
              v4  := x13
              v5  := x21
              v6  := 2
              v7  := x23
              v8  := x32
              v9  := 3
              v10 := v3 * v5
              v11 := v6 - v10
              v12 := v4 * v5
              v13 := v7 - v12
              v14 := v13 / v11
              v15 := v14 * v3
              v16 := v4 - v15
              v17 := v14 * v8
              v18 := v9 - v17
              v19 := v18 * v11
```

The above 19 instruction program computes the determinant of d1 in the variable v19. It should be pointed out that in the internal representation the variables are pointers to the corresponding instructions as in a directed acyclic graph.

It is not obvious at all that the GCD and factorization problems are feasible for polynomials in straight-line representation. To prove our point, consider a seemingly easier operation, that of computing partial derivatives. Letting

$$f(x_{1,1},\ldots,x_{n,n},y_1,\ldots,y_n) = \prod_{i=1}^{n}(\sum_{j=1}^{n} x_{i,j} y_j)$$

Valiant [87] observes that

$$\frac{\partial^n f}{\partial y_1 \cdots \partial y_n} = \text{permanent}(\begin{bmatrix} x_{1,1} & \cdots & x_{1,n} \\ \vdots & & \vdots \\ x_{n,1} & \cdots & x_{n,n} \end{bmatrix})$$

Clearly, f can be computed by a straight-line program of length $O(n^2)$, whereas the computation of the permanent is by Valiant's results #P-hard. Therefore it is believed that no straight-line program of length $n^{O(1)}$ exists that computes the permanent, and hence the intermediate expression swell for iterated partial derivatives is inherent even for the straight-line representation.

Aside from the just mentioned negative result, several efficient straight-line program transformations have been developed in the context of computational algebraic complexity. Most notably are the method by Strassen [83] for eliminating divisions from computations for polynomials, the method by Baur and Strassen [9] for computing all first partial derivatives, and the probabilistic equivalence test of straight-line programs [40]. One of the first results for polynomials represented by straight-line programs is the efficient computation of their factor degree pattern by von zur Gathen [28].

We now give the I/O specifications for the two main algorithms connected to straight-line factorization [51].

Algorithm *Factorization*

Input: $f \in F[x_1, \ldots, x_n]$, F a field, given by a straight-line program P of length l, a bound $d \geq \deg(f)$, and an allowed failure probability $\varepsilon \ll 1$.

Output: Either "failure", that with probability $< \varepsilon$, or $e_i \geq 1$ and irreducible $h_i \in F[x_1, \ldots, x_n]$, $1 \leq i \leq r$, given by a straight-line program Q of length

$$\text{len}(Q) = O(d^2 l + d^6)$$

such that with probability $> 1 - \varepsilon$, $f = \prod_{i=1}^{r} h_i^{e_i}$. In case $p = \text{char}(F)$ divides any e_i, that is $e_i = p^{\hat{e}_i} \overline{e}_i$ with $\overline{e}_i$ not divisible by p, we return $\overline{e}_i$ in place of e_i and Q will compute $h_i^{p^{\hat{e}_i}}$.

Algorithm *Sparse Conversion*

Input: $f \in F[x_1, \ldots, x_n]$ given by a straight-line program P of length l. Furthermore, a bound $d_0 \geq \max_{1 \leq i \leq n} \{\deg_{x_i}(f)\}$, the allowed failure probability $\varepsilon \ll 1$, and an upper bound $t \leq (d_0+1)^n$ for the number of monomials permitted in the answer.

Output: Either "failure" (that with probability $< \varepsilon$), or the representation of a sparse polynomial with no more than t monomials, or the message "f has (probably) more than t monomials." The latter two outputs are correct with probability $> 1 - \varepsilon$.

Both algorithms work over an abstract field and are randomized. Probabilistic choices are interpreted as picking random elements from a sufficiently large subset of the field. A major theoretical fact is that both algorithms have polynomial complexity in a certain natural and precise sense [51]. In particular, for $F = \mathbf{Q}$ and $F = \mathbf{F}_q$ the algorithms have polynomial running time in bit steps. We therefore get the following theorem, which resolves all problems with sparse lifting mentioned above.

Theorem: If in addition to the input parameters of the Factorization algorithm we are given t > 0, for $F = \mathbf{Q}$ or $F = \mathbf{F}_q$ we can find in polynomially many binary steps and random bit choices in

$$l, \ d, \ \log(\frac{1}{\varepsilon}), \ \text{el–size}(P), \ \text{cc–size}(f), \ \text{and } t$$

sparse polynomials that with probability $> 1 - \varepsilon$ constitute all irreducible factors of f with no more than t monomials. Here el–size(P) is the binary size of the scalars in P, and cc–size(f) is the binary size of the coefficients of f, which in the case of the rationals are considered with a common denominator.

We mention that the degree bound d can be probabilistically determined [49], §5, and that d can be exponential in the length of the input programs. As it turns out, the length of the shortest straight-line program for a factor can then become exponential in the input length (or the input degree in binary) [69]. We conclude this section with a non-trivial straight-line factoring example, executed on our system.

```
(c1) p : 'determinant(matrix([w+x+y+z,a+b+c,u+v,0],
                             [(a-x-y-z)^2,(u-b-c)^2,(d-w)^2,0],
                             [(a+b+c+d)^3,(x+y+z)^3,(u+v)^3,0],
                             [(u+z)^5,(x+d)^5,(a+w)^5,x^2+y^2+z^2]));
Time= 250.0 msecs.
```

```
             [   z + y + x + w      c + b + a     v + u        0      ]
             [                                                        ]
             [                 2             2           2            ]
             [(- z - y - x + a)  (u - c - b)   (d - w)      0         ]
(d1)determinant([                                                    ])
             [                 3             3           3            ]
             [ (d + c + b + a)   (z + y + x)   (v + u)      0         ]
             [                                                        ]
             [                 5             5           5  2   2    2]
             [      (z + u)        (x + d)      (w + a)  z + y + x   ]
```

```
(c2) sf : straightfactor(polytostraight(p),1000)$
Time= 37100.0 msecs.
```

Determine length of straight-line program for first factor
```
(c3) straightlength(sf[1][1]);
Time= 100.0 msecs.
(d3)                              11565
```

Optimize the straight-line program for first factor
```
(c4)  sfo : straightopt3(sf[1][1])$
1811(15%) instructions saved.
Time= 110000.0 msecs.
```

Convert first factor to sparse
```
(c5)  straighttosparse(sfo,10,terms=3);
Time= 111000.0 msecs.
```
$$\text{(d5)} \qquad\qquad z^2 + y^2 + x^2$$

Convert second factor to sparse unless it has more than 3 terms
```
(c6)  straighttosparse(sf[2][1],10,terms=3);
Term bound exceeded.
Time= 28900.0 msecs.
```
(d6)   false

Use sparse lifting algorithm to obtain factorization (takes 1.7 hours)
```
(c7)  factor(d1)$
Time= 6120000.0 msec.
```

6. Conclusion

Uni- and multivariate polynomial factorization is not only a classical problem, but efficient procedures also have important applications. An also classical one is that for determining the Galois group of a polynomial [60], [72], [13], etc. Multivariate polynomial factorization can help speed up the ubiquitous Gröbner basis construction [33], and some of the largest test cases have been successfully factored in this setting. Factorization over algebraic extensions is a key subroutine in the Cylindrical Algebraic Decomposition algorithm [7], and the references there, in integration in closed form [86], and takes part in Chou's method for geometrical theorem proving [19].

The question arises what major unresolved problems in the subject of polynomial factorization remain. One theoretical question is to remove the necessity of random choices from any of the problems known to lie within probabilistic polynomial-time, say factorization of univariate polynomials over large finite fields. Another problem is to investigate the parallel complexity of polynomial factorization, say for the NC model [20]. Kronecker's reduction from algebraic number coefficients [85], [59], Berlekamp's factorization algorithm over small finite fields [27], Kaltofen's deterministic Hilbert irreducibility theorem [47], §7, and Weinberger's irreducibility test for $Q[x]$ [90] all lead to NC solutions by simply applying known NC methods for linear algebra problems. It is open whether factoring in $Q[x]$ or irreducibility testing in $F_p[x]$, p large, or in $Q[x, y]$ can be accomplished in NC. We remark that testing a rational dense multivariate polynomial for absolute irreducibility can be shown to be in NC [44].

In connection with the Factorization algorithm presented in §5, we mention an open question. Assume that a straight-line program computes a polynomial whose degree is exponential in the length of the program. Do then at least the factors of polynomially bounded degree have feasible straight-line computations? A positive answer to this question would show that testing a polynomial for zero in a suitable decision-tree model is polynomial-time related to computing that polynomial. In general the theory of straight-line manipulation of polynomials may be extendable in part to unbounded input degrees, but even for the elimination of divisions problem [83] the answer is not known.

References

1. J. A. Abbot, R. J. Bradford, and J. H. Davenport, "The Bath algebraic number package," *Proc. 1986 ACM Symp. Symbolic Algebraic Comp.*, pp. 250-253, 1986.

2. L. M. Adleman, "Two theorems on random polynomial time," *Proc. 19th IEEE Symp. Foundations Comp. Sci.*, pp. 75-83, 1978.

3. L. M. Adleman, "On breaking generalized knapsack public key crypto systems," *Proc. 15th Annual ACM Symp. Theory Comp.*, pp. 402-412, 1983.

4. L. M. Adleman and H. W. Lenstra, "Finding irreducible polynomials over finite fields," *Proc. 18th ACM Symp. Theory Comp.*, pp. 350-355, 1986.

5. L. M. Adleman, K. Manders, and G. L. Miller, "On taking roots in finite fields," *Proc. 18th IEEE Symp. Foundations Comp. Sci.*, pp. 175-178, 1977.

6. L. M. Adleman and A. M. Odlyzko, "Irreducibility testing and factorization of polynomials," *Math. Comp.*, vol. 41, pp. 699-709, 1983.

7. D. S. Arnon, G. E. Collins, and S. McCallum, "Cylindrical algebraic decomposition I: The basic algorithm," *SIAM J. Comp.*, vol. 13, pp. 865-877, 1984.

8. L. Babai, "On Lovász' lattice reduction and the nearest lattice point problem," *Combinatorica*, vol. 6, pp. 1-13, 1986.

9. W. Baur and V. Strassen, "The complexity of partial derivatives," *Theoretical Comp. Sci.*, vol. 22, pp. 317-330, 1983.

10. M. Ben-Or, "Probabilistic algorithms in finite fields," *Proc. 22nd IEEE Symp. Foundations Comp. Sci.*, pp. 394-398, 1981.

11. M. Blum and S. Micali, "How to generate cryptographically strong sequences of pseudo-random bits," *SIAM J. Comp.*, vol. 13, pp. 850-864, 1984.

12. W. S. Brown and J. F. Traub, "On Euclid's algorithm and the theory of subresultants," *J. ACM*, vol. 18, pp. 505-514, 1971.

13. A. A. Bruen, C. U. Jensen, and N. Yui, "Polynomials with Frobenius groups of prime degree as Galois groups II," *J. Number Theory*, vol. 24, pp. 305-359, 1986.

14. M. C. R. Butler, "On the reducibility of polynomials over a finite field," *Quart. J. Math., Oxford Ser. (2)*, vol. 5, pp. 102-107, 1954.

15. J. Calmet and R. Loos, "An improvement of Rabin's probabilistic algorithm for generating irreducible polynomials over GF(p)," *Inf. Proc. Lett.*, vol. 11, pp. 94-95, 1980.

16. P. Camion, "A deterministic algorithm for factorizing polynomials of $F_p[x]$," *Ann. Discrete Math.*, vol. 17, pp. 149-157, 1983.

17. D. G. Cantor and H. Zassenhaus, "A new algorithm for factoring polynomials over finite fields," *Math. Comp.*, vol. 36, pp. 587-592, 1981.

18. A. L. Chistov and D. Yu. Grigoryev, "Polynomial-time factoring of multivariable polynomials over a global field," *LOMI preprint E-5-82*, Steklov Institute, Leningrad, 1982.

19. S.-C. Chou, "Proving elementary geometry theorems using Wu's algorithm," in *Theorem Proving: After 25 Years*, ed. Bledsoe, W. W. & Loveland, D. W., Contemporary Mathematics, vol. 29, pp. 243-286, AMS, Providence, RI, 1984.

20. S. A. Cook, "A taxonomy of problems with fast parallel algorithms," *Inf. Control*, vol. 64, pp. 2-22, 1985.

21. D. Coppersmith and S. Winograd, "Matrix multiplication via arithmetic progressions," *Proc. 19th Annual ACM Symp. Theory Comp.*, pp. 1-6, 1987.

22. U. Dieter, "How to calculate the shortest vector in a lattice," *Math. Comp.*, vol. 29, pp. 827-833, 1975.

23. D. Duval, "Une méthode géométrique de factorisation des polynômes en deux indéterminées," Tech. Report, Institut Fourier, Université de Grenoble I, 1983.

24. R. P. Ferguson and R. W. Forcade, "Multidimensional Euclidean algorithms," *J. reine angew. Math.*, vol. 334, pp. 171-181, 1982.

25. T. S. Freeman, G. Imirzian, and E. Kaltofen, "A system for manipulating polynomials given by straight-line programs," *Proc. 1986 ACM Symp. Symbolic Algebraic Comp.*, pp. 169-175, 1986.

26. A. Fröhlich and J. C. Shepherdson, "Effective procedures in field theory," *Phil. Trans. Roy. Soc., Ser. A*, vol. 248, pp. 407-432, 1955/56.

27. J. von zur Gathen, "Parallel algorithms for algebraic problems," *SIAM J. Comp.*, vol. 13, pp. 802-824, 1984.

28. J. von zur Gathen, "Irreducibility of multivariate polynomials," *J. Comp. System Sci.*, vol. 31, pp. 225-264, 1985.

29. J. von zur Gathen, "Factoring polynomials and primitive elements for special primes," Manuscript, April 1985.

30. J. von zur Gathen, "Irreducible polynomials over finite fields," Manuscript, 1986.

31. J. von zur Gathen and E. Kaltofen, "Factoring multivariate polynomials over finite fields," *Math. Comp.*, vol. 45, pp. 251-261, 1985.

32. J. von zur Gathen and E. Kaltofen, "Factoring sparse multivariate polynomials," *J. Comp. System Sci.*, vol. 31, pp. 265-287, 1985.

33. R. Gebauer, Private communication, April 1986.

34. A. O. Gelfond, *Transcendental and Algebraic Numbers*, Dover Publ., New York, 1960.

35. H. Gunji and D. Arnon, "On polynomial factorization over finite fields," *Math. Comp.*, vol. 36, pp. 281-287, 1981.

36. J. Hastad, B. Just, J. C. Lagarias, and C. P. Schnorr, "Polynomial time algorithms for finding integer rela-
     tions among reals," *Proc. STACS '86, Springer Lec. Notes Comp. Sci.*, vol. 210, pp. 105-118, 1986.

37. J. Heintz and M. Sieveking, "Absolute primality of polynomials is decidable in random polynomial-time
     in the number of variables," *Proc. ICALP '81, Springer Lec. Notes Comp. Sci.*, vol. 115, pp. 16-28, 1981.

38. M.-D. A. Huang, "Riemann hypothesis and finding roots over finite fields," *Proc. 17th ACM Symp.
     Theory Comp.*, pp. 121-130, 1985.

39. M.-P. van der Hulst and A. K. Lenstra, "Factorization of polynomials by transcendental evaluation,"
     Proc. EUROCAL '85, Vol. 2, Springer Lec. Notes Comp. Sci., vol. 204, pp. 138-145, 1985.

40. O. H. Ibarra and S. Moran, "Probabilistic algorithms for deciding equivalence of straight-line programs,"
     J. ACM, vol. 30, pp. 217-228, 1983.

41. E. Kaltofen, "Polynomial factorization," in *Computer Algebra, 2nd ed.*, ed. B. Buchberger et al., pp. 95-
     113, Springer Verlag, Vienna, 1982.

42. E. Kaltofen, "A polynomial reduction from multivariate to bivariate integral polynomial factorization,"
     Proc. 14th Annual ACM Symp. Theory Comp., pp. 261-266, 1982.

43. E. Kaltofen, "On the complexity of finding short vectors in integer lattices," *Proc. EUROCAL '83,
     Springer Lec. Notes Comp. Sci.*, vol. 162, pp. 236-244, 1983.

44. E. Kaltofen, "Fast parallel absolute irreducibility testing," *J. Symbolic Computation*, vol. 1, pp. 57-67,
     1985.

45. E. Kaltofen, "Effective Hilbert irreducibility," *Information and Control*, vol. 66, pp. 123-137, 1985.

46. E. Kaltofen, "Computing with polynomials given by straight-line programs II; Sparse factorization," *Proc.
     26th IEEE Symp. Foundations Comp. Sci.*, pp. 451-458, 1985.

47. E. Kaltofen, "Polynomial-time reductions from multivariate to bi- and univariate integral polynomial fac-
     torization," *SIAM J. Comp.*, vol. 14, pp. 469-489, 1985.

48. E. Kaltofen, "Sparse Hensel lifting," *Proc. EUROCAL '85, Vol. 2, Springer Lec. Notes Comp. Sci.*, vol.
     204, pp. 4-17, 1985.

49. E. Kaltofen, "Greatest common divisors of polynomials given by straight-line programs," *Math. Sci.
     Research Inst. Preprint*, vol. 01918-86, Berkeley, CA, 1986. Expanded version to appear in *J. ACM*. Prel-
     iminary version under the title "Computing with polynomials given by straight-line programs I: Greatest
     common divisors" in *Proc. 17th ACM Symp. Theory Comp.*, pp. 131-142, 1985.

50. E. Kaltofen, "Uniform closure properties of p-computable functions," *Proc. 18th ACM Symp. Theory
     Comp.*, pp. 330-337, 1986.

51. E. Kaltofen, "Factorization of polynomials given by straight-line programs," *Math. Sci. Research Inst.
     Preprint*, vol. 02018-86, Berkeley, CA, 1986. To appear in: "Randomness in Computation," *Advances in
     Computing Research*, S. Micali ed., JAI Press Inc., Greenwich, CT, January 1987.

52. E. Kaltofen, "Deterministic irreducibility testing of polynomials over large finite fields," *J. Symbolic
     Comp.*, vol. 3, 1987.

53. E. Kaltofen, D. R. Musser, and B. D. Saunders, "A generalized class of polynomials that are hard to fac-
     tor," *SIAM J. Comp.*, vol. 12, pp. 473-485, 1983.

54. E. Kaltofen and N. Yui, "Explicit construction of the Hilbert class field of imaginary quadratic fields with
     class number 7 and 11," *Proc. EUROSAM '84, Springer Lec. Notes Comp. Sci.*, vol. 174, pp. 310-320,
     1984.

55. R. Kannan, A. K. Lenstra, and L. Lovász, "Polynomial factorization and nonrandomness of bits of alge-
     braic and some transcendental numbers," *Proc. 16th Annual Symp. Theory Comp.*, pp. 191-200, 1984.

56. D. E. Knuth, *The Art of Programming, vol. 2, Semi-Numerical Algorithms, ed. 2*, Addison Wesley, Read-
     ing, MA, 1981.

57. J. C. Lagarias, "The computational complexity of simultaneous Diophantine approximation problems,"
     SIAM J. Comp., vol. 14, pp. 196-209, 1985.

58. J. C. Lagarias and A. M. Odlyzko, "Solving low-density subset sum problems," *J. ACM*, vol. 32, pp.
     229-246, 1985.

59. S. Landau, "Factoring polynomials over algebraic number fields," *SIAM J. Comp.*, vol. 14, pp. 184-195,
     1985.

60. S. Landau and G. L. Miller, "Sovability by radicals," *J. Comp. System Sci.*, vol. 30, pp. 179-208, 1985.

61. D. Lazard, "On polynomial factorization," *Proc. EUROCAM '82, Springer Lec. Notes Comp. Sci.*, vol.
     144, pp. 126-134, 1982.

62. A. K. Lenstra, "Lattices and factorization of polynomials over algebraic number fields," *Proc. EURO-
     CAM '82, Springer Lec. Notes Comp. Sci.*, vol. 144, pp. 32-39, 1982.

63. A. K. Lenstra, "Factoring polynomials over algebraic number fields," *Proc. EUROCAL '83, Springer Lec.
     Notes Comp. Sci.*, vol. 162, pp. 245-254, 1983.

64. A. K. Lenstra, "Factoring multivariate integral polynomials," *Theoretical Comp. Sci.*, vol. 34, pp. 207-
     213, 1984.

65. A. K. Lenstra, "Factoring multivariate polynomials over algebraic number fields," *Proc. MFCS '84,
     Springer Lec. Notes Comp. Sci.*, vol. 176, pp. 389-396, 1984.

66. A. K. Lenstra, "Factoring multivariate polynomials over finite fields," *J. Comput. System Sci.*, vol. 30, pp.
     235-248, 1985.

67. A. K. Lenstra, H. W. Lenstra Jr., and L. Lovász, "Factoring polynomials with rational coefficients,"
     Math. Ann., vol. 261, pp. 515-534, 1982.

68. J. Lipson, *Elements of Algebra and Algebraic Computing*, Addison-Wesley Publ., Reading, Mass., 1981.

69. R. Lipton and L. Stockmeyer, "Evaluations of polynomials with superpreconditioning," *Proc. 8th ACM
     Symp. Theory Comp.*, pp. 174-180, 1976.

70. M. Lucks, "A fast implementation of polynomial factorization," *Proc. 1986 ACM Symp. Symbolic Alge-
     braic Comp.*, pp. 228-232, 1986.

71. D. Lugiez, "A new lifting process for multivariate polynomial factorization," *Proc. EUROSAM '84,
     Springer Lec. Notes Comp. Sci.*, vol. 174, pp. 297-309, 1984.

72. J. McKay, "Some remarks on computing Galois groups," *SIAM J. Comp.*, vol. 8, pp. 344-347, 1979.

73. R. T. Moenck, "On the efficiency of algorithms for polynomial factoring," *Math. Comp.*, vol. 31, pp.
     235-250, 1977.

74. M. B. Monagan, "A heuristic irreducibility test for univariate polynomials," *J. Symbolic Comp.*, vol. sub-
     mitted, 1986.

75. A. M. Odlyzko, "Cryptoanalytic attacks on the multiplicative knapsack cryptosystem and on Shamir's fast
     signature scheme," *IEEE Trans. Inf Theory*, vol. IT-30/4, pp. 584-601, 1984.

76. M. O. Rabin, "Probabilistic algorithms in finite fields," *SIAM J. Comp.*, vol. 9, pp. 273-280, 1980.

77. C. P. Schnorr, "A more efficient approach for lattice basis reduction," *Proc. ICALP '86, Springer Lec. Notes Comp. Sci.*, vol. 226, pp. 359-369, 1986.

78. C. P. Schnorr, "A hierarchy of polynomial time basis reduction algorithms," in *Theory of Algebra*, ed. L. Lovász and E. Semerédi, Coll. Math. Soc. Janos Bolyai, vol. 44, pp. 375-386, North Holland Publ., Amsterdam, 1986.

79. R. J. Schoof, "Elliptic curves over finite fields and the computation of square roots mod p," *Math. Comp.*, vol. 44, pp. 483-494, 1985.

80. J. T. Schwartz, "Fast probabilistic algorithms for verification of polynomial identities," *J. ACM*, vol. 27, pp. 701-717, 1980.

81. A. Schönhage, "The fundamental theorem of algebra in terms of computational complexity," Tech. Report, Univ. Tübingen, 1982.

82. A. Schönhage, "Factorization of univariate integer polynomials by diophantine approximation and an improved basis reduction algorithm," *Proc. ICALP '84, Springer Lec. Notes Comp. Sci.*, vol. 172, pp. 436-447, 1984.

83. V. Strassen, "Vermeidung von Divisionen," *J. reine u. angew. Math.*, vol. 264, pp. 182-202, 1973. (In German).

84. V. Strassen, "Die Berechnungskomplexität von elementarsymmetrischen Funktionen und von Interpolationskoeffizienten," *Numer. Math.*, vol. 20, pp. 238-251, 1973. (In German).

85. B. M. Trager, "Algebraic factoring and rational function integration," *Proc. 1976 ACM Symp. Symbolic Algebraic Comp.*, pp. 219-228, 1976.

86. B. M. Trager, "Integration of algebraic functions," Ph.D. Thesis, MIT, 1984.

87. L. Valiant, "Reducibility by algebraic projections," *L'Enseignement mathématique*, vol. 28, pp. 253-268, 1982.

88. B. L. van der Waerden, *Modern Algebra*, F. Ungar Publ. Co., New York, 1953.

89. P. S. Wang, "Early detection of true factors in univariate polynomial factorization," *Proc. EUROCAL '83, Springer Lec. Notes Comp. Sci.*, vol. 162, pp. 225-235, 1983.

90. P. J. Weinberger, "Finding the number of factors of a polynomial," *J. Algorithms*, vol. 5, pp. 180-186, 1984.

91. P. J. Weinberger and L. P. Rothschild, "Factoring polynomials over algebraic number fields," *ACM Trans. Math. Software*, vol. 2, pp. 335-350, 1976.

92. K. Yokoyama and T. Takeshima, "Factorization of univariate polynomials over finite fields," Manuscript, 1986.

93. H. Zassenhaus, "Polynomial time factoring of integral polynomials," *SIGSAM Bulletin*, vol. 15, no. 2, pp. 6-7, 1981.

94. R. E. Zippel, "Probabilistic algorithms for sparse polynomials," *Proc. EUROSAM '79, Springer Lec. Notes Comp. Sci.*, vol. 72, pp. 216-226, 1979.

95. R. E. Zippel, "Newton's iteration and the sparse Hensel algorithm," *Proc. '81 ACM Symp. Symbolic Algebraic Comp.*, pp. 68-72, 1981.

Factorization Then and Now

D. H. LEHMER* University of California at Berkeley, Berkeley, California

It is a pleasure to discuss with you a remarkable development that extends from the 1920s to the present time and has to do with automation and the ancient problem of factorization.

This development began much earlier than 1920. One has to go back to Gauss and Legendre in the early nineteenth century to ground the development properly. It was Legendre who introduced the continued fraction of the square root of an integer as a vehicle for factoring that integer and it was Gauss who introduced us to the composition of binary quadratic forms of the same discriminant.

Throughout the nineteenth century a favorite method of factorization was to use the factor table, a sort of phonebook solution in which we look up a factor of N that has been prepared in advance by a wholesale process. More than half of the entries printed will never be read. By 1909 my father had brought out his factor table for the first ten million (1). This proved to be the factor table to end all factor tables.

The semimechanical method used to make factor tables was the so-called stencil method or movable strip method, which is applicable to many other problems and which developed into the sieve machines of the mid-1920s.

If one had no usable factor table, one could use trial divisions to search for small factors of N, the number to be factored. Ideally one would use a list of primes as a source of trial divisors. Sometimes one would use simple arithmetical progressions instead. An alternative procedure was to use a previously prepared product P of small primes and then compute the greatest common divisor of P and N. However it was done, the process was painful to carry out by hand, even on a desk calculator. The method of trial division is still used today, the pain borne by a computer instead of a human being.

The Legendre (2) method was quite different. It depended on the notion of quadratic residues. As the name implies, these are integers R such that there exists another integer x for which the congruence

$$x^2 \equiv R(\text{mod } N)$$

holds. The success of the Legendre method depends on the fact that if R is a quadratic residue of N, it is also a quadratic residue of every prime factor of N. By the law of quadratic reciprocity, the primes that have R for a quadratic residue are represented by a set of linear forms

$$p = 4Rn + a_i \qquad (n = 0, 1, 2, \ldots)$$

For example, the primes having 2 for a quadratic residue are of one of the forms

$$8n + 1, \qquad 8n + 7$$

So if you know that 2 is a quadratic residue of N, the unknown smallest divisor of N is restricted to one-half of all the primes less than $\sqrt{N}$. If one knows k independent nontrivial quadratic residues of N, then the number of trial divisors is reduced to not more than

$$2^{-k}\pi(\sqrt{N})$$

which is a mere handful for k large enough.

How are we to find 10 or 20 nontrivial quadratic residues of N? This is where the continued fraction for $\sqrt{N}$ comes in. If

$$\sqrt{N} = a_0 + \cfrac{1}{a_1 + \cfrac{1}{a_2 + \cdots + \cfrac{1}{a_{n-1} + \cfrac{1}{\frac{\sqrt{N} + P_n}{Q_n}}}}}$$

then $(-1)^n Q_n$ is a quadratic residue of N and is less than $2\sqrt{N}$. It is relatively cheap, you say, but not small enough. But the product of two quadratic residues is again a quadratic residue. So, by multiplying the Q_n together and canceling unwanted square factors, one is able to obtain a set of small quadratic residues of N.

Thus, the method of Legendre has three parts:

Step 1: Find a suitable set of quadratic residues of N.

Step 2: Select the corresponding linear forms from an existing table.

Step 3: Combine the linear forms to exclude almost all the primes less than $\sqrt{N}$.

My father and I explored this method and in 1926 we hit upon a scheme for mechanizing steps 2 and 3. We took the first 5000 primes, those less than 48,594, on the first page of Lehmer's list of primes (1) and made them correspond to a matrix M of 100 rows and 50 columns. For each square-free R less than 250 in absolute value we punched a hole in the matrix M at the spot where a prime represented the linear forms that go with this value of R. This produced a "factor stencil" for R. Step 3 merely consisted in stacking the selected stencils on top of one another. The row and column numbers of the few places where a light showed through the stack of stencils indicated the only primes that have a chance to divide N. [See (3).]

The punching of the factor stencils was a big job, comparable to the production of a good-sized factor table, but the stencils could be used to factor numbers up to $48593^2 = 2361279649$.

In 1939 J. D. Elder produced a set of 2000 Hollerith cards equivalent to a set of factor stencils. These were distributed to libraries by the Carnegie Institution. When you return home, see if your library has a set (4).

This leaves us only step 1. There is even today a controversy over how best to obtain quadratic residues on a big computer. The continued fraction method is recursive and it provides relatively small quadratic residues. However many modifications of this method have been used. For example, Kraitchik (5) in 1925 used the following device. If for a fixed value of x we have

$$x^2 = N + mp$$

where p is a prime too large to be useful, then

$$(x - p)^2 = N + m_1 p \qquad (m_1 = m + p - 2x)$$

Since mp and m_1p are both quadratic residues of N, so is their product; that is, mm_1 is a quadratic residue of N. It may be that the prime factors of m and m_1 are smaller than p. If not, we repeat the trick.

It has been suggested that a Monte Carlo procedure be introduced. This is okay. Unlike the companion problem of identifying prime numbers, where the Monte Carlo procedures are questionable, there can be no objection to them in factorization. After all, the factorization of N, no matter how obtained, can always be easily verified.

The following variant of the Legendre method was published by me and R. E. Powers (7) in 1931. It developed into the continued fraction method of Morrison and Brillhart (6) in 1970. If we write the truncated continued fraction

$$a_0 + \cfrac{1}{a_1} + \cdots + \cfrac{1}{a_{n-1}}$$

as a rational number A_{n-1}/B_{n-1}, then it transpires that

$$A_{n-1}^2 - NB_{n-1}^2 = (-1)^n Q_n$$

Taking this equation modulo N, we have

$$A_{n-1}^2 \equiv (-1)^n Q_n \quad (\bmod\ N)$$

This not only proves that $(-1)^n Q_n$ is always a quadratic residue of N, but also gives explicitly the A_{n-1} itself.

If we can find a set of n's,

$$n = n_1, n_2, n_3, \ldots, n_k$$

for which the product

$$(-1)^{n_1} \ldots (-1)^{n_k} Q_{n_1} \ldots Q_{n_k} = U^2$$

is a perfect square, and we write

$$A_{n_1-1} A_{n_2-1} \cdots A_{n_k-1} \equiv V \quad (\bmod\ N)$$

then

$$U^2 \equiv V^2 \quad (\bmod\ N)$$

Thus, N divides (U - V) · (U + V). If N fails to divide each factor, we
are in luck because the greatest common divisor of N and U - V is a nontrivial factor of N. (For N a prime we shall always be unlucky.) If we
are unlucky, we have to round up another set of n's and try, try again.
With a hand computer this could be demoralizing, but with a machine it
would be routine.

In 1925 A. J. C. Cunningham (8), a British Army officer retired from
India, published a slim volume giving the factorization of $b^n \pm 1$ for

b = 2, 3, 5, 6, 7, 10, 11, 12

and my father and I started filling in the blank entries. This was a slow
start of what was to become known as the "Cunningham project" (15). As
Cunningham had tested all the numbers in his tables for factors less than
100000, the factor stencils were not much help and I began to think about
how to perform step 3 in general. In 1927 (9) I constructed an automatic
electromechanical device for combining sets of linear forms that sifted
for answers at the rate of 3600 numbers a minute and ran without attention.
This was the first of half a dozen sieves and sieve programs constructed
over the next 50 years (10). With this device we could tackle a great
variety of problems, some of them having to do with factorization—for
example, representing N by quadratic forms due to Fermat, Euler, and Cheby-
shev. The Fermat method uses the form

$$N = x^2 - y^2 = (x + y)(x - y)$$

To find x and hence

$$y = \sqrt{x^2 - N}$$

we take a small excluding prime modulus q and ask what values of x (mod q)
make $x^2 - N$ a quadratic residue of q. Thus x (mod q) is restricted to
(q - 1)/2 cases, so x belongs to one of (q - 1)/2 linear forms

$$x = qn + a_i \qquad [i = 1(1)(q - 1)/2]$$

Choosing 10 or 20 values of q, we have a nice sieve problem.

For x we have the inequalities

$$\sqrt{N} < x < \frac{1}{2}\left(L + \frac{N}{L}\right)$$

where L is the limit to which we have searched for small factors of N.
At L = $\sqrt[3]{N}$, for example, we have

$$x = O(N^{2/3})$$

According to complexity theory this method is laughably inefficient when compared with the direct search for a prime factor of N which costs $O(N^{1/2})$. This shows the lack of reality of complexity theory. If one says that the cost of a method is $O(N^S)$, one means that there exists a positive constant C for which the running time could be as much as CN^S seconds. To compare the costs of two methods M_1 and M_2 for factoring a given N, one has to know the values of the respective constants C_1 and C_2 implies by the O symbols. The Fermat method applied to a number from the Cunningham project is vastly superior to the crude search for prime factors of N. This method got the Cunningham project off the ground.

A nineteenth-century method of factorization is based on the old Brahmagupta identity

$$(x_1^2 - Dy_1^2)(x_2^2 - Dy_2^2) = (x_1x_2 \pm Dy_1y_2)^2 - D(x_1y_2 \mp x_2y_1)^2$$

which shows that numbers of the form $x^2 - Dy^2$ are closed under multiplication. If a number N has two really different representations of the form

$$mN = t_1^2 - Du_1^2 = t_2^2 - Du_2^2$$

where m is a small multiplier, then $|t_1u_2 - t_2u_1|$ has a factor in common with N that can be revealed by the Euclidean algorithm. To represent mN by $x^2 - Dy^2$, one uses a good fast sieve as in the Fermat method. The number D is chosen square-free and may be positive or negative and $x < O(N^{1/2})$.

The only catch to this method is that one can choose a D for which there is no representation at all. If this disappointment occurs, one has to choose a different D and begin again. The psychological effect of the possibility of this happening made this method unpopular among most arithmeticians of the nineteenth century. To restore the reputation of this method to its rightful place, my wife and I used Steiner triple systems to solve the disappointment problem (11). Given N we give in return three forms

$$x^2 - D_1y^2, \quad x^2 - D_2y^2, \quad x^2 - D_3y^2$$

at least one of which is guaranteed to give two representations of mN in case N is composite. Thus, there can be two disappointments at most. With a high-speed sieve available this method is indeed a competitive one.

My father and I used to daydream of better methods of factorization. One of these dreams was related to Euler's totient function, the function $\phi(N)$ which gives the number of numbers not exceeding N and relatively prime to N. If p is a prime, then $\phi(p) = p - 1$. If N is a product of two distinct primes p and q, since

$$\phi(pq) = \phi(p)\ (q) = (p - 1)(q - 1) = pq - p - q + 1$$

we have

$$N + 1 - \phi(N) = N + 1 - \phi(pq) = p + q$$

Hence, p and q are the roots of the quadratic equation

$$X^2 - (N + 1 - \phi(N))X + N = 0$$

whose coefficients are known. Solving this equation we obtain p and q.

But how do we get the value of $\phi(N)$ short of factoring N? If only we could approximate $\phi(N)$ sufficiently closely, we could use a theorem of Euler

$$2^{\phi(N)} \equiv 1 \quad (\text{mod } N)$$

to get the exact value of $\phi(N)$. Despite its forbidding aspect, the problem of determining c in

$$2^b \equiv c \quad (\text{mod } N)$$

is simple and it costs only $O(\log b)$. If A is a known approximation to $\phi(N)$, so that

$$A - \phi(N) = \varepsilon, \qquad |\varepsilon| < K$$

then

$$2^A \equiv 2^{A-\phi(N)} \equiv 2^\varepsilon \quad (\text{mod } N)$$

Now ε is determined from a prepared table of 2^n (mod N) for $|n| < K$. Once we have ε, we can find the exact value of $\phi(N)$ and we can factor N.

I mention this foolish dream because it came true in a different context in 1971 when Dan Shanks discovered a new method of factorization. We have just been talking about the set of numbers less than or equal to N and relatively prime to N. These numbers constitute an abelian group closed under ordinary multiplication. The order of this group is $\phi(N)$. Shanks uses another abelian group of reduced binary quadratic forms

$$ax^2 + bxy + cy^2 \quad \text{with } b^2 - 4ac = -N$$

first discovered by Gauss. It is closed under a binary operation called *composition* and its order is denoted by h(-N). If F is any one of these forms, then, in the sense of group operations, we have

$$F^{h(-N)} = U$$

where U is the unit of the group. For example, if N is congruent to 3 modulo 4, then

$$U = x^2 + xy + (N + 1)/4$$

If N is composite, then h(-N) is even. Let

$$h(-N) = 2^k m$$

where m is odd. We first compute $F^m = V$. Then we start computing

$$V^2, V^4, V^8, \ldots$$

until we find an n such that $V^{2^n} = U$. This number n cannot exceed k. The preceding form $V^{2^{n-1}}$ is special. It is

$$ax^2 + bxt + cy^2$$

in which two of its coefficients are equal. But its discriminant is -N; that is,

$$-N = b^2 - 4ac = \begin{cases} a(a - 4c) & \text{if } a = b \\ (b - 2a)(b + 2a) & \text{if } a = c \\ b(b - 4a) & \text{if } b = c \end{cases}$$

In all cases N has been factored.

Hence, if we know the order h(-N) of this group, we can factor N in about log N steps. Unlike $\phi(N)$, the function h(-N) can be approximated by its conditionally convergent product formula

$$h(-N) = \frac{N^{1/2}}{\pi} \prod_p \frac{p}{p - \left(\frac{-N}{p}\right)}$$

where (-N/P) = ±1 is Kronecker's symbol and where the product is taken over all primes. If we take the product over just the primes less than x, we obtain an approximation A(x) to h(-N). A little fiddling involving $F^{A(x)}$ and the early powers of F determines h(-N) exactly.

Shanks' discovery spurred a search for methods depending on groups other than the usual set of integers prime to N. We have heard Professor Lenstra explain the impressive results obtained from the group of integer points of the cubic curve

$$y^2 = x^3 + ax^2 + bx + c$$

Who knows what algebraic varieties the future has in store for us.

In conclusion, I do not want to leave you with the impression that the Lehmer family initiated the recent development in factorization methods. A glance at the chapter entitled Methods of Factoring in the first volume of Dickson's *History* will give a truer picture of what was done before 1918. Then there was Maurice Kraitchik (5) with whom we kept up a lively correspondence from 1925 to 1947 and whose books contain fresh ideas on the subject. And we must mention J. M. Pollard with his two powerful methods, the "p - 1 method" (13) and the "rho method" (14), which changed the direction of development in the 1970s. But the greatest share of responsibility is borne by the electronic digital computer which since 1947 has made all this possible. We have seen how an inefficient hand method becomes a highly efficient machine method (6) simply because the machine is so much faster. Also, the machine does not resent being told to try, try again if at first it doesn't succeed.

REFERENCES

1. D. N. Lehmer, *Factor Tables for the First Ten Million*, Carnegie Inst. no. 105, Washington, D.C., 1909.

2. A. M. Legendre, *Essai Sur la Theorie des Nombres*, Paris, 1798.

3. D. N. Lehmer, *Factor Stencils*, Carnegie Inst., Washington, D.C., 1929.

4. J. D. Elder, *Factor Stencils*, Carnegie Inst., Washington, D.C., 1939.

5. M. Kraitchik, *Theorie des Nombres*, vol. 1, Paris, 1924; vol. 2, Paris, 1926; *Recherches sur la Theorie des Nombres*, vol. 1, Paris, 1924; vol. 2, Paris, 1929.

6. M. A. Morrison and J. D. Brillhart, A method of factoring and the factorization of F_7, *Math. Comp.* 29:183-205 (1975).

7. D. H. Lehmer and R. E. Powers, On factoring large numbers, *Bull. Amer. Math. Soc.* 37:770-776 (1931).

8. A. J. C. Cunningham and H. J. Woodall, Factorisation of $y^n \pm 1$, y = 2, 3, 5, 6, 7, 10, 11, 12 up to high powers (n). Hodgson & Sons, London, 1925.

9. D. H. Lehmer, The mechanical combination of linear forms, *Amer. Math. Monthly* 35:114-121 (1928).

10. D. H. Lehmer, A photo-electric number sieve, *Amer. Math. Monthly* 40:401-406 (1933).

11. D. H. and Emma Lehmer, A new factorization technique using quadratic forms, *Math. Comp.* 28:625-635 (1976).

12. Daniel Shanks, Class number, a theory of factorization and genera, *Amer. Math. Soc. Proc. Symposia in Pure Math.* 20:415-440 (1971).

13. J. M. Pollard, Theorem on factorization and primality testing, *Proc. Cambridge Phil. Soc.* 76:521-528 (1974).

14. J. M. Pollard, A Monte Carlo method of factorization, *Nordisk Todskrift for Inf. (BIT)* 15:331-334 (1975).

15. J. Brillhart, D. H. Lehmer, J. L. Selfridge, Bryant Tuckerman, and S. S. Wagstaff, Jr., Factorizations of $b^n \pm 1$, b = 2, 3, 5, 6, 7, 10, 11, 12 up to high powers, *Contemp. Math.* 22 (1983).

Computer Animation in Mathematics, Science, and Art

NELSON L. MAX   Lawrence Livermore National Laboratory, Livermore, California

At the conference whose proceedings make up this volume, my presentation consisted of a collection of excerpts from computer animated films. Since I showed only my own films, a description of my personal experiences in computer animation seems appropriate for this paper, with indications of the films shown at the conference, and where they can be obtained.

I started my career as a topologist and got involved in computer graphics because I wanted to produce an animated film showing how to evert a sphere, i.e., how to turn it inside out, using a regular homotopy. A regular homotopy is a continuous deformation of a smooth surface, during which the surface may intersect itself, but must remain continuously differentiable. A smooth, possibly self-intersecting surface is called an immersion, so a regular homotopy is a continuous family of immersions.

Steve Smale had given a constructive proof [1] that such a sphere eversion was possible, and Anthony Phillips had written an article in Scientific American [2], with illustrations showing several stages in the regular homotopy, sliced into ribbons by parallel cross section planes. However, it was difficult to visualize the continuous motion of the sphere, either from Smale's inductive proof or Phillips' ribbon diagrams. I hoped to show the motion of the surface in a computer animated movie, and proposed that the National Science Foundation (NSF) fund its production. My first proposal was written in 1969, and I had no idea how difficult a computer modelling, graphics rendering, and film production project I was attempting.

At the recommendation of the NSF, I worked with the Education Development Center (EDC) in Newton , Massachusetts, to revise the proposal and budget, and added a steering committee to advise on the film content. The executive producer at EDC suggested that it would be as easy for the NSF to fund several films as to fund one, so we proposed four films: "Space Filling Curves," "Regular Homotopies in the Plane, Parts I and II", and "Turning a Sphere Inside Out." The Topology Films Project was funded in 1970.

At the Battelle Rencontres held in Seattle in the summer of 1967, Bryce DeWitt [3] presented
another way of visualizing the regular homotopy in Scientific American [2], and Marcel Froissart
gave an illuminating talk on his own variant, which I found easier to understand. He and Bernard
Morin refined this eversion and made models of clay. I have tried to illustrate and describe the
history of this eversion in [4]. Later Bernard Morin was able to understand this eversion in terms
of its non-generic or singular positions, where the topological character of the immersion changes.
This sphere eversion requires 14 such singular positions, as outlined in [5]. Another newer
eversion of Bernard Morin [6] also requires 14. It seems unlikely that the sphere can be everted
with fewer singularities, since it has been shown that a quadruple point is required somewhere
during the eversion. The first proof of this fact [7] was based on the geometry of Morin's
eversion. Subsequently, John Hughes has given a shorter, more algebraic proof [8].

I planned to model key stages of the regular homotopy as the unions of bicubic polynomial
parametric surface patches, constrained to match in a continuously differentiable manner. I traveled
to Cambridge University to use the MultiObject system [9] of Andrew Armit, which was designed
to interactively create such models. However, after over a week with the system, I was unable to
model even the very symmetrical stage halfway through the eversion. I learned that the system's
successful computer models all were interactively modified from first approximations, created by
measuring actual physical models. Therefore, I was very pleased to discover that Charles Pugh
had created beautiful nickel-plated chicken-wire models of eleven key stages of the eversion, based
on the drawings in my script. I was able to lay out the patch boundaries on these models with
masking tape, measure their 3-D coordinates, and input the data to a program similar to
MultiObject, which I had written at Carnegie Mellon University.

Vector and dot images were created at Carnegie Mellon University and at MIT Lincoln Laboratory,
and color stills were contributed by Jim Blinn at the University of Utah. (See figure 1). Smooth
shaded color animation, showing the inside surface of the sphere in blue and the outside surface in
red, was produced on a special purpose shaded graphics system at Case Western Reserve
University. Further details of the modelling and rendering algorithms are given in [10].

The film starts with a discussion with Steve Smale about the mathematical context of his theorem,
and continues with a tour of the Charles Pugh's models. The final 10 minutes consist of computer
animation with a musical sound track, and this animated excerpt was shown at the conference.
When the film was completed in 1976, it represented the state of the art in specifying, animating,
and rendering smooth moving surfaces. It could be appreciated as art as well as mathematics.
However, I was disappointed to find that it left most mathematicians unsatisfied. Although, they
were convinced that they had seen a valid eversion of the sphere, they were unable to reconstruct
the motion from memory afterwards, because it was not structured as a sequence of key steps. I
hope that the discussion of the 14 singularities in [5] provides the needed structure.

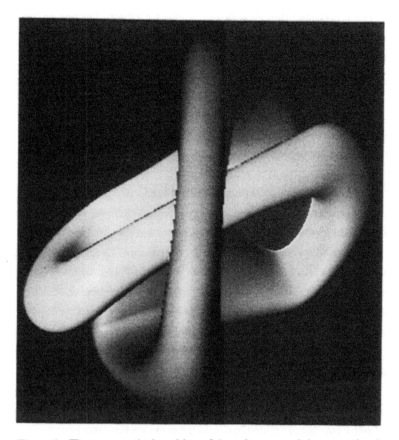

Figure 1. The symmetrical position of the sphere containing a quadruple point, halfway through the sphere eversion, rendered by James Blinn as a union of bicubic surface patches.

The other three films in the Topology Films Project were completed earlier. The one on "Space Filling Curves" shows three nowhere differentiable curves. The first is Koch's "snowflake" curve. The other two are continuous curves whose image is a complete square: a variant of Peano's original example, and one due to Sierpinski. The curves are studied as a sequence of approximations, animated to move continuously as they approach their limit curve. Individual points are followed as they define their limit points, leaving tracks behind. In "zoom" scenes, the approximations are magnified at the same rate as they grow new appendages, so that the self-similar shape of the limit curve becomes evident.

A subsequent film, "Limit Surfaces and Space Filling Curves" shows two three-dimensional self-similar surfaces: a volume filling surface analogous to Sierpinski's curve, and the Alexander Horned Sphere, a sphere famous to topologists because its exterior is not simply connected. These are also constructed as limits, and reviewed by zooms. On the advice of Gregg Edwards of the NSF, the various zooms in these two films were collected into one "art" film with a musical sound track, called "Zooms on Self Similar Figures," which was shown at the conference. Such self-similar shapes were popularized as "fractals" by Benoit Mandelbrot [11] and the film can be viewed as an introduction to precisely self-similar fractals. Statistically self-similar fractals have now become an important method of generating realism in complex natural objects such as clouds, trees, and mountains. (See [12], [13], and [14] respectively.)

The two films on "Regular Homotopies in the Plane" prove the "only if" and the "if" parts of the Whitney-Graustein theorem: two regular closed curves in the plane are regularly homotopic if and only if they have the same winding numbers. The original proof used derivatives to define the tangent vector and the winding number of a curve, and integrals to reconstruct a curve from the variation of its tangent vector. The steering committee for our project invented a geometric symbol with the tangent vector winding up a spiral to record the winding number. We were able to give an entirely visual proof of the theorem, without any formulas.

Another important collection of mathematical computer animation has been produced by Tom Banchoff and his colleagues at Brown University. A particularly striking film shows the four dimensional hypercube rotating around various planes, and sliced with colored sections by various families of parallel three-dimensional hyperplanes.

Another area where computer graphics is important in science is in illustrating the structure of molecules. For many years, the structures elucidated by x-ray crystallography have been presented with "ball-and-stick" line drawings produced by the ORTEP program [15]. More recently, smoothly shaded colored images of the space-filling [16, 17] and the solvent-accessible [18] molecular surfaces have been produced. (See [19] for a review of many molecular surface representation systems and algorithms.) At the conference, I showed excerpts from films on

DNA-drug interaction, on the structure of a virus, and on inertial-confinement fusion. These were produced at Lawrence Livermore National Laboratory, using the ATOMLLL program described in [17] and [20]. This program renders images of "space-filling" molecular models, formed as the union of colored spheres centered at the atomic nuclei, with radii equal to the atom's Van der Waals radius. The spheres for two covalently bonded atoms will overlap, while for non-bonded Van-der-Waals contacts, they will just touch. [See figure 2]. These images simulate the plastic CPK models invented by Corey, Pauling, and Koltun to study geometric relationships within and between molecules.

The essential core of ATOMLLL comes from the ATOMS program [21], written by Ken Knowlton and Lorinda Cherry at Bell Telephone Laboratories. They used a pseudo-perspective projection, in which the spheres project as circles, instead of the ellipses which would result from true perspective. They also approximated the elliptical projection of the intersection arc of two overlapping spheres by a circle. Thus the visible portion of an atom sphere (or a bond cylinder for ball-and-stick models) is one or more regions bounded by line segments or circular arcs. The visible surface algorithm works by successively modifying these visible regions to remove the parts hidden by occlusion or intersection with other spheres or cylinders. The outlines of the remaining regions are written to tape, and later filled in by a minicomputer, controlling a color film recorder. This is an efficient division of labor between a large mainframe, which computes the visible surfaces from the three-dimensional model using floating point arithmetic, and the minicomputer, which needs only fixed-point arithmetic to color in the resulting two dimensional regions with shading and highlights. It also decreases the IO burden on the mainframe, since the outlines are described much more compactly than the millions of color shading values needed to specify the final image.

When I added shading and highlights to the ATOMS program, I used a forward difference iteration to generate quadratic polynomial shading values. For speed, I implemented this in microcode, using the writable control store in our Varian V-75 minicomputer. I was later able to apply this same microcode to do quadratic interpolation between an array of values computed by the Cray 1 for a Hartree Fock solution to the time varying densities of two colliding nuclei. The results of this interpolation were sent through a color-look-up table before being recorded on film, producing smooth colored contour bands of nuclear density. An animation showing nuclei colliding with varying angular momenta was shown at the conference.

From November 1983 through December 1984, I was involved in the production of a red-blue stereo film called "We are Born of Stars," produced for the Fujitsu Pavillion at Expo '85, in Tsukuba, Japan (See [22].) This film was projected through a fisheye lens onto a tilted dome

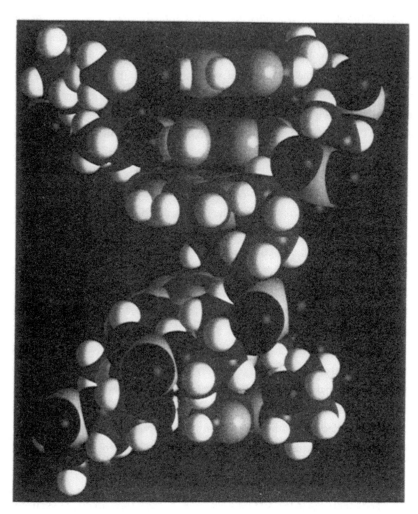

Figure 2. Twenty base pairs of DNA, drawn with shading, highlights, and soft shadows by the
ATOMLLL program.

Omnimax theater, so the image filled the viewers' peripheral vision. Because the screen edge was not visible to destroy the stereo illusion, the audience felt they were in the middle of the animated universe we created.

It was necessary to compensate in the computer for the distortion of the fisheye lens, as described in [17] and [23]. Also, to accommodate rapid motion across such a wide field of view without strobing or doubling due to the frame and flicker rate, we added motion blur to the images, using a "2 1/2 D" compositing algorithm [24]. Three non-stereo scenes from this film were projected at the conference, showing nuclear reactions inside a star, water freezing into ice, and DNA coiling into a chromosome. The DNA scene contained smooth transitions between four levels of detail, starting with individual atoms, and ending with a twisted tube representing the chromation fibre. (See [25]).

In addition to producing animation of mathematical examples and chemical simulations, I have been interested in producing realistic renderings of natural phenomena such as water, clouds, terrain, trees, and haze. The conference presentation included a film called "Carla's Island," showing an ocean scene during calm, a storm, sunset, and moonlight. The images were produced by ray tracing, so that the islands, clouds, sun, and moon reflected realistically in the water waves, which moved according to the laws of hydrodynamics. (See figure 3). For efficiency on the Cray-1, the ray tracing computations were vectorized, and color table animation was used to extend one cycle of the wave motion through the color changes during sunset and moonset. (See [26]).

Most recently, I have been working on algorithms simulating the scattering of light through clouds or haze. (See [27] and [28].) Animation illustrating these algorithms at the conference showed sunlight filtering through clouds lighting up the haze below, and candlelight shining beams through the holes in a jack-o-lantern. These algorithms assume a single-scatttering model [29] for the light diffusion by the haze. More accurate, but more expensive, models are described in [30] and [31].

It is difficult to convey here in words the content and appearance of my computer animation. Therefore, I append below a list of my films, and where they can be obtained. Those marked with an asterisk contain scenes shown at the conference.

Available in 16mm or video from

International Film Bureau
332 South Michegan Avenue
Chicago, Illinois 60604
(312) 427-4545

Figure 3. A frame from the film "Carla's Island," showing the reflection of the sunset in the moving water waves, rendered by ray tracing.

Topology Film Series

* 1.	Space Filling Curves	25 1/2 min	sound
2.	Regular Homotopies in the Plane, Part 1	14 min	sound
3.	Regular Homotopies in the Plane, Part II	18 1/2 min	sound
* 4.	Turning a Sphere Inside Out	23 min	sound

Topology Short Film Series

5.	Limit Curves and Curves of Infinite Length	14 min	silent
* 6.	Sphere Eversions	7 1/2 min	silent
* 7.	Limit Surfaces and Space Filling Curves	10 1/2 min	silent
8.	Sierpinski's Curve Fills Space	4 1/2 min	sound
* 9.	Zooms on Self-Similar Figures	8 min	sound
10.	The Butterfly Catastrophe	4 1/2 min	silent

The International Film Bureau also distributes Tom Banchoff's film "The Hypercube."

Available in U-Matic or VHS video from

SIGGRAPH Video Review
ACM Order Dept.
P.O. Box 64145
Baltimore, MD 21264
(800) 526-0359 ext. 75

*11.	DNA with Ethidium	Issue 1 # 7	silent
12.	Doxorubicin with DNA	Issue 2 # 5	silent
*13.	Carla's Island	Issue 5 # 3	sound
*14.	Light Beams	Issue 25 # 9	sound

Available in 16mm from

John Blunden
Lawrence Livermore National Lab
P.O. Box 808
Livermore, CA 94550

*15. DNA excerpt from "We are Born of Stars"   3 min   silent

*16. Computer animated excerpt from "Inertial
      Confinement Fusion"   2 min   silent

*17. Hartree Fock Calculation   10 min   silent
*18. Tomato Bushy Stunt Virus   4 min   silent

The Topology Films Project was supported by the National Science Foundation grants GY-7699, HES70-3128 A03, and SED75-15112 A01. "We are Born of Stars" was supported by Fujitsu Limited. The rest of the films were produced under the auspices of the U.S. Department of Energy by Lawrence Livermore National Laboratory under contract number W-7405-ENG-48.

References

[1] Smale, Stephen, "A classification of immersions of the two-sphere" Trans. Amer. Math. Soc., Vol. 90 (1959), p. 281.

[2] Phillips, Anthony, "Turning a Sphere Inside Out" Scientific American, Vol. 214 (May 1966), p. 112.

[3] DeWitt, Bryce, "Eversion of the 2-sphere," in "Battelle Rencontres," Cecile DeWitt and John Wheeler, eds., W.A. Benjamin, Inc., New York (1968), p. 546.

[4] Max, Nelson, "Le retournement de la sphère et le cinéma informatique" La Recherche, No. 122 (May 1981), p. 630 .

[5] Max, Nelson, "Turning a Sphere Inside Out," a guide to the 16mm color film "International Film Bureau, Chicago (1976). This volume, p. 334.

[6] Morin, Bernard and Petit, Jean Pierre, "Le Retournement de la Sphère" Pour la Science, Vol. 15 (1979), p. 34.

[7] Max, Nelson and Banchoff, Tom, "Every Sphere Eversion has a Quadruple Point" in "Contributions to Analysis and Geometry," D. N. Clark, G. Pecelli, and R. Sacksteder eds., John Hopkins Univ. Press, Baltimore (1982), p. 191.

[8] Hughes, John, "Another proof that every eversion of the sphere has a quadruple point." Amer. Journ of Math., Vol. 107 (1975), p. 501.

[9] Armit, Andrew, "Multipatch and MultiObject Design Systems" Proc. Royal Soc. London, Vol. A 321 (1971), p. 235.

[10] Max, Nelson and Clifford, William, "Computer Animation of the Sphere Eversion," Computer Graphics, Vol. 9, No. 1 (1975), p. 32.

[11] Mandelbrot, Benoit , "The Fractal Geometry of Nature" W.H. Freeman Press, San Francisco (1982).

[12] Voss, Richard, "Fourier synthesis of gaussian fractals: 1/f noises, landscapes, and flakes" in Tutorial on State of the Art Image Synthesis, ACM SIGGRAPH 83 course notes, Vol. 10.

[13] Oppenheimer, Peter, "Real Time Design and Animation of Fractal Plants and Trees" Computer Graphics, Vol. 20, No. 4 (1986), p. 55.

[14] Fournier, A., Fussel, D., and Carpenter, L., "Computer Rendering of Stochastic Models, " Communication of the ACM, Vol. 25 (1982), p. 38.

[15] Johnson, C. K., "ORTEP-II, a Fortran Thermal-Ellipsoid Plot Program for Crystal Structure Illustrations," ORNL-5138, third revision, Oak Ridge National Laboratory, Tenn., (1976).

[16] Porter, Tom, "Spherical Shading," Computer Graphics, Vol. 12, No. 3 (1978), p. 282.

[17] Max, Nelson, "ATOMLLL--ATOMS with Shading and Highlights," Computer Graphics, Vol. 13, No. 2 (1979), p. 165.

[18] Connolley, Michael, "Solvent-accessible surfaces of proteins and nuclei acids," Science, Vol. 221, (1983), p. 709.

[19] Max, Nelson, "Computer Representation of Molecular Surfaces," Molecular Graphics, Vol. 2, No. 1 (1984), p. 8.

[20] Max, Nelson , "Atoms with Transparency and Shadows" Computer Vision, Graphics, and Image Processing, Vol. 27 (1984), p. 46.

[21] Knowlton, Ken and Cherry, Lorinda "ATOMS, A Three-d Opaque Molecule System for color pictures of space-filling or ball-and-stick models," Computer and Chemistry, Vol. 1, (1977), p. 161.

[22] Max, Nelson, "We are Born of Stars," IEEE Computer Graphics and Applications, Vol. 5, No. 11, (November 1985), p. 4.

[23] Max, Nelson, "Computer Graphics Distortion for IMAX and Omnimax Projection," Nicograph '83 Proceedings, Nihon Keizai Shimbun, Inc., Tokyo (1983), p. 137.

[24] Max, Nelson, and Lerner, Douglas, "A two-and-one-half-D Motion Blur
      Algorithm,"Computer Graphics, Vol. 19, No. 3 (July 1985), p. 85.

[25] Max, Nelson, "DNA animation, from atom to chromosome," Molecular Graphics,
      Vol. 3, No. 2 (1985), p. 69, with correct figures in Vol. 3, No. 3.

[26] Max, Nelson, "Vectorized Procedural Models for Natural Terrain: Waves and Islands
      in the Sunset" Computer Graphics, Vol. 15, No. 3 (1981), p. 317.

[27] Max, Nelson, "Light Diffusion through Clouds and Haze," Computer Vision,
      Graphics, and Image Processing, Vol. 13 (1986), p. 280.

[28] Max, Nelson "Atmospheric Illumination and Shadows," Computer Graphics, Vol. 20,
      No. 4 (1986), p. 117.

[29] Blinn, James "Light Reflection Functions for Simulation of Clouds and Dusty
      Surfaces," Computer Graphics, Vol. 16, No. 3 (1982), p. 21.

[30] Kajiya, James, and Von Herzen, Brian, "Ray Tracing Volume Densities," Computer
      Graphics, Vol. 18, No. 3 (1984).

[31] Kushmeier, Holly and Torrance, Kenneth, "The Zonal Method for Calculating Light
      Intensities in the Presence of a Participating Medium," submitted to Computer
      Graphics, for SIGGRAPH '87 proceedings.

APPENDIX

Turning a Sphere Inside Out: A Guide to the Film

distributed by

‾INTERNATIONAL FILM BUREAU INC. 332 South Michigan Avenue, Chicago, Illinois **60604**

Material for this guide was prepared by Nelson L. Max, Case Western Reserve University, and Project Director of the Topology Films Project, Education Development Center, Newton, Massachusetts.

Other films in THE TOPOLOGY SERIES include:
SPACE FILLING CURVES
REGULAR HOMOTOPIES IN THE PLANE:
    PARTS I & II

INTRODUCTION

TURNING A SPHERE INSIDE OUT opens with a discussion of the problem of turning a sphere inside out by passing the surface through itself without making any holes or creases. Mathematicians believed that the problem was insoluble until 1958 when Stephen Smale proved otherwise. However, no one could visualize the motion, called a regular homotopy. The homotopy in this film was developed by Bernard Morin, a blind mathematician. It is illustrated with a sequence of chicken-wire models, built by Charles Pugh, showing the crucial stages in the motion. Mathematicians Nelson L. Max, Stephen Smale, and Charles Pugh, and physicist Judith Bregmann provide

the commentary. The film closes with several different sequences of computer animation revealing the continuous motion of the sphere.

Crucial stages in this motion are given in the discussion following, with figures placed above the text referring to them. The inner surface of the sphere has been shaded.

TURNING A SPHERE INSIDE OUT

Imagine that the surface of the sphere is made out of rubber like a hollow ball. If we make a hole in the ball, we can turn it inside out through the hole, but we have destroyed the surface. At the intermediate stages, it is no longer a complete sphere.

If we do not allow any holes, the problem becomes impossible; what is inside must remain inside. Therefore, instead of allowing holes, we allow the surface to cross itself. A physical sheet of rubber can never cross itself, but a mathematical surface in space can—just as a curve drawn on a piece of paper can cross itself. A position of the surface is merely a function from a round sphere into three-dimensional space; if the surface crosses itself, this only means that the function is not one-to-one.

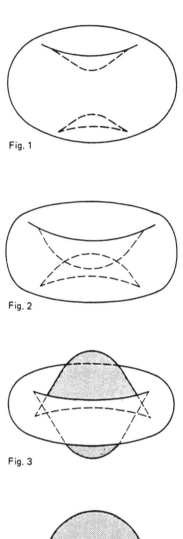

Fig. 1

Fig. 2

Fig. 3

Fig. 4

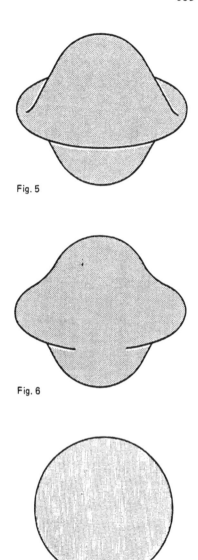

Fig. 5

Fig. 6

Fig. 7

The north pole can be pushed down past the south pole so that the sphere intersects itself along a circle and the inside of the surface near the south pole becomes visible from above. As the surfaces are pushed through each other, the circle of intersection grows.

Eventually the intersection circle reaches the equator and forms a crease there. Then the surface opens up into a round inside out sphere.

The crease is disagreeable because the surface is not entirely smooth; the function defining the surface is not differentiable. So we add another rule stating that every surface during the motion must be smooth. Mathematically this means that the partial derivative vectors must be continuous and linearly independent. Geometrically it means there is a tangent plane at every point on the surface. There is no tangent plane to a surface at a crease.

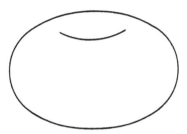

Fig. 8

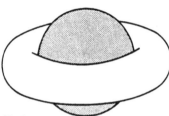

Fig. 9

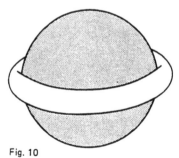

Fig. 10

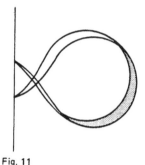

Fig. 11

The motions we have described so far all involved surfaces of revolution, obtained by revolving half of a cross section curve around a vertical axis. None gave regular homotopies, since each violated one of the rules.

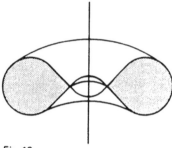

Fig. 12

Is there a regular homotopy which turns the sphere inside out through surfaces of revolution? If there were, each cross section through the axis would be a smooth curve in the cross section plane, and we would have a regular homotopy in the plane which turns a circle inside out.

The Whitney-Graustein theorem, discussed in the films REGULAR HOMOTOPIES IN THE PLANE: PARTS I and II, proves that the circle cannot be turned inside out by a regular homotopy. Therefore, the sphere cannot be turned inside out in space by a regular homotopy using only surfaces of revolution. This fact led mathematicians to believe that it was impossible to turn a sphere inside out by any regular homotopy at all. However, in 1958, Stephen Smale proved that in fact it was possible. His proof used a complicated sort of mathematical induction which pushed the surface wildly back and forth at each step. Although the construction of a regular homotopy was, in principle, contained in his proof, no one could visualize it.

Once they knew it was possible, a number of people invented homotopies; among them was Arnold Shapiro, René Thom, Anthony Phillips, Marcel Froissart and Bernard Morin. Anthony Phillips, wrote an article in the May, 1966 issue of *Scientific American* giving a history of the problem and a sequence of illustrations for the regular homotopy. The one illustrated here is somewhat simpler and was invented by Bernard Morin, a blind French mathematician who described it using the ideas of generic surfaces and singularities.

There is a tricky way around this rule. Near the equator we make a smaller and smaller round tube which disappears at the last instant. At every stage we have a smooth surface, even when the tube disappears in a "kink." This is the three dimensional version of the method of turning a figure eight into a circle by making a little loop disappear, as shown in REGULAR HOMOTOPIES IN THE PLANE: PART I. We prohibit this too.

A motion which satisfies all the rules—no holes, creases, or "kinks"—is called a regular homotopy. Our problem is to describe a regular homotopy which turns the sphere inside out.

GENERIC AND SINGULAR POSITIONS

A generic surface is one which will look basically the same if it is moved slightly. For example, if a surface intersects itself in a circle, and we move it to a slightly different position, it will still intersect itself in a nearby curve, which may not be exactly circular. If we imagine the surface imbedded in solid rubber or gelatin, we could move the gelatin from one position to the other and carry the surface along. Two positions of a surface are called topologically equivalent if they can be moved one to the other by such a pushing or twisting of all of space. So a position of a surface is called generic if all small motions give nearby surfaces which are topologically equivalent to it.

For example, the positions shown in Figures 2, 3, and 4 are all generic, as are all the stages between them. As a consequence, the positions in Figures 2 and 4 are topologically equivalent. Similarly all the positions before the north and south poles touch are topologically equivalent.

However, at the instant the north and south poles touch, the topology changes. If we pull the north and south poles apart, there will be no self-intersection, and if we push them past each other, there will be a circle of self-intersection. These are not equivalent to the position with one point of tangency, which is therefore not generic. We call the point of tangency a singular point, the position a singular position or singularity.

In a regular homotopy, the singular positions will in general occur at a finite number of special instants, at which the nature of the self-intersection changes. Between these instants, the surfaces are merely pushed or twisted by topological equivalences. Thus, a regular homotopy can be described in terms of a list of its singular positions, such as the one given at the end of these notes. Before we start on this list, let us look at standard models for the four types of singular points which occur in it.

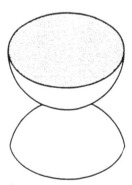

Fig. 13

We have already seen one type, a tangency at which a circle of self-intersection is created. A standard model for this is two curved bowls, tangent at one point.

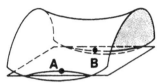

Fig. 14

The second sort of singular point occurs when an isthmus of land is flooded by a rising tide. In the standard model above, the isthmus is represented by the saddle shaped surface, and the sea level by the horizontal plane. Points A and B are on the shoreline of the isthmus at its narrowest part.

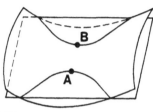

Fig. 15

Here is a view from the top, showing the shoreline as two arcs passing from left to right through the points A and B.

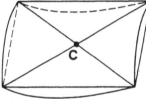

Fig. 16

When the sea rises to the low point, C, of the pass on the isthmus, it is tangent to the saddle shaped surface there. The two shorelines through A and B have met at C. This position is not topologically equivalent to those just preceeding it, so it is singular, and C is the singular point.

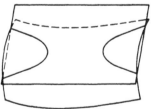

Fig. 17

As the water level rises still more, the two regions of land are separated by a shallow channel of water. There are again two arcs of shoreline representing the surface intersections, but they connect differently, and now run from top to bottom.

For our next singularity, we consider a triple point where three surfaces intersect. If the surfaces intersect like the two walls and ceiling at the corner of a room, then the triple point is not singular, because the surfaces would intersect in a topologially equivalent way if they were moved slightly.

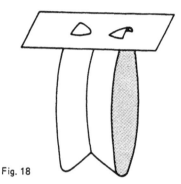

Fig. 18

But consider the picture above, showing two bowls with a vertical intersection circle, and a horizontal plane just above this circle.

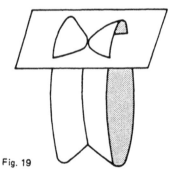

Fig. 19

If this plane is lowered until it is tangent to this circle, we get a singular triple point, where three intersection curves are tangent.

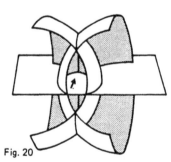

Fig. 20

If the plane is lowered still more, this singular triple point separates into two non-singular triple points, where the three self-intersection curves cross. One of them is shown at the arrow in the cut away picture above.

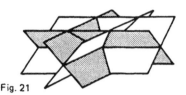

Fig. 21

The final sort of singularity is represented by a quadruple point, where four surfaces intersect, as in the picture above.

This quadruple point is singular, since if any of the planes were moved a little, it would separate into four triple points, forming the vertices of a small tetrahedron.

MORIN'S REGULAR HOMOTOPY

Now we are ready to study Morin's regular homotopy in terms of these singularities. Here are the first few stages, shown in cross section.

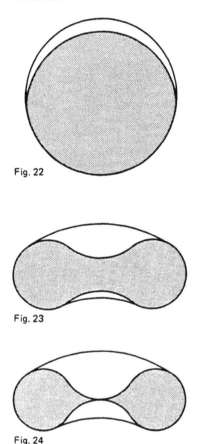

Fig. 22

Fig. 23

Fig. 24

The first singular point occurs when the north and south poles touch, and has Figure 13 as its standard model.

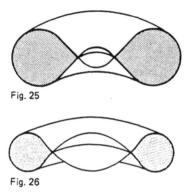

Fig. 25

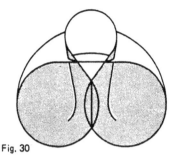

Fig. 30

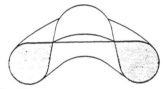

Fig. 26

and pass through each other, forming a vertical circle of self intersection. This is the same sort of singular point, represented by the model in Figure 13 turned sideways.

A horizontal self-intersection curve is created at this singular point, and then grows larger.

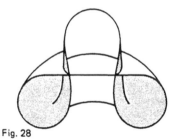

Fig. 27

The next stage in the homotopy involves twisting the surface so that it looks like a hat,

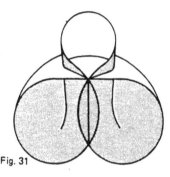

Fig. 31

In the next few stages, the second vertical circle of self-intersection continues to expand as the sides are pushed in. It soon touches the horizontal plane at a singular triple point like the one in Figure 19.

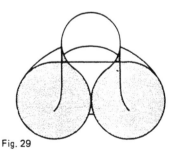

Fig. 28

and bringing the sides inward

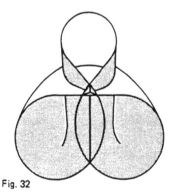

Fig. 32

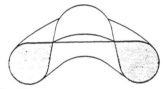

Fig. 29

so that they touch,

Pushing in the sides has the same effect as lowering the horizontal plane: Two non-singular triple points are created, as in Figure 20.

Fig. 33

Here is a perspective view of the surface so far, viewed from the side.

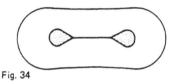

Fig. 34

Here is the same surface viewed from below. The two shaded areas are part of the back of the horizontal surface. We call them "eardrums" and call the tubes around them "ears" or "earlobes." There are two mirror planes of reflection symmetry, perpendicular to the page in this view. The surface is cut by them into four symmetrical quarters. The horizontal segment shows an arc of the "vertical" intersection circle.

Fig. 35

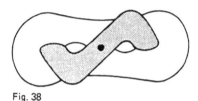

Fig. 36

The next stage involves twisting of the surface. The tube across the top grows longer, and the horizontal section becomes twisted so that the two "ears" on the sides lie in different planes. The twisted shapes are all topologically equivalent.

Fig. 37

Now that the twisting starts, the mirror symmetry is destroyed. Instead, there is a vertical axis of two-fold rotation symmetry; the surface is made up of two congruent halves, differing by a 180° rotation.

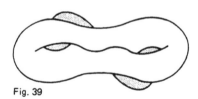

Fig. 38

Here is a top view, showing the twisted tube. The dot shows the position of the two-fold rotation axis. A rotation of 180° about this axis will take the set of points on the surface into the same set.

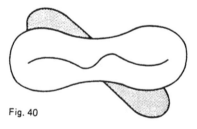

Fig. 39

Here is a view from the bottom, showing the two "ears" being twisted.

Fig. 40

Here the "ears" are twisted still more. You can no longer see the "eardrums." The bottom arc of the vertical self-intersection circle has become quite curved.

Fig. 41

Here is another side view. Notice that point A from the originally vertical intersection circle has moved close to point B which is from the originally horizontal intersection circle.

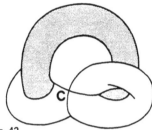

Fig. 42

Here is the next stage, an isthmus singularity as in Figure 16. The surfaces have become tangent at point C where A and B come together. A similar situation takes place on the other side of the model.

Fig. 43

If we go slightly past this singularity, the surface which previously separated points A and B is now partly hidden, and the other surface is now continuously visible.

The next stages are all topologically equivalent.

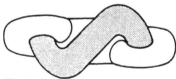

Fig. 44

In the top view we can see that the tube where the inside surface shows has been lengthened and further twisted.

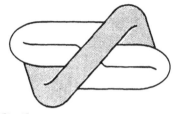

Fig. 45

Now the ears have been twisted more than 180°, and we can see a little into the holes from the top view, instead of from the bottom view. Also, the two earlobes line up to form a new tube, running under the old tube and back out again.

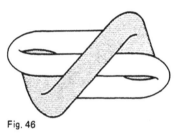

Fig. 46

Here is still further twisting. Small pieces of the eardrums are visible. Also the new tube now runs straight across, and even intersects, the old tube. We have passed a new singularity, which can be seen better from the side.

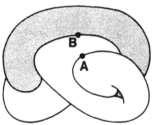

Fig. 47

Here is a side view again, slightly twisted from Figure 43. The top of the earlobe A continues into the surface and comes back out as the corresponding earlobe on the back side. Together they form another tube which has existed in a twisted form ever since the "isthmus" singularities were passed.

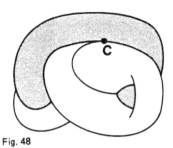

Fig. 48

This new tube rises and the old tube sinks, so that the points A and B meet in a new triple point C. The horizontal line running from left to right through C is tangent to all three surfaces at C; this is another singular triple point.

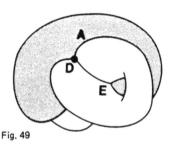

Fig. 49

As the new tube rises still further, C separates into two triple points. One is labeled D in the picture above. The other is hidden behind the earlobe. There are now four triple points which lie at the vertices of a curved tetrahedron. One edge of this tetrahedron is the arc DE.

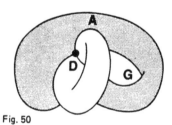

Fig. 50

As the tube containing A rises still further, this tetrahedron decreases in size and becomes proportionately less curved. In addition, further twisting takes place so that we can no longer see the eardrum containing E. However, we see a new eardrum G formed as the tube A cuts off a piece of the surface below B.

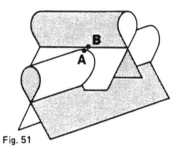

Fig. 51

The singularity may be easier to see if we just consider the pieces of tube near A and B and straighten them so that they are only curved in one direction. They could then be rolled from pieces of paper, if paper could pass through itself. Here is the stage just before the singularity.

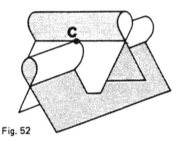

Fig. 52

Here the top of tube A has risen to become tangent to the straight line of intersection of the tube B. Thus all three surfaces are tangent to this horizontal line at the singular triple point C.

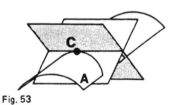

Fig. 53

Here is a close-up of the pieces of surface near C. This does not look much like the standard model in Figure 19 for such a triple point,

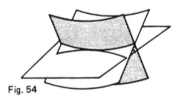

Fig. 54

but if we twist the picture by flattening the curved surface A and therefore curving the other two surfaces, we get a picture like the one above. This now looks more like an upside down version of the piece of our standard model near the triple point, in Figure 19.

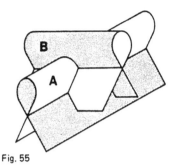

Fig. 55

If tube A rises still farther with respect to tube B, two new triple points are created. The four triple points form a tetrahedron.

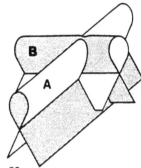

Fig. 56

As the tube A rises, this tetrahedron gets smaller. Also, as the top of tube A gets higher, more of it is visible. The place at the top of the tubes where their two intersection curves come together looks like the coastline of the isthmus. In fact, if we curl up the picture so the top of tube A flattens out and the top of tube B bends into a saddle shape, we can then get the standard model in Figure 14. This is similar to what we did with the triple point in Figure 53.

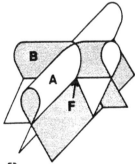

Fig. 57

When the two tubes are at the same level, they intersect as in the vault where two semicircular arches cross in Roman architecture. At the same time, the four triple points meet at a quadruple point F, where four planes intersect.

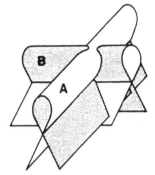

Fig. 58

When tube A (the sea) rises above tube B (the land), the intersection curves (the shoreline) near the top are joined in the opposite way. The quadruple point also opens again into a tetrahedron of the opposite orientation.

Fig. 59

Here is the same thing taking place on the whole sphere. The two tubes are now curved instead of straight, but the singularities take place the same way.

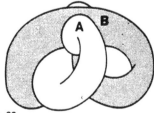

Fig. 60

As the two tubes approach each other, their top lines of intersection also approach until they join, forming the vaulted arch.

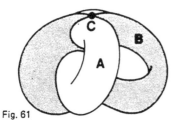

Fig. 61

The top point C is then a place where both surfaces have the same tangent plane. The model at this stage has four-fold symmetry, because a rotation of 90° about the vertical axis through C takes the model into the same shape.

However, this rotation reverses the inside and outside surfaces, because tube A (which came from the brim of the hat) was the original outside surface and tube B (which came from the top of the hat) had the inside surface visible. The sphere is now halfway inside out, because just as much of the inside surface is showing as the outside.

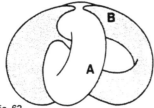

Fig. 62

When tube A rises still farther, it lies above tube B, which now intersects it in the same way tube A intersected tube B previously. If it continues to rise, we can see the rest of the process in reverse, with two of the triple points disappearing.

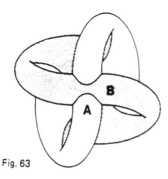

Fig. 63

The four-fold symmetry is easier to understand if the surfaces are viewed from above. Here is the surface just before the halfway stage, with the tube A below the tube B. You can see little pieces of the eardrums.

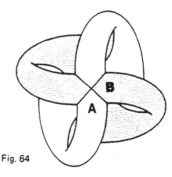

Fig. 64

Here is the halfway stage. A 90° rotation will interchange tubes A and B, and therefore the inside and outside surfaces. There are four equivalent ears.

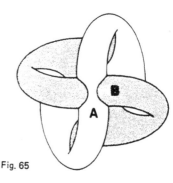

Fig. 65

Here is the surface just after the halfway stage. A 90° rotation of this picture produces the picture preceeding the halfway stage, with the outside tube A in the position of B, where the inside surface shows.

Thus the rest of the regular homotopy continues from the halfway stage in reverse, and the homotopy is symmetrical in time, by switching inside and outside. When all the steps are repeated, we will get a round, inside-out sphere.

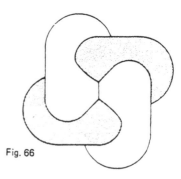

Fig. 66

Here are the same three stages, viewed from the bottom. When tube A is below tube B, a short segment of its intersection curve is visible from the bottom. It appears vertical in this view. This segment is the one visible edge of the tetrahedron formed by the four planes which slant inward toward the center. Its four vertices are the triple points where three of the planes meet.

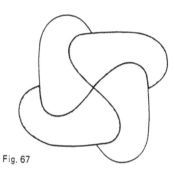

Fig. 67

When the two tubes are at the same level, the vertices of the tetrahedron come together in a single point, the quadruple point. This surface now has four-fold symmetry.

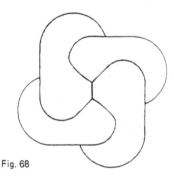

Fig. 68

When tube A is above tube B, B is closer to us as we view the model from below, and the small horizontal segment is a piece of its intersection curve. The quadruple point has split into four triple points again.

The motion then continues through the rest of the stages in reverse.

Altogether there are 14 singularities:

1) the creation of the first intersection circle,

2) the creation of the second intersection circle,

3) the creation of the first two triple points,

4&5) the simultaneous flooding of the two isthmuses,

6) the creation of the second two triple points

7&8) the simultaneous quadruple point and isthmus at the halfway stage,

9) the merging and disappearance of two triple points,

10&11) the exposing of two isthmuses,

12) the merging and disappearance of the last two triple points,

13) the disappearance of one circle of intersection, and

14) the disappearance of the last intersection circle.

No regular homotopy has been discovered with less than 14 singularities, so this one is among the simplest known.

References:

Smale's paper[1] first proved, using induction, that the sphere could be turned inside out in three dimensional space, and in fact, classified the positions of the ordinary sphere in a euclidean space of any dimension. The ideas in this paper were derived from an earlier one,[2] which studied regular homotopies of curves on an arbitrary surface, and were generalized in a later paper,[3] which considered spheres of arbitrary dimension. These three papers are rather technical, and the basic idea is explicated in the French mimeographed notes[4] of René Thom.

A more popular presentation[5] by Anthony Phillips gives a series of illustrations of a different regular homotopy.

[1] Stephen Smale, "A Classification of Immersions of the Two-sphere," *Transactions of the American Mathematical Society*, 90 (1959), 281.

[2] Stephen Smale, "Regular Curves on Riemannian Manifolds," *Transactions of the American Mathematical Society* 87 (1958), 492.

[3] Stephen Smale, "The Classification of Immersions of Spheres in Euclidean Space," *Annals of Mathematics* (2) 69 (1959), 327.

[4] René Thom, "La Classification des Immersions," Seminaire Bourbaki Exposé 157 (December 1957), Secrétariat Mathématique, Paris (1958).

[5] Anthony Phillips, "Turning a Surface Inside Out," *Scientific American* 214 (May 1966), 112.

The figures used in this guide were executed by Michael Hemby and William Urian.

To rent or borrow films, apply to your audiovisual director or your nearest media center.

To preview with a view to purchase, and to purchase, write:

INTERNATIONAL FILM BUREAU INC. 332 South Michigan Avenue, Chicago, Illinois 60604

Prices quoted in this guide apply in U.S. only. In Canada, apply to: **EDUCATIONAL FILM DISTRIBUTORS LTD.**
285 Lesmill Road, Don Mills, Ontario M3B 2V1

In the United Kingdom, apply to: **EDWARD PATTERSON ASSOCIATES LIMITED**
68 Copers Cope Road, Beckenham, Kent, England

Physicists and Computer Algebra

TULLIO E. REGGE Institute of Theoretical Physics, Torrino, Italy

I have now listened to many talks on computer algebra and related applications to many outstanding problems in science. My personal experience with computer algebra is rather limited but I have fairly clear ideas about my particular needs in research and about the needs of my colleagues. My talk is therefore essentially a shopping list; I have little to offer and much to ask.

After these preliminaries let me offer my first comments about what I have seen and listened to. When I came as a student to the United States, some firm was making an awful amount of money by selling ballpoint pens with the catchword "It writes underwater." Over many years I have never seen anybody writing underwater, and it reminds me anyway of Samuel Johnson and his remarks about dogs walking on their hindlegs.

How much of this computer algebra is "writing underwater?" Calculations involving the manipulation of enormous polynomials are certainly amazing but this does not prove that they are useful or needed. Only long experience, many disappointing failures, but also success will provide enough judgment to rule out the possibility that we are writing underwater. A second impression is, of course, that much programming is oriented toward pure mathematics. I have nothing against mathematics; I have always enjoyed it and envied those who can master it at the highest levels. Here,

however, I am speaking as a physicist and I must confess that I find
physics conspicuously absent at this meeting. For instance, the so-called
Schooship system, which proved itself to be very useful in physical applications and which scored many successes in quantum electrodynamics, is not
represented here.

This last remark brings me to the central theme of my talk. What do
we really want from computer algebra? I will list a few requests but it
should be clear that different physicists may have different needs.

If you read any textbook in theoretical physics you shall immediately
be struck by the large amount of information concerning the rotation group
SO(3) or rather SU(2). In particular, we have now detailed formulae concerning the Clebsch-Gordan decomposition, related tables, and all sorts of
useful machinery to deal with the most exhaustive computational needs (see,
for instance, the 6-j or Racah coefficients). Nothing comparable exists
for other groups, even finite groups. One reason for this is clear: SU(2)
is the smallest nonabelian simple Lie group, it is simply reducible, and
physics without SU(2) could not be imagined. For instance, SU(3) is enormously more complicated and not so useful. Yet the achievement of similar
total control and user friendly facility in another group would be highly
desirable. When I prepared this talk I did not know of any sizable effort
in this direction but now I have learned of many groups that are active in
Lie groups and deserve careful attention.

Related to this subject is the wasteland of special functions. I con-
sult regularly the now famous five books of the Bateman project. They do
not contain any deep mathematics but they are nevertheless extremely useful.
It is a fact that many special functions quoted and used in the literature
owe their special properties to their role in the theory of group representations (see, for instance, Vilenkin's book on this subject).

The Bateman project is the end product of the random growth of the
subject with a gradual superimposition of uncorrelated notations and as
such it is highly inefficient. Computer algebra should be used to simplify
the notation, generalize the final results, and make it easier to discover
old results which are potentially useful. I want an optical disk version
of the Bateman project, not a copy but a highly revised and reorganized
version, with information retrieval, and display of the whole subject of
special functions, and its relation to other chapters of modern mathematics.

Let me go now into the realm of tensor calculus. I am aware that
there are many implementations of it in computer algebra; some of them are

already quite old. Any tensor package should have, of course, facilities
to implement spinor calculus, including gamma matrices algebra in any dimension. A major stumbling block in dealing with spinors is, however, the
occurrence of the so-called Fierz identities involving polynomials of degree at least 3 or 4 in spinor fields defined in higher dimensional spaces.
These identities have a distinct group-theoretical flavor and in fact are
related to the Clebsch-Gordan series for spin groups. They are usually
very complicated and messy. The construction of the first supergravity
theory was long delayed by the need to prove just one such Fierz identity.
I want a facility with a Fierz identity chapter.

But also I would like a tensor package in which the dimension not only
is arbitrary but does not need to be declared beforehand. Many computations,
such as traces or raising or lowering of indices, do not need an a priori
knowledge of the dimension and should be excised in a separate package.
The advantage would be enormous. For instance, the techniques of dimensional renormalization need a dimension which is a complex variable and
have been extensively used in particle theory.

A related package should include a broad spectrum of computational
tools in analytical mechanics (Poisson brackets) and in quantum mechanics
where the need to manipulate long expressions involving noncommuting variables and field components is often a major stumbling block in discussing
theories of great physical relevance. One example is the ordering problem
in quantum gravity, yet unsolved.

I finish with a question which is no longer very relevant to physics
but was inspired by a physical problem. Let S be a simplex in four-dimensional euclidean space, having therefore five vertices, ten sides, ten
faces or plaquettes, five tetrahedral cells. The shape of S is completely
determined, modulo congruences, by the ten sides. An old formula by Cayley
or just the use of Gram determinants allows us to calculate from these
sides all areas, volumes, and angles that we can possibly need. It stands
to reason that it should be possible to express the same quantities directly
in terms of the ten areas of the plaquettes, whose squares are polynomials
in the square sides. The actual expression can be extremely complicated
and is certainly beyond manual computation and certainly beyond the primitive PC computer algebra packages available to me. I am not satisfied by
an existence theorem; if computer algebra has any use it lies in the actual
computation and display of formulae. Can one write directly the (square)
measure of S in terms of the ten areas of the plaquettes? In principle

this is certainly possible by going through a monstrous sequence of algebraic manipulation of polynomials and resolvents. But the final answer
may be relatively simple.

Computer algebra is here to stay and there is no way to go back; in a
few years new monstrously large and cheap memories will contain an immense
variety of computational tools and will make it accessible to a personal
computer. The problem is how to manage it efficiently. Quite possibly
most manuals on this subject have been written by secret agents of the competition and as a rule they are remarkably unclear, user unfriendly, and
smartproof. I cannot conceive spending one week in learning the ropes on
a frustrating manual in order to save a couple of days hard work. I prefer
indoctrinating a deserving student. Easy access is a must in this type of
endeavor. Most probably in the near future computer algebra will be a
routine facility implemented directly into the gigabyte ROM of a small PC.
At the moment it is still in the pioneer stage, relatively expensive or
rather still a steep function of investment. From this point of view, it
may introduce an additional discrimination between rich and poor institu-
tions. I find this intolerable and an inefficient and immoral use of our
human and intellectual resources. Also I fear a proliferation of different
standards and a useless duplication of efforts. The simultaneous existence
of a West Coast and an East Coast metric tensor is already too much to
bear as a source of endless confusion in theoretical physics. Please save
me the horror of a Mexican theta function versus a Canadian one. Let us
call an international meeting where we try to agree on some consistent use
of mathematical symbols before it is too late.

Symbolic Computation: The Early Days (1950-1971)

JEAN E. SAMMET   Federal System Division, IBM, Bethesda, Maryland

1.   INTRODUCTION

This is a brief and informal introduction to some of the early work that
occurred in symbolic computation. It is an indication of some historical
highlights as perceived by the author, and pointers to some early references that are probably more obscure by now. It is not meant to be a detailed description of the history of even the early days of this field.
The choice of the end date of 1971 is to permit inclusion of material on
the second major conference held on this topic.

Section 2 covers some examples of the earliest work in this field
during the 1950s. In Section 3, which is devoted to a description of the
field during the early and mid-1960s, a number of problems and technical
issues are delineated. The first *language* to be widely used—namely,
FORMAC—is described in Section 4, to provide some idea of the capabilities
of one significant early system. Since one of the early problems in this
field was the dissemination of information, Section 5 discusses this problem and describes some of the early conferences and working meetings.
Section 6 lists the systems being used by 1971. The last section, Section
7, draws a few obvious conclusions.

One of the ironic aspects of this activity is that the terminology
keeps changing; that is of course quite common in the computer field but

less common in mathematics, and this work is certainly a combination and
bridge between them. The early terms that were generally used were "formula manipulation" or "formal algebraic manipulation." At later periods
equivalent terms were "nonnumerical mathematics," "formal mathematical
computation," "symbolic mathematics," and even "symbol manipulation."
The 1986 term in vogue seems to be "symbolic computation." But regardless
of what term is used, the intent is to consider those activities in which
some aspect of *formal* mathematics is being done on a digital computer.
The word *formal* is emphasized to distinguish this work from numerical
analysis.

A very detailed annotated bibliography covering work up until 1966
is in Sammet (1968).

2. VERY EARLY PROGRAMS

I cite a few examples of what appear to be the earliest programs in this
general area. According to Cannon (1969), the earliest use of computers
for group theory was for "coset enumeration and construction of character
tables of symmetric groups." This was done as early as 1953 by Haselgrove
on the EDSAC-1 according to Cannon, who also states that "Probably.the
first reference to computing with groups is Newman (1951)."

In 1954, there were two Master's theses written to do formal differentiation. One was by J. Nolan at MIT on the Whirlwind, and the other by
H. G. Kahrimanian at Temple University on UNIVAC I. From the perspective
of today's work, people may be aghast to realize that people indeed obtained a Master's thesis from good universities for work that is now sometimes given to freshman students as a homework exercise. But remember
that in 1953 there was no LISP, and very little software of any kind. Doing formal mathematics on a computer was truly unheard of!

Clearly, one of the milestones in the broad area of symbolic mathematics was the work on proving theorems in the propositional calculus
(Newell and Shaw, 1957). These people literally took theorems from the
Russell and Whitehead *Principia Mathematica* and, using their new programming language IPL-V, coded the axioms and rules of deduction and had the
program proving theorems, and even learning a little about how to prove
them more efficiently.

One of the areas that the computer was used for in the early days
was to look at varying conjectures in number theory. A conjecture of

that period was that the following terms were all primes:

$$2 + 1, \quad 2^2 + 1, \quad 2^{2^2} + 1, \quad 2^{2^{2^2}} + 1$$

This was proved false on the SWAC in 1954 (Lehmer, 1963). The fifth term is indeed factorable.

3. EARLY TO MID-1960s

The most commonly used broad term during the first half of the 1960s was "nonnumerical mathematics." There were normally three topics included under that scope: formula manipulation, "pure" mathematics, and theorem proving.

Formula manipulation referred to the use of the computer to do the type of formal mathematical computations frequently done by high school and college students in math classes, and by engineers working on real problems. These included formal differentiation and integration, expansion and factoring expressions, simplification, etc. The *"pure" mathematics* referred to attempts to use the computer for number theory and group theory. The *theorem proving* area included both theorems in pure logic (e.g., the propositional calculus work referred to above) but also involved proving theorems in mathematics per se.

3.1 Problems in Symbolic Computation Work

There were a number of problems in those early days in trying to deal with what is now called symbolic computation. This section will attempt to discuss some of them.

Probably the most difficult problem was to explain to individuals what was meant by formula manipulation and/or nonnumeric mathematics. Only by giving examples of high school algebra or college calculus could the point be gotten across, and it usually was extremely difficult to explain, even to sophisticated audiences. Along with the problem in explaining what was meant, it was difficult to show engineers and scientists how such facilities could be useful to them. Certainly by the late 1950s and mid-1960s, engineers and scientists were accustomed to using the computer to do numerical computations and relieve them of laborious numerical calculations that had previously been done using either slide rules or mechanical desk calculators. But it was extremely difficult to convince them that the tedious algebra that they frequently had to do by hand could also be done on a computer—and more accurately!

Another difficulty was to determine what types of problems were reasonable to use for experimentation. Although differentiation and integration were normally taught in the same college calculus course, naturally all mathematicians, and even their students, understand that there is a fundamental difference between the ease of differentiation and integration. I refer to the fact that for formal differentiation there is an established set of simple rules which will always work if applied properly. The same statement could *not* be made for integration. A similar illustration is the difference between expanding an expression—no matter how complicated it is at the start—and factoring one. Expansion will always work whereas factoring requires insight.

It is undoubtedly shocking to a mid-1980s audience to learn that one of the biggest difficulties until the mid-1960s was for one group or person to find out if and where other people were doing work in this specific technical area. Up until 1965, there was no infrastructure for nonnumerical mathematics by which people could communicate among each other, except for the few tendrils of an invisible college. There were no conferences, and there was no SIGSAM. I guarantee the accuracy of that later statement, because I founded SIGSAM (then called SICSAM) in 1965, under ACM auspices. The mere fact that individuals, whether in universities or in industry, had difficulty in knowing what type of similar work might be done elsewhere certainly inhibited moré rapid progress.

3.2 Some Technical Issues in Developing
     "Formula Manipulation" Systems

In examining this particular aspect of the field, a number of technical issues kept coming up over and over again in creating the best design for formula manipulation systems. I have been led to believe that even today there is a dispute about the best solution for some of these!

1. There were many debates about the relative merits of

a language and system integrated together

an integrated package of subroutines

a set of nonintegrated routines

(Note that in this context, the word *integrated* is used in its nonmathematical meaning.)

The use of nonintegrated routines occurred in the early days using LISP, where routines for differentiation, simplification, and so on, were developed independently and generally could not be used together. The systems ALPAK and PM were examples of an integrated package of subroutines. These permitted the user to access individual routines via calling sequences in a manner which was coordinated but not necessarily easy for the user. The first example of a useful, practical, integrated language and system was FORMAC, shortly followed by ALTRAN (a language superimposed on ALPAK). Several years later, REDUCE and MACSYMA followed the philosophy of defining a specific language.

2. There were major disputes as to whether a language—if one was to be developed—should be a unique stand-alone language developed with the needs of the users (e.g., engineers and mathematicians) in mind, or whether some existing language (e.g., FORTRAN, ALGOL) should be extended with subroutines and/or language facilities. There are examples of both from that period. An example of a separate stand-alone language was the original REDUCE and an example of an extended existing language was the original FORMAC (based on FORTRAN).

3. There was an open question as to the amount and level of functionality that should be made available in any particular system. It was at least one order of magnitude—if not more—easier to handle just polynomials, rather than full expressions which could contain transcendental functions.

Another functionality debate involved simplification. People disagreed on whether all simplification should be done by the user, or whether it should all be done by the system, or whether there should be something in between. (See a further discussion of this issue in point 5 below.)

The question of providing the functionality for integration was certainly open to question, because by the mid-1960s integration was still considered extremely difficult. Therefore the early systems certainly did *not* contain that functional capability. However, the programs SAINT and SIN, written as Ph.D. theses at MIT by J. Slagle and J. Moses, respectively, in 1961 and 1967 opened the door to eventual inclusion in useful systems. A description of this problem is in Moses (1971a).

4. The format of the output was a significant problem, but more as a result of equipment limitations than individual desires. Almost everybody could agree that the most desirable output was notation as close to what one would see in a normal mathematical textbook as possible. Unfortunately,

there was the major limitation that the primary source of fast output was
a high-speed printer that had only uppercase letters and could only print
characters in a single line unless very clever programming was done. The
use of terminals—which were only beginning to come into significant use
in the mid-1960s—made it a little bit easier because at least upper- and
lowercase letters were possible, but of course they were very slow. Figures 1 and 2 are examples of early output using just the high-speed printer;
Figure 1 is pure linear output whereas Figure 2 shows what can be done even
with just an uppercase high speed printer. The later availability of terminals provided much better results, e.g., Millen (1968).

   5. An open question was the real meaning of simplification. Not
only does the question arise as to whether the user or the system should
control it, but it is very unclear just what is really meant. Figure 3
shows a number of expressions in two forms and asks the question, "Which
is simpler?" The first few can be judged fairly readily but they get more
difficult and in reality there is no proper answer. Thus the question of
whether

$$(X + 1)^2 - X^2$$

is simpler than

$$2X + 1$$

really does *not* have an answer. It simply depends on what one is trying
to accomplish at a particular point in the problem. Similarly the question of whether the expression

$$AB + AC$$

is simpler than

$$A(B + C)$$

is entirely dependent on what one is trying to do at that stage. A discussion of the general problem is in Moses (1971b).

   6. There was an open question about batch versus interactive systems,
but again this was a function of equipment rather than desire. Not until
the mid-1960s did terminals become fairly commonplace, and screens became
readily available even later; hence if anybody was going to do work in
formula manipulation prior to then, it had to be done in batch. The first
few people that began to develop interactive systems immediately claimed
that you could not do anything productive in batch. I guarantee that you

```
-(F+F+(G+FMCSIN(I)++2.0+H+FMCCOS(I)++2.0)+T+(F+(G+FMCSIN(I))
++2.0+F+FMCCOS(I)++2.0)+2.0+F++2.0+C++2.C+FMCSIN(I)++2.0+H++
2.C+FMCCCS(I)++2.C)++5.CE-1+((G+FPCSIN(I)++2.0+H+FMCCOS(I)++
2.C)+T+2.C+(G++2.C+FPCSIN(I)++2.0+F++2.C+FPCCCS(I)++2.C)+T++
2.C+1.C)++5.CE-1+G+FPCSIN(I)++2.0+(G++2.0+FPCSIN(I)++2.C+++
2.C+FPCCCS(I)++2.C)+T++FPCCCS(I)++2.C)+(F+(F+(G+FPCSIN(I)++
2.C+++FPCCCS(I)++2.C)+2.C+F++2.C+G++2.0+FPCSIN(I)++2.C+++
2.C+FPCCCS(I)++2.C)++5.CE-1+G+FPCSIN(I)++2.C+++FPCCCS(I)++
2.C)++(-2.0)+(-F+T+1.C)+(-1.C)+((F+(C+FPCSIN(I)+FPCCCS(I)++
2.C+++FPCSIN(I)+FPCCCS(I))+(-2.C))+2.C+G++2.C+FPCSIN(I)++
FPCCCS(I)++2.C+++2.C+FPCSIN(I)+FPCCCS(I)+(-2.C))+(F+(G+
FPCSIN(I)++2.C+H+FPCCCS(I)++2.C)+2.C+F++2.C+C++++2.C+FPCSIN(I)
++2.C++++2.C+FPCCCS(I)++2.C)++(-5.0E-1)+5.CE-1+G+FPCSIN(I)++
FPCCCS(I)++2.C+++FPGSIN(I)+FPCCCS(I)+(-2.C))+(F+(F+(G+FPCSIN(
I)++2.C+++FPCCCS(I)++2.C)+2.C+F++2.C+C+C++2.C+FPCSIN(I)++2.C++
++2.0+FPCCCS(I)++2.0)++5.CE-1+G+FPCSIN(I)++2.C+++FPCCCS(I)++
2.0)++(-1.0)+(F+(C+FPCSIN(I)+FPCCCS(I)++2.C+++FPCSIN(I)++
FPCCCS(I)+(-2.C))+T+(F+(C+FPCSIN(I)+FPCCCS(I)++2.C+++FPCSIN(I
)+FPCCCS(I)+(-2.C))+2.C+C++2.C+FPCSIN(I)+FPCCCS(I)++2.C+++
2.C+FPCSIN(I)+FPCCCS(I)+(-2.C))+(F+(C+FPCSIN(I)++2.C+++
FPCCCS(I)++2.0)+2.C+F++2.C+(++2.C+FPCSIN(I)++2.C+++2.C+
FPCCCS(I)++2.C)++(-5.CE-1)+((G+FPCSIN(I)++2.C+++FPCCCS(I)++
2.C)+T+2.C+(G++2.C+FPCSIN(I)++2.C+++2.C+FPCCCS(I)++2.C)+T++
2.C+1.C)++5.CE-1++(+((C+FPCSIN(I)++2.C+++FPCSIN(I)++
2.C)+2.C+F++2.C+C++2.C+FPCSIN(I)++2.C+++2.C+FPCCCS(I)++2.C)
++5.CE-1+((C+FPCSIN(I)+FPCCCS(I)++2.C+++FPCSIN(I)+FPCCCS(I)+(
-2.C))+T+2.C+(G++2.C+FPCSIN(I)+FPCCCS(I)++2.C+++2.C+FPCSIN(I
)+FPCCCS(I)+(-2.0))+T++2.C)+((G+FPCSIN(I)++2.C+++FPCCCS(I)+
2.C)+T+2.0+(C++2.C+FPCSIN(I)++2.0+++2.C+FPCCCS(I)++2.C)+T++
2.C+1.C)++(-5.CE-1)+5.CE-1+C+FPCSIN(I)+FPCCCS(I)++2.C+(C++2.C
+FPCSIN(I)+FPCCCS(I)++2.C+++2.C+FPCSIN(I)+FPCCCS(I)+(-2.C)+
T+H+FPCSIN(I)+FPCCCS(I)+(-2.C))+(-F+T+1.C)++(-1.C)
```

Figure 1   Linear high-speed printer output.   (From *Communications of the ACM*, Vol. 9, No. 8 (August 1966), p. 595.)

```
                                              B
                                         - ----X
                                           2A
            (Y(0)A + (B + C)Y(0) - 1)E
         *
                   1             2
            COS(----SQRT(4C*A - B )X)
                  2A
      +
          X
          E
   *
       SQRT(4C*A - B )
                     2
   +
              2
       2DY(0)A  + (B*Y(0) + (2B + 2C)DY(0) - 2)A
   +
          2
       (B  + C*B)Y(0) - B
   *
        B
      - ----X
        2A         1            2
      E       SIN(----SQRT(4C*A - B )X)
                   2A

               2
(A + B + C)SQRT(4C*A - B )
```

Figure 2   Two-dimensional high-speed printer output.   (From Klerer and Reinfelds, 1968.)

1A	=	A
A + OB	=	A
X^2X^3	=	X^5
$\dfrac{(X - 1)^2}{(X - 1)(X + 1)}$	=	$\dfrac{X - 1}{X + 1}$
AB + AC	=	A(B + C)
$C\dfrac{2A}{D(2D + C)}$	=	$\dfrac{C}{D} \dfrac{A}{1 + (1/2)(C/D)}$
$(X + 1)^2 - X^2$	=	2X + 1

Figure 3   Which side is simpler?

could achieve important results using batch processing because many people
did. However, the way of thinking about the problem was entirely different
between batch and interactive, and in fact entirely different from what was
done numerically. By the mid-1960s scientists and engineers were quite
used to having to program all the steps ahead of time for the calculations
they wanted performed. But it was much more difficult to think ahead in
that level of detail when dealing with expressions whose form you can't
predict. In the numerical case you can state with some certainty that the
resulting number might be positive or negative, between 0 and 100 or 101
and 1000, and so one could chop up the range of potential answers and write
programs to deal with each of them if that was appropriate. But often when
two expressions are multiplied together the user has no idea how many terms
the result will have, nor what the expression will look like then; similarly
one might take the derivative of a very complicated expression—and the result might be a constant! The real fact of life was that while interactive
systems were certainly easier to use, there were a lot of very good results
obtained from batch systems.

   7. Performance was a major concern and the emphasis was as much on
storage requirements as on the computer time used. After all, in a batch
environment one could at least conceivably run the problem overnight. But
if the problem stopped because there was not enough core memory (and paging
systems and virtual memory were not common) that was a much more difficult
limitation to get around. It took a lot of experimentation and trial and
error to find techniques which enabled people to successfully run problems

which generated expressions which would normally exceed the available memory.

The best example of one of these problems and its solution is the "intermediate swell" problem. In this case, using a straightforward approach, expanding an expression could generate something which exceeded the memory capacity, even though the final result of the expansion could collapse into a tiny expression. (Note that this was before the days of virtual memory.) The solution generally was to avoid a brute force approach and instead perform simplification periodically throughout the expansion process.

8. The observant reader will have noticed the *lack* of a particular topic. I have not referred to anything pertaining to the development of mathematical algorithms or the use of mathematics per se in the development of the systems. Up until the mid-1960s there was very little of that being done. Most of the emphasis and attention was on the actual development of the software systems. In some cases it was a real tour de force just to get something to really run and help a person produce useful results. The mathematical rigor and better algorithms lay somewhat in the future.

9. There was then, as there still is now, a debate between the proponents of artificial intelligence and those who claimed that algorithmic approaches would suffice. And there was then, as I believe there is now, a debate and a failure to distinguish between the tools used to develop these systems and the claims of artificial intelligence. I know of two systems developed at around the same time with rather similar capabilities. One was developed using LISP and the other was developed using assembly language. The one developed using LISP was claimed to be an "artificial intelligence" formula manipulation system whereas the one developed in assembly language was simply a "system." The amount of artificial intelligence (whatever that term means) in both systems was identical, namely, zero. Nevertheless that debate frequently persisted.

4. FORMAC: FIRST LANGUAGE IN PRACTICAL USE (1964-ff)

Admitting to my own bias, since the development of FORMAC was my idea and I managed its development, I still feel it is a milestone and a landmark which is worth discussing in some detail, because it was really the first *language* in *practical* use. Prior to that, and in parallel with it, there

were many powerful *systems of subroutines*, but they were not languages.
Apparently the first language of any kind was ALGY, developed prior to
1961. A brief secondary description is in Sammet (1969), but it never
had much use or publicity.

FORMAC was a language extension to FORTRAN on the IBM 7090. It was
originally developed as an experiment, but because of great interest from
potential users we made it available externally—but without any official
support. Technically, FORMAC added to the normal facilities of FORTRAN
what was called a "formal algebraic variable" and defined some operations
on that. The motivation behind that approach was our belief that the user
needed the logical, arithmetic, and input/output facilities of FORTRAN
combined with the formula manipulation facilities that were the new feature.
FORMAC was used successfully on a number of practical problems in a batch
mode. Just one example of such usage was Howard (1967). One of the amusing things is that one of the tests that we used to help test the original
system (known as the "f and g Series") has to some extent become a classical
benchmark in measuring system performance.

The functional capabilities of FORMAC included the following:

Substitution of an expression for a variable
Expansion of an expression to remove parentheses
Obtaining the coefficient of a variable
Separation of expressions into terms, factors, etc.
Comparison of expressions for equivalence or identity
Automatic simplification
Differentiation
Evaluation of an expression numerically (using the normal FORTRAN
    facilities)

More details can be found in Sammet (1969) and Sammet and Bond (1964).

The original FORMAC was in use for several years on the 7090, as an
experimental system. With the advent of the IBM System/360, and the IBM
intent to have PL/I become its major language, a second version was developed using PL/I as the base in the same sense that the first version
used FORTRAN as the base. Based on the experience from the 7090 FORMAC,
a number of improvements were made in the PL/I-FORMAC, but the basic concepts remained the same. Both systems eventually faded from use, although
I believe that there are still people or organizations using the PL/I-FOR-
MAC, or the later version modified by others and known as FORMAC 73.

5. CONFERENCES AND WORKSHOPS IN THE 1960s

I said earlier that one of the difficulties in the very early days was that essentially no practical means of communication existed—no newsletters, no conferences, etc. The first conference that purported to use the term *symbolic computation* was in 1960, but it really did not deal with any mathematics as we later defined the scope of ACM SIGSAM to be. However, there were several important papers presented and published in the May 1960 issue of the *Communications of the ACM*, including two of interest to this community (McCarthy, 1960; Wang, 1960).

I may be biased, since I founded ACM SIGSAM and was the General and Program chairman for the first SIGSAM conference held March 1966 in Washington. Nevertheless, I believe that SIGSAM in general, and that conference in particular, were milestones that began to build the infrastructure that permitted many of the conferences and groups to grow and flourish. (More details on this conference are given in Section 5.1.) The Computers and Mathematics Conference held at Stanford in August 1986 is merely the most recent example.

5.1 1966 SIGSAM Conference

The SIGSAM conference was the first one that involved algebraic manipulation in any significant way. When the program committee met to generate a call for papers, we thought we knew where most of the work was occurring, even if we did not know the details. We were surprised to receive papers from people and organizations that were completely unknown to us, and we were delighted to be able to put many of them on the program. A subsidiary result of that SIGSAM conference was to determine and define to a significant extent the field which today we are calling symbolic computation. A summary of the topics covered is as follows, with the author shown. The full program and many of the papers are in *Communications of the ACM* (1966).

Topic	Authors
Survey of Formula Manipulation	Sammet
Major Languages/Systems	
ALTRAN/ALPAK	Brown
FORMAC	Bond; Tobey
Formula Algol	Perlis et al.

Topic	Authors
Minor Languages	
AMBIT	Christensen
L^6	Knowlton
PANON-1B	Caracciolo di Forino et al.
Packages and Systems	
ALGEM	Gotlieb and Novak
AUTOMAST	Ball and Berns
CADET	Wolman
CONVERT	Guzman and McIntosh
GRAD Assistant	Fletcher
MANIP	Bender
PM	Collins
Unnamed packages	Bobrow; Teitelman
Applications	
Boundary value problems	Cuthill
Physics	Hearn
Poisson series	Danby et al.
Miscellaneous Topics	
Algebraic representations	Lapidus et al.
Analytic processing of mathematical functions	Clapp
Automated mathematical assistant	Korsvold
Associative algebras	Morris
Finite algebra	Maurer
Logical statements in real number theory	Hodes
Solving polynomial equations	Moses
Polynomial factoring	Jordan et al.
Sequence predictions	Abrahams

5.2 1966 IFIP Conference

The second conference on algebraic manipulation was a small IFIP working conference held September 1966 in Pisa, Italy. The primary subject areas were string and list processing, and formula manipulation. The proceedings are in Bobrow (1968).

5.3 1968 IBM Summer Institute

In 1968 an eight-week Summer Institute on Symbolic Mathematical Computation was held at IBM's Boston Programming Center. The main objective was to have formal and informal lectures and discussions on PL/I-FORMAC and other aspects of symbolic mathematics. The participants had access to both the batch PL/I-FORMAC, and an experimental interactive version of it. An informal proceedings was prepared (Tobey, 1968). Although the main thrust involved PL/I-FORMAC, there were papers on REDUCE, by Risch on integration, and on computing time analyses for a greatest common divisor algorithm.

5.4 1971 SIGSAM Conference

The second major conference on symbolic and algebraic manipulation was held just five years after the first. By then, the field was better defined, and the topics and systems had improved considerably from the 1966 pioneering work. There were new systems described as well as improvements to existing ones; many of the systems were interactive. There were many papers about group theory, and several about the topics of algorithmic analysis, integration, simplification, and rational functions. A topic which appeared for the first time was two-dimensional input/output; practical experience with the early systems had shown the great necessity for this facility! The full proceedings are in Petrick (1971). Many of the papers appeared in one of two journals: *Communications of the ACM* (1971) and *Journal of the ACM* (1971).

Many symbolic math systems were described. Those which are modifications and improvements to systems discussed at the 1966 conference are:

ALTRAN, MATHLAB 68, PL/I-FORMAC, SAC-1

(PL/I-FORMAC was the new version of FORMAC based on PL/I, and SAC-1 was an improved and renamed version of the previous polynomial system PM.)

The systems new since the 1966 conference are:

ALADIN, CAMAL, IAM, MACSYMA, Reduce 2, SCRATCHPAD/1

(Some of these had been described at other conferences between 1966 and 1971.)

5.5 Other Conferences

There were, of course, other conferences pertaining to mathematics taking place during this period, although their emphasis was not on formal algebraic manipulation.

An ACM conference on Interactive Systems for Experimental Applied
Mathematics, held in Washington in August 1967, contained two relevant
papers: REDUCE (Hearn, 1968) and CHARYBDIS (Millen, 1968).

In April 1970 there was a conference on Mathematical Software held at
Purdue University. Regrettably, in spite of the title, out of 23 papers,
there was only one paper on *nonnumerical* software (Sammet, 1971). It was
certainly true then that the numerical analysts did not really want to pay
any attention to this upstart field of nonnumerical computation.

6. SYSTEMS IN USE BY 1971

By 1971 there were a number of "formula manipulation" systems in use on
various computers. Some were experimental (e.g., SCRATCHPAD) whereas
others were in their "second version" and were really used for practical
work (e.g., PL/I-FORMAC). Most were interactive to some extent. All except SAC-1 were languages and were generally implemented on only one computer. Of that list, the two in significant use (albeit greatly improved)
in the mid-1980s are MACSYMA and REDUCE-2.

Name	Computer
Aladin	IBM System/360
ALTRAN	Honeywell 6070
CAMAL	Titan (British)
IAM	DEC PDP-10
MACSYMA	DEC PDP-10
MATHLAB68	DEC PDP-6
PL/I-FORMAC	IBM System/360
Reduce 2	DEC PDP-10
SAC-1	Several
SCRATCHPAD	IBM System/360
Symbal	CDC 6600
Trigman	CDC 6600

7. CONCLUSIONS

The field of symbolic computation has moved a long way from the primitive
formal differentiation of 1954 to the sophisticated systems of the late
1980s. The interests and skills of systems programmers, mathematicians,
and users have all contributed to moving this field ahead.

REFERENCES

Bobrow, J. G. (ed.) (1968). *Symbol Manipulation Languages and Techniques*, North-Holland, Amsterdam.

Cannon, John J. (1969). Computers in group theory: A survey, *Communications of the ACM*, Vol. 12, No. 1 (Jan.).

Communications of the ACM (April 1960).

Communications of the ACM (August 1966).

Communications of the ACM (August 1971).

Hearn, Anthony C. (1968). REDUCE: A user-oriented interactive system for algebraic simplification, in *Interactive Systems for Experimental Applied Mathematics*, Melvin Klerer and Juris Reinfelds (eds.), Academic Press, Orlando.

Howard, James C. (1967). Computer formulation of the equations of motion using tensor notation, *Communications of the ACM*, Vol. 10, No. 9 (Sept.).

Journal of the ACM (October 1971).

Klerer, M., and J. Reinfelds (1968). *Interactive Systems for Experimental Applied Mathematics*, Academic Press, Orlando, p. 158.

Lehmer, D. H. (1963). Automation and pure mathematics, in *Applications of Digital Computers*, W. Freiburger and W. Prager (eds.), Ginn, Waltham, Mass.

McCarthy, John (1960). Recursive functions of symbolic expressions and their computation by machine, Pt. 1, *Communications of the ACM*, Vol. 3, No. 4 (April).

Millen, Jonathan K. (1968). CHARYBDIS: A LISP program to display mathematical expressions on typewriter-like devices, in *Interactive Systems for Experimental Applied Mathematics*, Melvin Klerer and Juris Reinfelds (eds.), Academic Press, Orlando.

Moses, Joel (1971a). Symbolic integration: The stormy decade, *Communications of the ACM* (Aug.).

Moses, Joel (1971b). Algebraic simplification: A guide for the perplexed, *Communications of the ACM* (Aug.).

Newell, A., and J. C. Shaw (1957). Empirical explorations of the logic theory machine, *Proc. Western Joint Computer Conference* (Feb.).

Newman, M. H. A. (1951). The influence of automatic computers on mathematical methods, *Proceedings of the Inaugural Conference of the Manchester University Computer*.

Petrick, S. R. (ed.) (1971). *Proceedings of the Second Symposium on Symbolic and Algebraic Manipulation*, ACM, New York (March).

Sammet, Jean (1968). Revised annotated descriptor based bibliography on the use of computers for non-numerical mathematics, in *Symbol Manipulation Languages and Techniques*, J. G. Bobrow (ed.), North-Holland, Amsterdam.

Sammet, Jean (1969). *Programming Languages: History and Fundamentals*, Prentice-Hall, Englewood Cliffs, N.J.

Sammet, Jean (1971). Software for non-numerical mathematics, in *Mathematical Software*, John Rice (ed.), Academic Press, Orlando.

Sammet, Jean, and Elaine Bond (1964). Introduction to FORMAC, *IEEE Transactions on Electronic Computers*, Vol. EC-13, No. 4 (Aug.).

Tobey, Robert G. (ed.) (1969). *Proceedings of the 1968 Summer Institute on Symbolic Mathematical Computation*, IBM Federal Systems Division Programming Laboratory Report # FSC69-0312 (June).

Wang, H. (1960). Proving theorems by pattern recognition, I, *Communications of the ACM*, Vol. 3, No. 4 (April).

What Computer Algebra Systems Can and Cannot Do

DAVID R. STOUTEMYER   University of Hawaii, Honolulu, Hawaii

1. INTRODUCTION

For a family of related problems, the amount of time and memory required
to complete a computation generally grows with the size of the example.
Typical measures of size are the degrees or the number of digits, variables, factors, terms or matrix rows and columns in the inputs and outputs.
For a given class of problems, manual computation is often practical for
small sizes, but computer algebra systems are practical for larger sizes.
However, systems generally exhaust memory or consume unacceptable time
beyond a certain problem size. Figure 1 qualitatively summarizes the
situation.

Attempting a problem on a computer algebra system involves an investment of time and perhaps also money. Having a sense of these boundaries en-
ables an estimate of the cost and the likelihood of success, thereby encour-
aging promising use of computer algebra while avoiding futile investments.

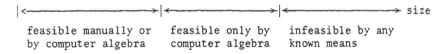

Figure 1 Feasibility ranges.

If average computing time or space is a high-degree polynomial function of problem size, then there may be only a small range of problem sizes
for which only computer algebra is feasible. This range is likely to be
even smaller when the growth with size is exponential or worse, as is often
the case.

The lower boundary of this range depends on your threshold of pain for
manual computation, and the upper boundary depends on your system, computer and budget. However, when the growth of *output* size is rapid, it is
possible to indicate a rough upper bound irrespective of a particular computer, system, or future algorithmic improvements.

The following sections contain estimates of the current upper boundaries for several fundamental classes of problems. They should be taken
only as points of departure to be refined by your experience. However, in
order to refine my estimates, I would like to know if they differ from
your experience.

2. INFINITE PRECISION ARITHMETIC

The numerical coefficients in most computer algebra inputs are rarely more
than ten or twenty decimal digits, but the number of digits in result coefficients occasionally exceeds several hundred digits. Moreover, the number
of digits in intermediate results occasionally exceeds several thousand—
for example, this may happen with polynomial greatest common divisor routines.

Most computer algebra systems can perform exact rational arithmetic
on numbers having at least several thousand decimal digits. Although asymptotically faster algorithms are known, most computer-algebra systems use
integer multiplication algorithms whose time is asymptotically proportional
to the product of the number of digits in the two operands. The time complexity of the implemented integer division and greatest common divisor
routines are also typically quadratic despite known asymptotically faster
algorithms. Consequently, the implemented rational arithmetic also exhibits
this complexity.

Quadratic-growth algorithms tend to become unacceptably slow beyond
input sizes of a few thousand. Thus, the arithmetic algorithms in general-
purpose computer algebra systems are appropriate for the vast majority of
algebra problems, but not for extreme numeric projects such as determining
record Mersenne primes or record approximations to pi.

3. POLYNOMIAL EXPANSION

Polynomial expansion is often necessary in order to enable critical cancellations. And expansion is sometimes necessary in order to explicitly exhibit certain coefficients of interest.

Most computer algebra systems on most computers can represent polynomials having more than 1000 terms in several variables, provided the average number of digits per coefficient is not excessive. Systems with a large address space can easily represent such polynomials exceeding 10,000 terms or perhaps even 100,000 terms.

Let m be the number of terms in the multiplier and n be the number of terms in the multiplicand. Although asymptotically faster polynomial multiplication algorithms are known, systems tend to use algorithms that require a number of coefficient multiplications asymptotically proportional to mn and a number of exponent comparisons asymptotically proportional to about $m^2 n$. This is proportional to m^3 when m is proportional to n, and cubic-growth algorithms tend to become unacceptably slow beyond input sizes of a few hundred. Thus, the polynomial expansion algorithms in general-purpose computer algebra systems are adequate for the vast majority of results that humans care to view, but perhaps not for certain specialized problems where it is worth implementing asymptotically faster algorithms.

Expansion of a polynomial raised to an integer power can be done by repeated multiplication, but it is usually faster to use a balanced recursive variant of the binomial theorem. However, unlike general products of polynomials, such multinomial expansions are rarely an explicit part of any other algorithm. Thus, the main advantage of implementing this more complicated algorithm is to dazzle spectators with the speed of expanding contrived inputs such as $(x + 1)^{200}$. Thus, do not be misled by speed comparisons on examples of this sort.

4. RATIONAL FUNCTIONS

Simplification of rational functions generally includes computation of polynomial greatest common divisors followed by division in order to reduce a numerator and denominator to lowest terms. Polynomial division is about as efficient as polynomial expansion. However, the best known algorithms for computing polynomial gcds are significantly more costly than polynomial expansion, with the time and space growingly rapidly with the number of variables and their degrees as well as the number of terms.

It is not unusual for a computer algebra system to exhaust space or your
patience while computing the gcd of two polynomials having only several
hundred terms in four or five variables. Thus, polynomial expansion can
easily generate numerators and denominators that it is impractical to
reduce. For this reason, it is often worth rephrasing a rational function
problem as a polynomial problem if possible by techniques such as clearing
denominators or temporarily representing complicated denominators by the
reciprocals of unique new variables.

5. FACTORING AND DECOMPOSITION

Factoring is typically even harder than gcds, so computer algebra systems
do not tend to compute gcds by factoring.

In order to specify what is achievable, it is important to specify
the class of expressions that are allowed as inputs and the class of expressions that are allowed as results. A factor is *irreducible* with respect to a particular class if the factor cannot be further factored with
respect to that class.

For factoring polynomials, the input class is usually multivariate
polynomials having rational numbers as coefficients. Clearing denominators makes this class effectively equivalent to multivariate polynomials
having integers as coefficients.

Some possible allowable classes of the irreducible result factors are
multivariate polynomials having coefficients that are:

1. Exact rational numbers
2. Expressions that are (perhaps nested) radicals involving only rational
   numbers and any parametric variables that we do not wish to include in
   the factoring
3. Approximate rational or floating-point numbers, serving as approximations to real numbers
4. Approximate rational or floating-point numbers together with $(-1)^{1/2}$,
   serving as approximations to complex numbers

The more powerful computer algebra systems contain algorithms for
factorization into irreducibles of class 1 above, which is essentially
equivalent via clearing denominators to the class having exact integer
coefficients. It is important to realize that the resulting factors
might be of any degree up to that of the given polynomial. Consequently,

this class alone might not be sufficient for the important application of
solving equations, where we usually seek linear factors. Nonetheless,
factoring into irreducibles of class 1 is an important first step for factoring into irreducibles of larger classes. This step is often quite hard
for polynomials having only about 100 terms.

Factorization into linear factors involving radicals can always be
accomplished by the quadratic, cubic, or quartic formulas for polynomials
of degree 2, 3, or 4, respectively. Moreover, although such factoring is
not generally possible for polynomials of degree exceeding 4, there is a
known algorithm for accomplishing such factorization whenever it is possible. This algorithm, which is based on some rather advanced group theory,
is not currently implemented in any general-purpose computer algebra system. However, some computer algebra systems do implement polynomial decomposition algorithms that cover some important special cases.

There are known algorithms for factoring univariate polynomials into
irreducible linear factors with approximate complex coefficients or irreducible linear and quadratic factors with approximate real coefficients.
Most numerical methods libraries contain such routines. It is possible
to make these algorithms foolproof using the arbitrary precision arithmetic provided by most computer algebra systems. Some of those systems
contain such an algorithm for approximate real coefficients. A few of
these systems also contain such an algorithm for approximate complex
coefficients.

6. DETERMINANTS AND INVERSES

Gaussian elimination is a common method for computing determinants or
matrix inverses using approximate arithmetic on numeric entries. With
fixed precision arithmetic, Gaussian elimination requires time asymptotically proportional to n^3 for a dense n-by-n matrix, making it feasible to
treat such matrices with n being several hundred.

For exact rational arithmetic on dense numeric entries, the size of
intermediate and final numbers tends to grow with the number of steps,
making the computation time grow more rapidly than for finite precision.
Thus, $n \simeq 50$ is a more typical feasibility boundary for such problems.
Of course, the threshold increases as the density of nonzero entries
decreases.

The situation is worse for nonnumeric entries. For example, consider an n-by-n matrix having merely n^2 distinct letters as entries. The expanded determinant will have n! distinct terms, each containing a product of n letters preceded by a plus or a minus sign. Thus, it will take super-exponential time of at least (n + 1)! even to display the result.

Algorithms whose expected computing time is exponential or worse tend to become impractical for n somewhere between 4 and 10. For example, display of the above expanded determinant for n = 8 would entail about 362,000 nonblank characters to display, which is about 90 pages excluding all blanks!

Inverses are typically about n^2 times as bulky as the determinant, because each entry is a reduced ratio of a determinant of a minor to the determinant of the matrix, and dramatic gcd cancellation is atypical.

The situation is generally even worse when the matrix entries are complicated, such as nontrivial polynomials in many variables. However, the entries are often simpler instead: Often, many of the entries are 0, ±1, or a letter that occurs in several entries, perhaps with some symmetry. These characteristics tend to significantly extend the feasibility threshold. For example, determinants of this type are often feasible even for n ≃ 30 provided there are no more than about three or four nonzero entries per row.

7. SOLUTION OF ALGEBRAIC EQUATIONS

A *closed form* is an explicit finite composition of functions and operators from a given class. Implicit solutions or infinite series are not allowed. Thus the phrase "closed form" is meaningful only if we specify the allowable class of functions and operators. For example, a closed-form solution might be impossible in the class of rational expressions but possible if this class is extended by Bessel functions.

By inventing new functions, it is always possible to express a solution as a finite composition of functions. Consequently, if a solution exists, you can always claim a closed form representation in a sufficiently novel class. However, such inventions are of no use unless you can also discover useful identities, qualitative properties, and sufficiently accurate computable approximations to these functions.

Such facts are cataloged for classic special higher transcendental functions such as Bessel functions. Study of their properties was once commonplace in advanced mathematics curricula. However, a closed form is

of limited use to someone who has no familiarity with some of its components, and most contemporary engineers, scientists, and mathematicians are
unfamiliar with most named functions beyond the elementary functions—rational expressions extended by radicals, exponentials, logarithms, trigonometric functions, and their inverses.

Most computer algebra systems contain a SOLVE function that attempts
to provide the set of all closed-form solutions to an "algebraic" equation.
Usually, the allowable class of expressions in the equations and the allowable class of expressions in the solutions are the elementary functions.

The primary techniques are factorization and decomposition, together
with the use of known inverses. Using the quadratic through quartic formulas, the SOLVE function can always succeed on polynomial functions
through fourth degree. However, for an indecomposable cubic or quartic
that is irreducible over the rationals, the cubic and quartic formulas
usually yield incomprehensibly lengthy results, even if the coefficients
are small integers. There is very little insight to be gained from studying such exact results, and approximations to square roots and cube roots
must be made to reduce each solution to a numeral. Thus, it might be better
to use an approximate factorization algorithm in the first place.

For some systems, the SOLVE function can also accommodate simultaneous algebraic equations—usually only simultaneous polynomials equations.
Unlike the corresponding routines in approximate numerical methods libraries, the equations can contain extra nonnumeric parameters, and SOLVE
attempts to find *all* the solutions exactly or at least to an arbitrary
specified precision.

When the equations are all linear in the unknowns, then the feasibility boundary is usually between those for computing determinants and inverses of matrices having the corresponding coefficients.

It is harder to predict the difficulty of solving simultaneous polynomial equations. Impressive successes have been reported for as many as
30 unknowns, but disappointing failures have been reported for as few as
four.

8. TRUNCATED SERIES

Computer algebra systems can be of help even when a closed-form solution
is impossible or too bulky for use. Most systems provide functions that
can compute a truncated series approximation to most expressions. These

facilities are perhaps the most underutilized features of computer algebra systems.

In order of increasing generality, the various types of supported power series are:

1. Taylor series
2. Laurant series (Taylor plus negative integer powers)
3. Puiseaux series (Laurant plus fractional powers), perhaps together with logarithmic and essential singularities

Some systems provide support for truncated Poisson series, which can be regarded as truncated Fourier series with coefficients that are truncated power series. Moreover, at least one system (MACSYMA) provides a facility that attempts to compute an *untruncated* power series including the general form of the nth term. For example,

$$\text{Power series } (\ln((\sin x)/x),x,0) \longrightarrow \frac{1}{2} \sum_{n=1}^{\infty} \frac{(-1)^n 2^{2n} \text{Bern}(2n) x^{2n}}{n(2n)!}$$

where Bern(k) is the kth Bernouli number.

9. EIGENVALUES AND EIGENVECTORS

We have seen that computer algebra systems can:

1. Expand determinants of matrices having nonnumeric entries
2. Factor multivariate polynomials
3. Solve singular systems of linear algebraic equations having nonnumeric coefficients

Thus we have all the appropriate tools to determine closed-form expressions for the eigenvalues and eigenvectors of matrices having nonnumeric entries.

Given an n-by-n matrix A containing some nonnumeric entries, forming the characteristic equation $\det(A - \lambda I)$ is the easiest part. This is practical for n through 5 or 6 with a completely general A having a distinct letter as each element. More specialized matrices permit correspondingly larger values of n.

Regrettably, the resulting nth degree polynomial in λ almost never factors over the rationals or decomposes, making an exact closed form generally impossible for n > 4. Moreover, the presence of extra nonnumeric parameters usually makes approximate factorization inapplicable too. Thus, exact closed-form eigenvalues are generally impossible for n > 4.

Table 1. Time and Output Size for Exact Eigenvalues

Matrix	Form char. poly.	Solve for eigenvalues
general 4 × 4		
$\begin{vmatrix} a & b & c & d \\ e & f & g & h \\ i & j & k & l \\ m & n & o & p \end{vmatrix}$	2 seconds 5 lines	38 minutes >700 pages
symmetric		
$\begin{vmatrix} a & b & c & d \\ b & e & f & g \\ c & f & h & i \\ d & g & i & j \end{vmatrix}$	2 seconds 4 lines	35 minutes $>10^4$ pages
symmetric Toeplitz		
$\begin{vmatrix} a & b & c & d \\ b & a & b & c \\ c & b & a & b \\ d & c & b & a \end{vmatrix}$	2 seconds 4 lines	3 minutes 60 pages
symmetric tridiagonal Toeplitz		
$\begin{vmatrix} a & b & 0 & 0 \\ b & a & b & 0 \\ 0 & b & a & b \\ 0 & 0 & b & a \end{vmatrix}$	2 seconds 1 line	4 seconds 12 lines
general 3 × 3		
$\begin{vmatrix} a & b & c \\ d & e & f \\ g & h & i \end{vmatrix}$	2 seconds 1 line	1 minute 3 pages
symmetric		
$\begin{vmatrix} a & b & c \\ b & d & e \\ c & e & f \end{vmatrix}$	2 seconds 1 line	1 minute 3 pages
symmetric Toeplitz		
$\begin{vmatrix} a & b & c \\ b & a & b \\ c & b & a \end{vmatrix}$	2 seconds 1 line	24 seconds 12 lines
symmetric tridiagonal Toeplitz		
$\begin{vmatrix} a & b & 0 \\ b & a & b \\ 0 & b & a \end{vmatrix}$	1 second 1/2 line	22 seconds 5 lines

Even for n = 3 and n = 4, the coefficients of the powers of λ in the
characteristic equation are usually sufficiently complicated to make the
results of the cubic and quartic formulas horrendously unpalatable. Substituting each value into A - λI and solving the consistent singular system
for the corresponding eigenvector amplifies the total horror into truly
astronomical portions.

To illustrate the hopelessness of these problems, Table 1 lists the
mere eigenvalue computing times and output sizes for a sequence of successively easier 4-by-4 matrices and 3-by-3 matrices. Computing times are for
muMATH-83 running on an IBM-PC-compatible lap-top Toshiba T1100.

For the 4-by-4 general and symmetric matrices, the outputs were so
lengthy that a 720 kilobyte diskette was insufficient to hold even one
eigenvalue. Moreover, merely displaying one eigenvalue on the screen requires an unknown amount of time exceeding 36 hours, which would correspond to more than 10,000 printed pages. The extensive common subexpres-
sions generated by the quartic formula permitted storage of this astro-
nomically large expression despite the less than 512 kilobytes that was
available to muMATH.

The table illustrates that with these matrix families and n > 2, exact
parameterized eigenvalues are practical only for the symmetric tridiagonal
Toeplitz family. Eigenvectors are even less practical.

These examples are intended to prepare you for probable disappointment rather than to totally discourage you from trying to compute exact
eigenvalues. You may be lucky. For example, circulant matrices such as

$$\begin{bmatrix} a & b & c & d \\ d & a & b & c \\ c & d & a & b \\ b & c & d & a \end{bmatrix}$$

have a very simple eigenstructure. Moreover, computer algebra systems
have enjoyed success in perturbation series analysis of eigenstructure.

10.   INTEGRATION

You may have heard that the Risch algorithm has solved the integration
problem. This is untrue. Problems remain:

10.1 Incomplete Implementations

No system currently contains the entire Risch algorithm. The complete algorithm treats integrands that are arbitrary mixtures of algebraic, logarithmic and exponential extensions to rational functions, determining an antiderivative in this same class if one exists. Note that the class includes trigonometric extensions via complex exponentials and inverse trigonometric extensions via complex logarithms.

Several systems include the portion of the algorithm that allows arbitrary mixtures of logarithmic and exponential extensions. Several systems also include special cases of algebraic extensions alone. No implementation yet includes the entire algebraic case, let alone arbitrary mixtures of the three extension types.

A few inroads have been made into permitting certain higher transcendental functions such as error functions and logarithmic integrals in the result and perhaps also the integrand. However, most users will not be familiar with such functions, making them of questionable value for the majority of users.

10.2 The Constant Problem

The decidability of the Risch algorithm depends upon determining whether or not various portions of an integrand are algebraically independent, which can be accomplished by sufficient simplification. Surprisingly, it turns out that it is harder to simplify expressions containing transcendental constants than variables. For example, it is not known if e^e or e + pi are rational. Thus, given such an expression f(e,pi),

$$\int f(e,pi) \, \exp(-x^2) \, dx$$

is elementary iff f(e,pi) = 0, which is generally undecidable. Although you can compute bounds for $\dot{f}(e,pi)$ using interval analysis, if these bounds continue to include 0 for increasing precision, you may compute forever without being certain whether or not the integral is elementary.

This "constant problem" is a limitation—but more in theory than in practice. Moreover, the problem could be categorized as a problem with simplification rather than a problem with integration.

10.3 High-Degree Denominators

The Risch algorithm generally requires a certain amount of factorization of integrand denominators and/or of resultants computed from the denominators.

When explicit exact irreducible linear factors are unobtainable even with the help of radicals, an explicit exact antiderivative might be unobtainable. Provided we are willing to accept an implicit antiderivative, we can introduce new indeterminates representing the roots, then do all subsequent calculations in terms of these, using the irreducible polynomial as a side relation to help simplify the resulting expressions.

However, many users are dissatisfied with such an implicit representation, preferring an explicit representation that relies upon approximate factorization.

10.4 Definite Integration

The Risch algorithm addresses only indefinite integrals (antiderivatives), and if the numerical methods literature and subroutine libraries are any indication, applications usually require *definite* integration instead. The numerical methods literature and subroutine libraries devote little or no attention to robust and efficient methods for constructing a table or a formula approximating an antiderivative. If a computer algebra user is requesting an antiderivative merely to substitute integration limits, then the user might have forgotten the pitfalls of this approach. For example, an antiderivative of x^{-2} is $-1/x$, but

$$\int_{-3}^{2} x^{-2}\, dx \neq (-1/2) - (-1/-3) = -5/6$$

because the integrand is positive throughout the interval. What these users really need is a definite integration package that automatically determines singularities, chooses contour paths, computes residues, etc. Of the various general-purpose computer algebra systems, only MACSYMA is currently distributed with such a package.

11. SUMMATION

Many of the fundamental antiderivatives are simple and easy to remember. For example, even I remember that

$$\int x^{n}\, dx = \frac{x^{n+1}}{n + 1} \qquad \text{for } n \neq -1$$

and

$$\int a^x \, dx = \frac{a^x}{a} \qquad \text{for } a \neq 0$$

but I cannot remember that

$$\sum_{k=0}^{n} k^2 = \frac{n^3}{3} + \frac{n^2}{2} + \frac{n}{6}$$

or that

$$\sum_{k=0}^{n} a^k = \frac{a^n - 1}{a - 1}$$

Moreover, I can remember techniques such as integration by parts or by substitution, whereas I quickly forget any general-purpose techniques for antidifferences. Finally, with the increasing emphasis on discrete mathematics, closed-form summation is receiving attention that is long overdue. Thus, it is good news that there are summation analogs to the Risch algorithm and that some systems contain some of these algorithms.

12. ORDINARY DIFFERENTIAL EQUATIONS

Several systems include functions that attempt to determine closed-form solutions to ordinary differential equations. For first-order equations, the usual primary technique is to attempt matching the equation to one of the classic forms such as linear, separable, homogeneous, exact, Bernoulli, etc., then apply the corresponding formulas. The performance can be quite good. For example, some of the implementations can solve virtually all of the problems in the classic encyclopedic collections.

The likelihood of disappointment increases rapidly with the order of the equation. In theory, linear equations with constant coefficients are trivial. However, an exact solution involves solving an nth degree polynomial characteristic equation exactly, which is generally impossible for $n > 4$. Moreover, the solution is likely to be unpalatably complicated even for $n = 3$ and 4—especially if the equation involves nonnumeric parameters and a nontrivial inhomogeneous term.

There are also now Risch-like algorithms for higher-order linear equations having variable coefficients, but in practice closed-form solutions are rare beyond second order.

There are a few techniques applicable to higher order nonlinear equations, such as recognition of exactness or reduction to a succession of lower-order equations when certain terms are missing.

There are also a few techniques applicable to *systems* of ordinary differential equations. For example, Laplace transforms are applicable to
linear systems—particularly in the case of constant coefficients. However, an exact solution essentially entails determining the exact eigenstructure, and a previous section has outlined the difficulty of doing
that even for systems having total degree 3 or 4.

Overall, the best packages tend to do a much better job than humans,
but the expectation that a closed-form solution even exists decreases
rapidly with order, number of equations, and nonlinearity.

Fortunately, infinite or truncated series solutions are often obtainable even when closed forms are not. Applicable techniques include Taylor
series, Picard iteration, and perturbation series.

13. PARTIAL DIFFERENTIAL EQUATIONS

Newcomers to computer algebra often hope that computer algebra can determine closed-form solutions to their partial differential equations. Unfortunately, closed-form solutions usually do not exist for even the simplest
applications. For example, consider the two-dimensional Laplace's equation
on a square, with boundary value 0 along three edges and boundary value 1
along the other edge: The solution is a double Fourier series for which
no elementary closed form representation is known. For these reasons,
computer algebra systems do not generally provide built-in facilities for
such attempts.

This is not to say that computer algebra cannot *contribute* toward the
closed-form solution of PDEs. For example, computer algebra is often used
to *verify* alleged closed-form solutions to the PDEs of general relativity
or *to categorize* verified solutions or to *compare* verified solutions for
equivalence.

Since Einstein's theory involves ten simultaneous nonlinear equations,
you may wonder how humans could propose solutions for verification: Usually
only the homogeneous vacuum case is considered, often highly symmetric; and
the boundary conditions at infinity and the origin are often trivial.

Computer algebra has been more successful in determining truncated
series solutions to PDEs. For example, most of the highly successful applications in astrophysics and high-energy physics are of this nature. Computer algebra has also been quite successful as a preprocessing step for
approximate numerical solution of PDEs. For example, computer algebra has

often been used to do tedious exact integrations in support of finite element methods.

14. CONCLUSIONS

Computer algebra can accomplish amazing feats, but there are fundamental limitations, such as

1. The frequent nonexistence of closed-form solutions
2. The inherent rapid growth of computing time and space for many families of problems
3. The limited amount of output that users are able to digest

The other papers in this proceedings and their references provide a good point of departure for learning more detail about some of the applications and limitations mentioned in this paper.

Solution of Equations I:
Polynomial Ideals and Gröbner Bases

FRANZ WINKLER   University of Delaware, Newark, Delaware

0. Introduction

The idea of using Gröbner bases for solving equations which arise from a given polynomial ideal is relatively new. Gröbner bases have been introduced by B.Buchberger in his dissertation in 1965, but only about ten years ago has the computer algebra community become aware of them and packages for computing and using Gröbner bases have only recently been added to existing computer algebra systems.

The problems, however, which can be attacked via Gröbner bases have a long history. In [Hilbert 1890] D. Hilbert investigates the problem of computing a basis for the syzygies of a finite set of polynomials over a field. Actually he considers a more general problem, namely solving a system of equations

$$(1) \quad f_{i1}z_1 + ... + f_{is}z_s = 0 \quad , (i{=}1,...,t).$$

By induction on the number of variables he shows how to construct a basis for the space of all solutions of the system (1). In [Hermann 26] G. Hermann uses Hilbert's method to derive a bound for the degree of a basis for the solutions of (1). After a slight correction, this bound is

$$m(t,q,n) = 1/2 \sum_{i=0}^{n\text{-}1} (2qt)^{2^i},$$

where q is the maximal degree of the f_{ij}'s and n is the number of variables. Hermann goes on to show in Satz 4 that for every ideal I there exists a basis $f_1,...,f_t$ such that every g in I can be represented by a linear combination

(2) $g = g_1 f_1 + ... + g_t f_t$

such that the degree of every summand $g_i f_i$ is bounded by the degree of g. She also
gives an algorithm for constructing such a basis from an arbitrary basis of the ideal I.
The degree of such a distinguished basis can be bounded by $m(1,q,n)$, where q is the
maximal degree of the given basis polynomials. A basis for which (2) holds has been
called an H-basis in [Macauley 16]. Obviously such a basis immediately leads to an algorithm for deciding the ideal membership problem. In [Buchberger 65] Buchberger in-
troduces the concept of a Gröbner basis, as a basis which allows to reduce all polynomials in the ideal to 0. He also gives an algorithm for constructing such a basis. It turns
out that every Gröbner basis is an H-basis, but not vice versa [Buchberger 81].
Hironaka's definition of a standard basis [Hironaka 64] for an ideal in a regular local
ring is basically identical to Buchberger's definition of a Gröbner basis. However, Hironaka does not give an algorithm for computing such a basis.

Gröbner bases are useful in computer algebra systems at least in two different ways.
First, they allow to find solutions to a number of algebraic problems which are important in their own right. Some of these problems will be discussed in this paper. For a
more extensive list of applications we refer to [Buchberger 83], [Buchberger 85],
[Winkler et al 85], [Trinks 78], and [Möller/Mora 86]. Secondly, Gröbner bases can be
used to simplify complicated partial results with respect to polynomial side relations.
Simplification is one of the main problems in computer algebra.

Gröbner bases can be viewed in the context of equational theorem proving. Actually a Gröbner basis gives rise to a canonical reduction relation for the ideal congruence.
Buchberger's algorithm for constructing a Gröbner basis is essentially a completion algorithm for the associated equational theory [Llopis 83], [Kandri-Rody/Kapur 83],
[Winkler 84]. This correspondence has been helpful in tranferring computational improvements in the construction of a Gröbner basis to the general situation of completing
a reduction system for an equational theory [Winkler/Buchberger 83].

In this paper we concentrate on applying Gröbner bases to solving equations that
arise from a given set of polynomials. In chapter 1 we specify the problems considered
in this paper, in chapter 2 we give the definition of a Gröbner basis and an algorithm
for constructing one, and in chapter 3 we describe how a Gröbner basis can be used to
solve the problems stated in chapter 1.

1. Problems

In this and the following chapters we assume that K is a field. Whenever we talk about polynomials we mean polynomials over K in the indeterminates $x_1,...,x_n$.

1.1. System of algebraic equations

We suppose that we are given polynomials $f_1,...,f_s$. We are looking for the common zeros of the given polynomials in the algebraic closure $\bar{K}$ of the field K, i.e. we want to solve the following problem:

(P1)   given: a finite set $F=\{f_1(x_1,...,x_n),...,f_s(x_1,...,x_n)\}$ of polynomials,
         find: the set of points $(x_1,...,x_n)$ in $\bar{K}^n$, for which all the polynomials in F vanish.

1.2. The equational theory associated with a polynomial ideal

For given equations $f_1=0,...,f_s=0$ we ask whether another equation $f=0$ can be derived from them, using the following equational calculus:

$$\frac{}{\overline{f=g}} \qquad \text{for all } f=g \text{ in the set of axioms,}$$

$$\frac{}{\overline{f=f}}, \quad \frac{f=g}{\overline{g=f}}, \quad \frac{f=g, g=h}{\overline{f=h}} \qquad \text{for all polynomials } f,g,h$$

$$\frac{f=g}{\overline{fh=gh}}, \quad \frac{f=g}{\overline{f+h=g+h}} \qquad \text{for all polynomials } f,g,h,$$

$$\frac{f=f', g=g'}{\overline{f+g=f'+g'}}, \quad \frac{f=f', g=g'}{\overline{fg=f'g'}} \qquad \text{for all polynomials } f,f',g,g'.$$

$f=0$ can be proved in this equational calculus if and only if f is in the ideal generated by the polynomials $f_1,...,f_s$. The provability problem for the equational theory associated with a finite set of polynomials or the membership problem for the ideal generated by a finite set of polynomials is the following:

(P2)   given: polynomials $f_1,...,f_s$, and f,
         decide: whether $f=0$ can be proved in the equational calculus with the axioms $f_1=0,...,f_s=0$,
         or, stated as an ideal membership problem, whether $f \in ideal(f_1,...,f_s)$.

1.3. System of linear equations with coefficients in $K[x_1,...,x_n]$

Information about the structure of an ideal given by a basis $f_1,...,f_s$ is provided by the syzygies of this basis, i.e. the s-tuples $(z_1,...,z_s)$ in $K[x_1,...,x_n]^s$, for which
$$f_1 z_1 + ... + f_s z_s = 0.$$
More generally, we will consider the following problem:

(P3)   given: polynomials $f_{11},...,f_{1s}$, ... , $f_{t1},...,f_{ts}$, and $f_1,...,f_t$ in $K[x_1,...,x_n]$,
        find: (a description of) all s-tuples $(z_1,...,z_s)$ of polynomials in $K[x_1,...,x_n]$, such that

$$\begin{pmatrix} f_{11} & \cdots & f_{1s} \\ \cdot & & \cdot \\ \cdot & & \cdot \\ \cdot & & \cdot \\ f_{t1} & \cdots & f_{ts} \end{pmatrix} \cdot \begin{pmatrix} z_1 \\ \cdot \\ \cdot \\ \cdot \\ z_s \end{pmatrix} = \begin{pmatrix} f_1 \\ \cdot \\ \cdot \\ \cdot \\ f_t \end{pmatrix} .$$

1.4. Membership in the radical of an ideal

Again we are given a finite set of polynomials $f_1,...,f_s$, and an additional polynomial f. We ask whether the conditions $f_1(\bar{x})=0$,..., $f_s(\bar{x})=0$ for an n-tuple $\bar{x}$ in $\bar{K}^n$ imply $f(\bar{x})=0$. That means we want to decide whether the polynomial f vanishes on the variety of $ideal(f_1,...,f_s)$, i.e. whether f is contained in the radical of $ideal(f_1,...,f_s)$.

(P4)   Given: polynomials $f_1,...,f_s$, and f,
        decide: whether f is contained in the radical of $ideal(f_1,...,f_s)$,
        or in other words, whether $f_1(\bar{x})=0$,..., $f_s(\bar{x})=0 \Longrightarrow f(\bar{x})=0$ for all n-tuples $\bar{x}$ in $\bar{K}^n$.

2. What is and how can we construct a Gröbner basis?

Before we show how Gröbner bases can be used for solving various problems, let us just say a few words about what a Gröbner basis is and how we can construct one. For further details we refer to [Buchberger 76a,b] and [Buchberger 85].

Let $\gg$ be a linear ordering on the power products of $x_1,...,x_n$, such that $1=x_1^0...x_n^0$ is minimal under $\gg$ and multiplication by a power product preserves the ordering. Examples for such an ordering are the lexicographic ordering or the graduated lexicographic ordering. Once the ordering $\gg$ is fixed, every nonzero polynomial f has a greatest power product (with nonzero coefficient) occurring in it, the *leading power product* of f, $lpp(f)$. The coefficient of $lpp(f)$ in f is the *leading coefficient* of f, $lc(f)$. The polynomial which results from f by subtracting the leading power product multiplied by its coefficient is called the *reductum* of f, $red(f)$.

Every nonzero polynomial f gives rise to a *reduction relation* $\rightarrow_f$ on the ring of polynomials in the following way: $g_1 \rightarrow_f g_2$ if and only if there is a power product p with a nonzero coefficient a in g_1, i.e. $g_1 = ap + h$ for some polynomial h which does not contain p, such that $lpp(f)$ divides p, i.e. $p = lpp(f)q$ for some q, and $g_2 = -(a/lc(f)) \cdot q \cdot red(f) + h$. If F is a set of polynomials, the *reduction relation* modulo F is defined such that $g_1 \rightarrow_F g_2$ if and only if $g_1 \rightarrow_f g_2$ for some $f \in F$. In this case g_1 *is reducible to* g_2 by F. If there is no such g_2, g_1 *is irreducible modulo F*. By $\rightarrow^+$, $\rightarrow^*$, and $\longleftrightarrow^*$ we denote the transitive, reflexive-transitive, and symmetric-reflexive-transitive closure of $\rightarrow$, respectively. For every set of polynomials F the reduction relation $\rightarrow_F$ is Noetherian, i.e. every decreasing chain terminates. We say that g is a *normal form of f modulo F*, if $f \rightarrow_F^* g$ and g is irreducible modulo F. In general, normal forms are not unique.

If two polynomials f and g are given, we can reduce the least common multiple (lcm) of $lpp(f)$ and $lpp(g)$ both by $\rightarrow_f$ and by $\rightarrow_g$. The difference of the results is called the *S-polynomial (S-pol)* of f and g. I.e.

$$S\text{-}pol(f,g) = lc(g) \cdot (lcm(lpp(f),lpp(g))/lpp(f)) \cdot f - lc(f) \cdot (lcm(lpp(f),lpp(g))/lpp(f)) \cdot g.$$

A *Gröbner basis* for a polynomial ideal I is a finite set of polynomials G such that I is generated by G, $I = ideal(G)$, and every nonzero polynomial f in I is reducible modulo G. According to a theorem of Buchberger G is a Gröbner basis if and only if $S\text{-}pol(g_1, g_2) \rightarrow_G^* 0$ for all $g_1, g_2 \in G$. Based on this theorem we get (a first, primitive version of) an algorithm for computing a Gröbner basis. For more advanced versions we refer to [Buchberger 76a] and [Buchberger 85].

Gröbner basis algorithm:

input: F, a finite set of polynomials,
output: G, a finite set of polynomials such that G is a Gröbner basis for the ideal generated by F.

$G := F;$

for all pairs $(g_1, g_2) \in G \times G$ **do**
    $h :=$ a normal form of $S\text{-}pol(g_1, g_2)$ modulo G;
    if $h \neq 0$ **then** $G := G \cup \{h\}$ **endif**
endfor •

For every input F the Gröbner basis algorithm will terminate after a finite number of steps [Buchberger 70]. Once a Gröbner basis is computed, the polynomials in the basis can be reduced with respect to each other and the leading coefficients can be normalized to 1, leading to a uniquely defined *minimal reduced Gröbner basis*. Wheras in general normal forms modulo a basis F are not unique, every polynomial f has a unique normal form modulo a Gröbner basis G, i.e. the reduction relation $\rightarrow_G$ has the Church-Rosser property.

3. Application of the Gröbner basis method to the solution of equations

3.1. System of algebraic equations

Before we set out to generate the solutions to the problem (P1) for given polynomials $f_1, ..., f_s$, we might want to know whether the system of equations

$$f_i(x_1, ..., x_n) = 0 \quad (i=1,...,s)$$

has any solutions at all. This question can easily be answered once we have computed a Gröbner basis for the given polynomials.

Theorem 1: Let $F = \{f_1, ..., f_s\}$ be a set of polynomials and G the minimal reduced Gröbner basis for $ideal(F)$. Then the system of equations

$$(3) \qquad f_i(x_1, ..., x_n) = 0 \quad (i=1,...,s)$$

is unsolvable (in $\bar{K}$) if and only if $1 \in G$.

Proof: If $1 \in G$, then 1 is in $ideal(F)$. So every solution of (3) is a solution of $1 = 0$. Thus, there is no solution of (3).

On the other hand, assume that (3) is unsolvable. Then 1 vanishes for every root of (3). So by Hilbert's Nullstellensatz [Waerden 37] there exists a positive integer m such that $1^m = 1 \in ideal(F)$. So there has to be a polynomial in G which allows to reduce 1, i.e. $1 \in G$.

Now suppose that (3) is solvable. We might want to determine whether there are finitely or infinitely many solutions.

Theorem 2: Let F and G be as in Theorem 1. Then (3) has finitely many solutions if and only if for every i ($i=1,...,n$) there is a polynomial g_i in G such that $lpp(g_i)$ is a power of x_i.

Proof: The system of equations (3) has finitely many solutions if and only if the vector space $K[x_1,...,x_n]/ideal(F)$ has finite vector space dimension [Gröbner 49]. That is the case if and only if the number of irreducible power products modulo G is finite, see lemma 6.7 in [Buchberger 83]. From this condition, the theorem follows immediately.

For really carrying out the elimination process, we compute the Gröbner basis with respect to the lexicographic ordering. The following elimination property of a Gröbner basis w.r.t. the lexicographic ordering has been observed in [Trinks 78]. It means that the i-th elimination ideal of a Gröbner basis G is generated by the polynomials in G that depend only on the variables $x_1,...,x_i$.

Theorem 3: Let G be a Gröbner basis w.r.t. the lexicographic ordering (without loss of generality assume $x_n \gg x_{n-1} \gg ... \gg x_1$). Then
$$ideal(G) \cap K[x_1,...,x_i] = ideal(G \cap K[x_1,...,x_i]) \quad \text{for } i=1,...,n,$$
where the ideal on the right hand side is formed in $K[x_1,...,x_i]$.

Proof: Obviously the right hand side is contained in the left hand side.

On the other hand, assume that $f \in ideal(G) \cap K[x_1,...,x_i]$. Then f can be reduced to 0 modulo G w.r.t. the lexicographic ordering. So all the polynomials occurring in this reduction process depend only on the variables $x_1,...,x_i$, and we get a representation of f as a linear combination of polynomials in G, where all the summands in this representation depend only on the variables $x_1,...,x_i$.

Example 1: Let $F=\{f_1,f_2,f_3\} \subset \mathbf{Q}[x,y,z]$ be the set of polynomials
$$f_1 = xz - xy^2 - 4x^2 - \tfrac{1}{4},$$
$$f_2 = y^2z + 2x + \tfrac{1}{2},$$
$$f_3 = x^2z + y^2 + \tfrac{1}{2}x.$$
Let $\gg$ be the lexicographic ordering with $z \gg y \gg x$. The Gröbner basis algorithm

applied to F generates the minimal reduced Gröbner basis $G=\{g_1,g_2,g_3\}$, where

$$g_1 = z + \frac{64}{65}x^4 - \frac{432}{65}x^3 + \frac{168}{65}x^2 - \frac{354}{65}x + \frac{8}{5},$$

$$g_2 = y^2 - \frac{8}{13}x^4 + \frac{54}{13}x^3 - \frac{8}{13}x^2 + \frac{17}{26}x,$$

$$g_3 = x^5 - \frac{27}{4}x^4 + 2x^3 - \frac{21}{16}x^2 + x + \frac{5}{32}.$$

Applying theorem 1 we see that the system of equations $f_1=0$, $f_2=0$, $f_3=0$ is solvable in the algebraic closure of $\mathbf{Q}$ (the field of algebraic numbers). Furthermore, by theorem 2, this system of equations has finitely many solutions. The variables in the Gröbner basis G are totally separated. An approximation of a root of g_3 up to $\pm\frac{1}{100000}$ is -0.128475. This solution of $g_3=0$ can be continued to solutions of $g_2=0$ and $g_1=0$, yielding the approximation (-0.128475, 0.321145, -2.356718) for the original system of equations. •

Example 2: The same method can be applied to algebraic equations with symbolic coefficients. For example let $F=\{f_1,...,f_4\}\subset\mathbf{Q}(a,b,c,d)$ consist of the polynomials

$$f_1 = x_4 + (b\text{-}d),$$

$$f_2 = x_4 + x_3 + x_2 + x_1 + (\text{-}a\text{-}c\text{-}d),$$

$$f_3 = x_3x_4 + x_1x_4 + x_2x_3 + (\text{-}ad\text{-}ac\text{-}cd),$$

$$f_4 = x_1x_3x_4 + (\text{-}acd).$$

Let $\gg$ be the lexicographic ordering with $x_4 \gg x_3 \gg x_2 \gg x_1$. The minimal reduced Gröbner basis for $ideal(F)$ is $G=\{g_1,...,g_4\}$, where

$$g_1 = x_4 + (b\text{-}d),$$

$$g_2 = x_3 + (\text{-}b^2+2bd\text{-}d^2)/(acd)x_1^2 + (\text{-}abc\text{-}abd+acd+ad^2\text{-}bcd+cd^2)/(acd)x_1 + (\text{-}a\text{-}c\text{-}d),$$

$$g_3 = x_2 + (b^2\text{-}2bd+d^2)/(acd)x_1^2 + (abc+abd\text{-}ad^2+bcd\text{-}cd^2)/(acd)x_1 + (\text{-}b+d),$$

$$g_4 = x_1^3 + (ac+ad+cd)/(b\text{-}d)x_1^2 + (a^2cd+ac^2d+acd^2)/(b^2\text{-}2bd+d^2)x_1 + (a^2c^2d^2)/(b^3\text{-}3b^2d+3bd^2\text{-}d^3).$$

Thus, the system has finitely many solutions. A particular solution of $g_4=0$ is $(\text{-}ad)/(b\text{-}d)$, which can be continued to the solution

$$\left(\frac{\text{-}ad}{b\text{-}d}, \frac{ab+b^2\text{-}bd}{b\text{-}d}, c, \text{-}b+d\right).$$

3.2. The equational theory associated with a polynomial ideal

Once we have computed a Gröbner basis G for the ideal generated by $F=\{f_1,...,f_s\}$, the provability of an equation $f=0$ in the equational theory associated with F, or in other words, the membership of f in $ideal(F)$, can be decided by reducing f to the uniquely defined normal form modulo G.

Theorem 4: Let G be a Gröbner basis, f a polynomial. Then $f \in ideal(G)$ if and only if $f \to_G^* 0$.

Proof: If $f \to_G^* 0$, then obviously f can be expressed as a linear combination of the elements of G.

On the other hand, if $f \in ideal(G)$ and $f \neq 0$, by the definition of a Gröbner basis f can be reduced modulo G, $f \to_G f'$, where $f' \in ideal(G)$. f' is also reducible modulo G, leading to f'' and so on. Since $\to_G$ is Noetherian, this process has to stop, i.e. f has to be reducible to 0.   •

Example 3: Let $F=\{\ f_1,\ f_2,\ f_3\ \}$, where
$$f_1 = x_2x_3x_4 - x_1x_3x_4 + x_2^2x_4 - 2x_1x_2x_4 + x_2x_3^2,$$
$$f_2 = x_1x_3^2x_4 - x_2^2x_3x_4 + x_1^2x_3x_4 - x_2^3x_4 + 2x_1x_2^2x_4 + x_2^2x_3^2 - x_1x_2x_3^2,$$
$$f_3 = x_1x_3x_4 - x_1x_2x_4 - x_1^2x_4 + 2x_2^2x_3 - x_1x_2x_3.$$
We want to decide, whether for
$$f = 2x_1x_2^2x_4 - 2x_1^2x_2x_4 - x_1^3x_4 + x_1x_2^2x_3^3 - 2x_2^4x_3^2 + 2x_1x_2^3x_3^2 - 2x_1^2x_2^2x_3^2 + x_1x_2x_3^2 -$$
$$2x_2^5x_3 + 5x_1x_2^4x_3 - 2x_1^2x_2^3x_3 - 2x_2^3x_3 + 3x_1x_2^2x_3 - x_1^2x_2x_3$$
the equation $f=0$ is provable in the equational theory associated with F, or whether $f \in ideal(F)$.

A Gröbner basis for $ideal(F)$, w.r.t. the lexicographic ordering $x_4 \gg x_3 \gg x_2 \gg x_1$, is $G=\{\ g_1,\ g_2,\ g_3,\ g_4\ \}$, where $g_1=f_1$, $g_2=f_3$,
$$g_3 = x_1x_2^2x_4 - x_1^2x_2x_4 - \tfrac{1}{2}x_1^3x_4 + \tfrac{1}{2}x_1x_2x_3^2 - x_2^3x_3 + \tfrac{3}{2}x_1x_2^2x_3 - \tfrac{1}{2}x_1^2x_2x_3,$$
$$g_4 = x_1x_2x_3^3 - 2x_2^3x_3^2 + 2x_1x_2^2x_3^2 - 2x_1^2x_2x_3^2 - 2x_2^4x_3 + 5x_1x_2^3x_3 - 2x_1^2x_2^2x_3.$$
f is reducible to 0 modulo G, so by theorem 4 the equation $f=0$ is provable in the equational theory associated with F.   •

3.3. System of linear equations with coefficients in $K[x_1,...,x_n]$

Let us first consider the case $t=1$, i.e. only one inhomogeneous equation

$$(3) \qquad f_1 z_1 + ... + f_s z_s = f.$$

If $G=\{g_1,...,g_m\}$ is a Gröbner basis, then a generating set for the solutions of the homogeneous equation

$$(4) \qquad g_1 z_1 + ... + g_m z_m = 0$$

can be computed by reducing the S-polynomials of G to 0 and storing the multiples of the basis polynomials used in this process ([Buchberger 85]). We provide a correctness proof for this method.

Theorem 5: Let $G=\{g_1,...,g_m\}$ be a Gröbner basis. For all $1 \leq i < j \leq m$, let $p_{i,j}$, $q_{i,j}$, $k_{i,j}^1,...,k_{i,j}^m$ be such that

$$S\text{-}pol(g_i,g_j) = p_{i,j} g_i - q_{i,j} g_j = k_{i,j}^1 g_1 + ... + k_{i,j}^m g_m,$$

where $k_{i,j}^1, ..., k_{i,j}^m$ are the polynomials extracted from the reduction of $S\text{-}pol(g_i,g_j)$ to 0. For $i=1,...,m$ let e_i denote the vector $(0,...,0,1,0,...,0)$, where the 1 occurs at the i-th position. Then

$$S = \{ \underbrace{p_{i,j} e_i - q_{i,j} e_j - (k_{i,j}^1,...,k_{i,j}^m)}_{S_{i,j}} \mid 1 \leq i < j \leq m \}$$

generates all the syzygies of G, i.e. the solutions of the homogeneous equation

$$g_1 z_1 + ... + g_m z_m = 0 \ .$$

Proof: Obviously every element of S is a syzygy. On the other hand, let $H=(h_1,...,h_m) \neq (0,...,0)$ be an arbitrary syzygy, i.e.

$$(*) \qquad g_1 h_1 + ... + g_m h_m = 0.$$

Let $i_1 < ... < i_k$ be those indices such that $lpp(g_{i_j} h_{i_j})=p$, the maximal power product w.r.t. $\gg$ in $(*)$. We have $k \geq 2$. Suppose $k>2$. By subtracting a suitable multiple of S_{i_{k-1},i_k}, we can reduce the number of positions in H that contribute to the highest power product in $(*)$. Iterating this process $k-2$ times, we finally reach a situation, where only two positions in the syzygy contribute to the highest power p in $(*)$. Now the highest power product in $(*)$ can be decreased by subtracting a suitable multiple of S_{i_1,i_2}. Since $\gg$ is Noetherian, this process has to terminate, leading to an expression of H as a linear combination of elements of S. •

Having solved equation (3) for a Gröbner basis G for the ideal generated by F, we want to transform this solution to a solution of

$$(5) \qquad f_1 z_1 + \dots + f_s z_s = 0.$$

Such a transformation algorithm is given in [Buchberger 85]. We provide a correctness proof.

Theorem 6: Let $F = \{f_1, \dots, f_s\}$ be a set of polynomials and $G = \{g_1, \dots, g_m\}$ a Gröbner basis for $ideal(F)$. We think of F and G as column vectors, i.e. $F = (f_1, \dots, f_s)^T$, $G = (g_1, \dots, g_m)^T$. Let the r rows of the matrix R be a basis for the syzygies of G, and let the matrices X^T, Y^T be such that $G = X^T F$ and $F = Y^T G$. Then the rows of Q are a basis for the syzygies of F, where

$$Q = \begin{pmatrix} I_s - Y^T X^T \\ \cdots\cdots\cdots\cdots \\ R X^T \end{pmatrix} .$$

Proof: Let $b_1, \dots, b_{s+r}$ be polynomials, $b = (b_1, \dots, b_{s+r})$.
$(b \cdot Q) \cdot F =$
$((b_1, \dots, b_s) \cdot (I_s - Y^T X^T) + (b_{s+1}, \dots, b_{s+r}) \cdot R X^T) \cdot F =$
$(b_1, \dots, b_s) \cdot \underbrace{(F - Y^T X^T F)}_{F} + (b_{s+1}, \dots, b_{s+r}) \cdot R \underbrace{X^T F}_{G} = 0.$

So every linear combination of the rows of Q is a syzygy of F.

On the other hand, let $H = (h_1, \dots, h_s)$ be a syzygy of F. Then $H \cdot Y^T$ is a syzygy of G. So for some H', $H \cdot Y^T = H' \cdot R$, and therefore $H \cdot Y^T \cdot X^T = H' \cdot R \cdot X^T$. Thus,

$$H = H \cdot (I_s - Y^T \cdot X^T) + H' \cdot R X^T = (H, H') \cdot Q,$$

i.e. H is a linear combination of the rows of Q. •

What we still need is a particular solution of (3). (3) has a solution if and only if $f \in ideal(F) = ideal(G)$. If the reduction of f to normal form modulo G yields $f \neq 0$, then (3) has no solution (compare Section 3.2). Otherwise one can extract from this reduction polynomials $h_1', \dots, h_m'$ such that

$$g_1 h_1' + \dots + g_m h_m' = f.$$

So $h' X^T$ is a particular solution of (3).

LinSolve1:

input: $f_1, \dots, f_s$, f, polynomials,
output: a polynomial vector $\bar{a}$ of length s and a polynomial matrix A of dimension (s, m) for some m, such that the set of solutions of

$f_1 z_1 + ... + f_s z_s = f$ is $\{\bar{a} + A \cdot (c_1, ..., c_m)^T \mid c_1, ..., c_m \text{ polynomials}\}$
or **unsolvable**.

compute a Gröbner basis $(g_1, ..., g_m)$ for $(f_1, ..., f_s)$ along with the transformation matrices X^T and Y^T (as in Theorem 6);

let R be a basis for the syzygies of $(g_1, ..., g_m)$ as described in Theorem 5;

let Q be the basis for the syzygies of $(f_1, ..., f_s)$ constructed from X^T, Y^T and R as described in Theorem 6;

$A := Q^T$;

$f :=$ normal form of f modulo G;

if $f \neq 0$

then   return unsolvable

else   {let $\bar{a}' = (\bar{a}'_1, ..., \bar{a}'_m)^T$ be the polynomials used as multiplicands of
           $g_1, ..., g_m$ in the reduction of f to 0;
           $\bar{a} := (\bar{a}'^T \cdot X^T)^T$;

           return $(\bar{a}, A)$}   •

Example 4: Let F and G be as in example 1. We want to compute a basis for the syzygies of F, i.e. a basis for the solutions of the equation

$$f_1 h_1 + ... + f_3 h_3 = 0.$$

From the Gröbner basis algorithm we can extract the transformation matrix $X^T = (X_{i,j})_{i=1,...3, \ j=1,...,3}$.

$X_{1,1} = -\frac{32}{35} y^2 z^2 + \frac{16}{5} xz^2 + \frac{32}{35} y^4 z - \frac{96}{35} xy^2 z - \frac{8}{5} x^2 z - \frac{64}{35} xz + \frac{68}{35} z + \frac{16}{5} xy^4 + \frac{8}{5} x^2 y^2 + \frac{64}{35} xy^2 - \frac{744}{455} y^2 - \frac{32}{5} x^4 - \frac{8}{5} x^3 - \frac{192}{35} x^2 - \frac{204}{455} x - \frac{32}{5}$,

$X_{1,2} = \frac{32}{35} x^3 z^2 + \frac{32}{35} xz^2 - \frac{32}{35} x^3 y^2 z - \frac{32}{35} xy^2 z - \frac{16}{35} x^4 z - \frac{16}{35} x^2 z + \frac{104}{35} z - \frac{16}{5} x^4 y^2 - \frac{16}{5} x^2 y^2 - \frac{104}{35} y^2 - \frac{64}{5} x^5 - \frac{852}{65} x^4 - \frac{5548}{455} x$,

$X_{1,3} = -\frac{32}{35} xy^2 z^2 - \frac{16}{5} z^2 + \frac{32}{35} xy^4 z + \frac{16}{35} x^2 y^2 z + \frac{16}{5} y^2 z - \frac{64}{35} x^2 z + \frac{488}{35} xz + \frac{16}{5} x^2 y^4 + \frac{64}{5} x^3 y^2 + \frac{64}{35} x^2 y^2 - \frac{76}{91} xy^2 + \frac{256}{35} x^3 - \frac{32}{7} x^2 - \frac{14}{13}$,

$X_{2,1} = \frac{4}{13} y^2 - \frac{14}{13} x$,

$X_{2,2} = -\frac{4}{13} x^3 - \frac{4}{13} x$,

$X_{2,3} = \frac{4}{13} xy^2 + \frac{14}{13}$,

$$X_{3,1} = -\frac{1}{7}y^2 z + \frac{1}{2}xz - \frac{1}{2}xy^2 + \frac{7}{4}x^2 - \frac{2}{7}x + \frac{17}{56},$$

$$X_{3,2} = \frac{1}{7}x^3 z + \frac{1}{7}xz + \frac{1}{2}x^4 + \frac{1}{2}x^2 + \frac{13}{28},$$

$$X_{3,3} = -\frac{1}{7}xy^2 z - \frac{1}{2}z - \frac{1}{2}x^2 y^2 - \frac{2}{7}x^2 + \frac{5}{28}x.$$

Reduction of the polynomials of F to 0 modulo G yields the entries for the matrix

$$Y^T = \begin{pmatrix} x & -x & -\frac{8}{5} \\[2ex] \frac{8}{13}x^4 - \frac{54}{13}x^3 + \frac{8}{13}x^2 - \frac{17}{26}x & z & -\frac{512}{845}x^3 + \frac{3456}{845}x^2 - \frac{64}{65}x + \frac{16}{5} \\[2ex] x^2 & 1 & -\frac{64}{65}x \end{pmatrix}.$$

Reducing the S-polynomials of G to 0 we get the rows R_1, R_2, R_3 of R, which form a basis for the syzygies of G.

$$R_1 = (\, y^2 - \frac{8}{13}x^4 + \frac{54}{13}x^3 - \frac{8}{13}x^2 + \frac{17}{13}x \,,\, -z - \frac{64}{65}x^4 + \frac{432}{65}x^3 - \frac{168}{65}x^2 + \frac{354}{65}x - \frac{8}{5} \,,\, 0\,),$$

$$R_2 = (\, x^5 - \frac{27}{4}x^4 + 2x^3 - \frac{21}{16}x^2 + x + \frac{5}{32} \,,\, 0 \,,\, -z - \frac{64}{65}x^4 + \frac{432}{65}x^3 - \frac{168}{65}x^2 + \frac{354}{65}x - \frac{8}{5}\,),$$

$$R_3 = (\, 0 \,,\, x^5 - \frac{27}{4}x^4 + 2x^3 - \frac{21}{16}x^2 + x + \frac{5}{32} \,,\, -y^2 + \frac{8}{13}x^4 - \frac{54}{13}x^3 + \frac{8}{13}x^2 - \frac{17}{13}x\,).$$

Now according to theorem 6, the rows $Q_1,...,Q_6$ of

$$Q = \begin{pmatrix} I_3 - Y^T X^T \\ \hdotsfor{1} \\ R X^T \end{pmatrix}$$

form a basis for the syzygies of F.

$Q_1 =$

(1/35) (32 x y² z² − 112 x² z² − 32 x y² z + 96 x² y² z − 8 y² z + 56 x³ z

$\quad$ + 64 x² z − 40 x z − 112 x² y⁴ − 66 x³ y² − 64 x² y² + 40 x y² + 224 x⁵

$\quad$ + 56 x⁴ + 192 x³ + 76 x² + 208 x + 62,

$\quad$ − 32 x⁴ z² − 32 x² z² + 32 x⁴ y² z + 32 x² y² z + 16 x⁵ z + 24 x³ z − 96 x z

$\quad$ + 112 x⁵ y² + 112 x³ y² + 104 x y² + 448 x⁶ + 476 x⁴ + 444 x² + 26,

$\quad$ 32 x² y² z² + 112 x z² − 32 x² y⁴ z − 16 x³ y² z − 120 x y² z + 64 x³ z

$\quad$ − 466 x² z − 26 z − 112 x³ y⁴ − 448 x⁴ y² − 64 x³ y² + 12 x² y² − 256 x⁴

$\quad$ + 160 x³ − 16 x² + 10 x)

$$Q_2 = (1/5915)(3328\,x^4 y^2 z^2 - 22464\,x^3 y^2 z^2 + 3328\,x^2 y^2 z^2 - 3536\,x\,y^2 z^2$$
$$- 11648\,x^5 z^2 + 78624\,x^4 z^2 - 11648\,x^3 z^2 + 12376\,x^2 z^2 - 3328\,x^4 y^4 z$$
$$+ 22464\,x^3 y^4 z - 3328\,x^2 y^4 z + 3536\,x\,y^4 z + 9984\,x^5 y^2 z - 67392\,x^4 y^2 z$$
$$+ 9472\,x^3 y^2 z - 7152\,x^2 y^2 z - 832\,x\,y^2 z + 864\,y^2 z + 5824\,x^6 z - 32656\,x^5 z$$
$$- 44384\,x^4 z + 36108\,x^3 z - 11232\,x^2 z + 4420\,x\,z - 11648\,x^5 y^4 + 78624\,x^4 y^4$$
$$- 11648\,x^3 y^4 + 12376\,x^2 y^4 - 5824\,x^6 y^2 + 32656\,x^5 y^2 + 43264\,x^4 y^2$$
$$- 28548\,x^3 y^2 + 10112\,x^2 y^2 + 3140\,x\,y^2 + 23296\,x^8 - 151424\,x^7 + 3952\,x^6$$
$$- 145808\,x^5 - 17300\,x^4 - 158640\,x^3 - 20570\,x^2 - 17576\,x - 5746,$$
$$- 3328\,x^7 z^2 + 22464\,x^6 z^2 - 6656\,x^5 z^2 + 26000\,x^4 z^2 - 3328\,x^3 z^2 + 3536\,x^2 z^2$$
$$+ 3328\,x^7 y^2 z - 22464\,x^6 y^2 z + 6656\,x^5 y^2 z - 26000\,x^4 y^2 z + 3328\,x^3 y^2 z$$
$$- 3536\,x^2 y^2 z + 1664\,x^8 z - 11232\,x^7 z + 3840\,x^6 z - 16456\,x^5 z - 7808\,x^4 z$$
$$+ 66900\,x^3 z - 9984\,x^2 z + 10608\,x\,z + 11648\,x^8 y^2 - 78624\,x^7 y^2 + 23296\,x^6 y^2$$
$$- 91000\,x^5 y^2 + 22464\,x^4 y^2 - 85384\,x^3 y^2 + 10816\,x^2 y^2 - 11492\,x\,y^2$$
$$+ 46592\,x^9 - 314496\,x^8 + 96096\,x^7 - 383656\,x^6 + 96800\,x^5 - 371846\,x^4$$
$$+ 48960\,x^3 - 67854\,x^2 + 2704\,x - 2873,$$
$$3328\,x^5 y^2 z^2 - 22464\,x^4 y^2 z^2 + 3328\,x^3 y^2 z^2 - 3536\,x^2 y^2 z^2 + 11648\,x^4 z^2$$
$$- 78624\,x^3 z^2 + 11648\,x^2 z^2 - 12376\,x\,z^2 - 3328\,x^5 y^4 z + 22464\,x^4 y^4 z$$
$$- 3328\,x^3 y^4 z + 3536\,x^2 y^4 z - 1664\,x^6 y^2 z + 11232\,x^5 y^2 z - 13824\,x^4 y^2 z$$
$$+ 83848\,x^3 y^2 z - 12480\,x^2 y^2 z + 13260\,x\,y^2 z + 6656\,x^6 z - 95680\,x^5 z$$
$$+ 349232\,x^4 z - 59616\,x^3 z + 66020\,x^2 z - 2912\,x\,z + 3094\,z - 11648\,x^6 y^4$$
$$+ 78624\,x^5 y^4 - 11648\,x^4 y^4 + 12376\,x^3 y^4 - 46592\,x^7 y^2 + 307840\,x^6 y^2$$
$$- 416\,x^5 y^2 + 34424\,x^4 y^2 + 7200\,x^3 y^2 + 6234\,x^2 y^2 - 26624\,x^7 + 196352\,x^6$$
$$- 139968\,x^5 + 48560\,x^4 + 2796\,x^3 + 2528\,x^2 + 766\,x)$$

$$Q_3 = (1/455)(416\,x^2 y^2 z^2 - 1456\,x^3 z^2 - 416\,x^2 y^4 z + 1248\,x^3 y^2 z - 64\,x\,y^2 z$$
$$+ 728\,x^4 z + 832\,x^3 z - 660\,x^2 z - 1456\,x^3 y^4 - 728\,x^4 y^2 - 832\,x^3 y^2$$
$$+ 520\,x^2 y^2 - 140\,y^2 + 2912\,x^6 + 728\,x^5 + 2496\,x^4 + 988\,x^3 + 2784\,x^2 + 626\,x,$$
$$- 416\,x^5 z^2 - 416\,x^3 z^2 + 416\,x^5 y^2 z + 416\,x^3 y^2 z + 208\,x^6 z + 272\,x^4 z$$
$$- 1288\,x^2 z + 1456\,x^6 y^2 + 1456\,x^4 y^2 + 1352\,x^2 y^2 + 5824\,x^7 + 6188\,x^5$$

$$+ 5912 x^3 + 348 x, \quad 416 x^3 y^2 z^2 + 1456 x^2 z^2 - 416 x^2 y^2 z - 208 x^4 y^2 z$$

$$- 1520 x^2 y^2 z + 832 x^4 z - 6344 x^3 z - 224 x z - 1456 x^4 y^4 - 5824 x^5 y^2$$

$$- 832 x^4 y^2 + 156 x^3 y^2 - 140 x y^2 - 3328 x^5 + 2080 x^4 - 128 x^3 - 410 x^2 - 35)$$

$Q_4 =$

$$(1/465)\,(-416 y^4 z^2 + 256 x^4 y^2 z^2 - 1728 x^3 y^2 z^2 + 256 x^2 y^2 z^2 + 1184 x y^2 z^2$$

$$- 896 x^5 z^2 + 6048 x^4 z^2 - 896 x^3 z^2 + 952 x^2 z^2 + 416 y^6 z - 256 x^4 y^4$$

$$+ 1728 x^3 y^4 z - 256 x^2 y^4 z - 976 x y^4 z + 768 x^5 y^2 z - 5184 x^4 y^2 z$$

$$+ 768 x^3 y^2 z - 1544 x^2 y^2 z - 832 x y^2 z + 744 y^2 z + 448 x^6 z - 2512 x^5 z$$

$$- 3552 x^4 z + 3708 x^3 z - 1088 x^2 z + 1088 x z + 1456 x^6 y - 896 x^5 y^4$$

$$+ 6048 x^4 y^4 - 896 x^3 y^4 + 1680 x^2 y^4 + 832 x y^4 - 744 y^4 - 448 x^6 y^2$$

$$+ 2512 x^5 y^2 + 416 x^4 y^2 - 2924 x^3 y^2 - 1856 x^2 y^2 + 72 x y^2 - 3136 y^2$$

$$+ 1792 x^8 - 11648 x^7 + 304 x^6 - 11216 x^5 - 1252 x^4 - 12336 x^3 - 1010 x^2$$

$$- 1120 x, \quad 416 x^3 y^2 z^2 + 416 x y^2 z^2 - 256 x^7 z^2 + 1728 x^6 z^2 - 512 x^5 z^2$$

$$+ 2000 x^4 z^2 - 256 x^3 z^2 + 272 x^2 z^2 - 416 x^3 y^4 z - 416 x y^4 z + 256 x^7 y^2 z$$

$$- 1728 x^6 y^2 z + 512 x^5 y^2 z - 2208 x^4 y^2 z + 256 x^3 y^2 z - 480 x^2 y^2 z$$

$$+ 1352 y^2 z + 128 x^8 z - 864 x^7 z + 256 x^6 z - 1000 x^5 z - 704 x^4 z$$

$$+ 5620 x^3 z - 832 x^2 z + 1024 x z - 1456 x^4 y^4 - 1456 x^2 y^4 - 1352 y^4$$

$$+ 896 x^8 y^2 - 6048 x^7 y^2 + 1792 x^6 y^2 - 12824 x^5 y^2 + 1728 x^4 y^2 - 12532 x^3 y^2$$

$$+ 832 x^2 y^2 - 6432 x y^2 + 3584 x^9 - 24192 x^8 + 7392 x^7 - 29512 x^6 + 7584 x^5$$

$$- 28638 x^4 + 4000 x^3 - 4390 x^2 + 224 x,$$

$$- 416 x y^4 z^2 + 256 x^5 y^2 z^2 - 1728 x^4 y^2 z^2 + 256 x^3 y^2 z^2 - 272 x^2 y^2 z^2$$

$$- 1456 y^2 z^2 + 896 x^4 z^2 - 6048 x^3 z^2 + 896 x^2 z^2 - 952 x z^2 + 416 x y^6 z$$

$$- 256 x^5 y^4 z + 1728 x^4 y^4 z - 256 x^3 y^4 z + 480 x^2 y^4 z + 1456 y^4 z$$

$$- 128 x^6 y^2 z + 864 x^5 y^2 z - 1024 x^4 y^2 z + 6184 x^3 y^2 z - 1728 x^2 y^2 z$$

$$+ 7156 x y^2 z + 512 x^6 z - 7360 x^5 z + 26864 x^4 z - 4448 x^3 z + 4148 x^2 z$$

$$- 490 z^2 + 1456 x^2 y^6 - 896 x^6 y^4 + 6048 x^5 y^4 - 896 x^4 y^4 + 6776 x^3 y^4$$

$$+ 832 x^2 y^4 - 380 x y^4 - 3584 x^7 y^2 + 23680 x^6 y^2 - 32 x^5 y^2 + 2648 x^4 y^2$$

$$+ 3744 x^3 y^2 - 1566 x^2 y^2 - 224 x y^2 + 490 y^2 - 2048 x^7 + 15104 x^6 - 10688 x^5$$

$$+ 2672 x^4 + 3932 x^3 - 1566 x^2 + 2989 x - 784)$$

$$Q_5 = (1/7280)(-6656\,x^6 y^2 z^2 + 44928\,x^4 y^2 z^2 - 13312\,x^3 y^2 z^2 + 8736\,x^2 y^2 z^2$$

$$- 6656\,x^2 y^2 z^2 + 23296\,x^6 z^2 - 157248\,x^5 z^2 + 46592\,x^4 z^2 - 30576\,x^3 z^2$$

$$+ 23296\,x^2 z^2 + 6656\,x^5 y^4 z - 44928\,x^4 y^4 z + 13312\,x^3 y^4 z - 8736\,x^2 y^4 z$$

$$+ 6656\,x^4 y^4 z + 1040\,y^4 z - 19968\,x^6 y^2 z + 134784\,x^5 y^2 z - 38912\,x^4 y^2 z$$

$$+ 19296\,x^3 y^2 z - 17280\,x^2 y^2 z - 5144\,x^2 y^2 z + 1664\,y^2 z - 11648\,x^7 z$$

$$+ 65312\,x^6 z + 77120\,x^5 z - 82616\,x^4 z + 24704\,x^3 z - 26612\,x^2 z + 8320\,x z$$

$$+ 23296\,x^6 y^4 - 157248\,x^5 y^4 + 46592\,x^4 y^4 - 30576\,x^3 y^4 + 23296\,x^2 y^4$$

$$+ 3640\,x y^4 + 11648\,x^7 y^2 - 65312\,x^6 y^2 - 74880\,x^5 y^2 + 67496\,x^4 y^2$$

$$- 20224\,x^3 y^2 + 10932\,x^2 y^2 - 4000\,x y^2 - 1860\,y^2 - 46592\,x^9 + 302848\,x^8$$

$$- 54496\,x^7 + 291616\,x^6 - 49016\,x^5 + 292528\,x^4 - 41208\,x^3 + 14224\,x^2 - 31738\,x$$

$$- 10816,\ 6656\,x^8 z^2 - 44928\,x^7 z^2 + 19968\,x^6 z^2 - 53664\,x^5 z^2 + 19968\,x^4 z^2$$

$$- 8736\,x^3 z^2 + 6656\,x^2 z^2 - 6656\,x^8 y^2 z + 44928\,x^7 y^2 z - 19968\,x^6 y^2 z$$

$$+ 53664\,x^5 y^2 z - 19968\,x^4 y^2 z + 7696\,x^3 y^2 z - 6656\,x^2 y^2 z - 1040\,x y^2 z$$

$$- 3328\,x^9 z + 22464\,x^8 z - 11008\,x^7 z + 33744\,x^6 z + 7936\,x^5 z - 133232\,x^4 z$$

$$+ 35584\,x^3 z - 26888\,x^2 z + 19968\,x z - 23296\,x^9 y^2 + 157248\,x^8 y^2$$

$$- 69888\,x^7 y^2 + 187824\,x^6 y^2 - 91520\,x^5 y^2 + 172952\,x^4 y^2 - 66560\,x^3 y^2$$

$$+ 24752\,x^2 y^2 - 21632\,x y^2 - 3380\,y^2 - 93184\,x^{10} + 628992\,x^9 - 285376\,x^8$$

$$+ 790608\,x^7 - 385792\,x^6 + 753884\,x^5 - 291520\,x^4 + 143886\,x^3 - 103328\,x^2$$

$$+ 4538\,x - 5408,\ -6656\,x^6 y^2 z^2 + 44928\,x^5 y^2 z^2 - 13312\,x^4 y^2 z^2$$

$$+ 8736\,x^3 y^2 z^2 - 6656\,x^2 y^2 z^2 - 23296\,x^5 z^2 + 157248\,x^4 z^2 - 46592\,x^3 z^2$$

$$+ 30576\,x^2 z^2 - 23296\,x z^2 + 6656\,x^6 y^4 z - 44928\,x^5 y^4 z + 13312\,x^4 y^4 z$$

$$- 8736\,x^3 y^4 z + 6656\,x^2 y^4 z + 1040\,x y^4 z + 3328\,x^7 y^2 z - 22464\,x^6 y^2 z$$

$$+ 30976\,x^5 y^2 z - 168528\,x^4 y^2 z + 62608\,x^3 y^2 z - 32080\,x^2 y^2 z$$

$$+ 24960\,x y^2 z + 3640\,y^2 z - 13312\,x^7 z + 191360\,x^6 z - 711776\,x^5 z$$

$$+ 224064\,x^4 z - 170728\,x^3 z + 110912\,x^2 z - 6264\,x z + 6824\,z + 23296\,x^7 y^4$$

$$- 157248\,x^6 y^4 + 46592\,x^5 y^4 - 30576\,x^4 y^4 + 23296\,x^3 y^4 + 3640\,x^2 y^4$$

$$+ 93184\,x^8 y^2 - 615680\,x^7 y^2 + 94016\,x^6 y^2 - 78832\,x^5 y^2 + 72960\,x^4 y^2$$

$$+ 16028\,x^3 y^2 + 1824\,x^2 y^2 - 950\,x y^2 + 63248\,x^8 - 392704\,x^7 + 333184\,x^6$$

$$- 143712\,x^5 + 58024\,x^4 - 23968\,x^3 - 6082\,x^2 + 6760\,x + 1226)$$

$$Q_6 = (1/1456)\ (208\ y^4\ z - 128\ x^4\ y^2\ z + 864\ x^3\ y^2\ z - 128\ x^2\ y^2\ z - 592\ x\ y^2\ z$$
$$+ 448\ x^5\ z - 3024\ x^4\ z + 448\ x^3\ z - 476\ x^2\ z + 728\ x^4\ y^2 + 448\ x^3\ y^2$$
$$- 2660\ x^2\ y^2 + 864\ x\ y^2 - 372\ y^2 - 256\ x^5 + 432\ x^4 - 1700\ x^3 - 1024\ x^2$$
$$- 534\ x,\ - 208\ x^3\ y^2\ z - 208\ x\ y^2\ z + 128\ x^7 - 864\ x^6\ z + 256\ x^5\ z$$
$$- 1000\ x^4\ z + 128\ x^3\ z - 136\ x^2\ z - 728\ x^4\ y^2 - 728\ x^2\ y^2 - 676\ y^2 - 448\ x^8$$
$$+ 112\ x^5 - 480\ x^4 - 2766\ x^3 - 32\ x^2 - 512\ x,$$
$$208\ x\ y^4\ z - 128\ x^5\ y^2\ z + 864\ x^4\ y^2\ z - 128\ x^3\ y^2\ z + 136\ x^2\ y^2\ z + 728\ y^2\ z$$
$$- 448\ x^4\ z + 3024\ x^3\ z - 448\ x^2\ z + 476\ x\ z + 728\ x^2\ y^4 + 448\ x^4\ y^2$$
$$- 112\ x^3\ y^2 + 864\ x^2\ y^2 - 190\ x\ y^2 - 256\ x^6 + 3456\ x^5 - 11920\ x^4 + 3568\ x^3$$
$$- 2228\ x^2 + 1568\ x + 248)$$

Let us now deal with the general case where we have t linear inhomogeneous equations. The idea is to solve the first equation and substitute the solution into the second equation. So the number of equations has been reduced by one. Iterating this process, the problem of solving a system of equations can be reduced to the problem of solving a single equation.

Theorem 7: Let $f_{11},...,f_{1s},\ \cdots\ ,f_{t1},...,f_{ts},\ f_1,...,f_t$ be polynomials. Let $\bar{a}=(\bar{a}_1,...,\bar{a}_s)^{\mathrm{T}}$ be a polynomial vector and

$$A = \begin{pmatrix} a_{11} & \cdots & a_{m1} \\ \vdots & & \vdots \\ a_{1s} & \cdots & a_{ms} \end{pmatrix}$$

a polynomial matrix such that every solution $z=(z_1,...,z_s)^{\mathrm{T}}$ of

$$(*)\qquad \begin{pmatrix} f_{11} & \cdots & f_{1s} \\ \vdots & & \vdots \\ f_{t1} & \cdots & f_{t1\,s} \end{pmatrix} \cdot \begin{pmatrix} z_1 \\ \vdots \\ z_s \end{pmatrix} = \begin{pmatrix} f_1 \\ \vdots \\ f_{t1} \end{pmatrix}$$

is of the form

$$z = \bar{a} + A \cdot (c_1,...,c_m)^{\mathrm{T}}$$

for some polynomials $c_1,...,c_m$.

Let $\bar{b}=(\bar{b}_1,...,\bar{b}_m)^T$ be a polynomial vector and

$$B = \begin{pmatrix} b_{11} & \cdots & b_{m1} \\ \vdots & & \vdots \\ b_{1s} & \cdots & b_{ms} \end{pmatrix}$$

a polynomial matrix such that every solution $z=(z_1,...,z_m)^T$ of

$$(**) \quad (g_{11} \ \cdots \ g_{1s}) \cdot z = g,$$

where $g_i = \sum_{j=1}^0 f_{tj} \cdot a_{ij}$, $i=1,...,m$, and $g = f_t - \sum_{j=1}^0 f_{tj} \cdot \bar{a}_j$, is of the form

$$z = \bar{b} + B \cdot (d_1,...,d_k)^T$$

for some polynomials $d_1,...,d_k$.

Then every solution of

$$(***) \quad \begin{pmatrix} f_{11} & \cdots & f_{1s} \\ \vdots & & \vdots \\ f_{t1} & \cdots & f_{ts} \end{pmatrix} \cdot \begin{pmatrix} z_1 \\ \vdots \\ z_s \end{pmatrix} = \begin{pmatrix} f_1 \\ \vdots \\ f_t \end{pmatrix}$$

is of the form

$$z = (\bar{a} + A \cdot \bar{b}) + (A \cdot B) \cdot (e_1,...,e_k)^T$$

for some polynomials $e_1,...,e_k$.

Proof: Let z be of the form $z = (\bar{a} + A \cdot \bar{b}) + (A \cdot B) \cdot (e_1,...,e_k)^T$ for some polynomials $e_1,...,e_k$. Then $z = \bar{a} + A \cdot (\bar{b} + B \cdot (e_1,...,e_k)^T)$, so z is of the form $\bar{a} + A \cdot (c_1,...,c_m)^T$, and therefore z solves the first t-1 equations in $(***)$.

$(f_{t1},...,f_{ts}) \cdot z = (f_{t1},...,f_{ts}) \cdot ((\bar{a} + A \cdot \bar{b}) + (A \cdot B) \cdot (e_1,...,e_k)^T) =$
$(f_{t1},...,f_{ts}) \cdot \bar{a} + (f_{t1},...,f_{ts}) \cdot A \cdot (\bar{b} + B \cdot (e_1,...,e_k)^T) =$
$f_t - g + (g_1,...,g_m) \cdot (\bar{b} + B \cdot (e_1,...,e_k)^T) =$
$f_t - g + g = f_t.$

So z also solves the t-th equation in $(***)$.

On the other hand, let z be a solution of $(***)$. Then z is a solution of $(*)$, and therefore it is of the form

$$z = \bar{a} + A \cdot (c_1,...,c_m)^T, \text{ for some polynomials } c_1,...,c_m.$$

z is also a solution of the t-th equation in $(***)$, so

$$(f_{t1},...,f_{ts}) \cdot \bar{a} + (f_{t1},...,f_{ts}) \cdot A \cdot (c_1,...,c_m)^T = f_t$$

and therefore

$$(g_1,...,g_m)\cdot(c_1,...,c_m)^T = f_t - g.$$

So $(c_1,...,c_m)^T$ is a solution of (**), i.e.

$$(c_1,...,c_m)^T = \bar{b} + B\cdot(d_1,...,d_k)^T, \text{ for some polynomials } d_1,...,d_k.$$

Thus,

$$z = \bar{a} + A\cdot(\bar{b} + B\cdot(d_1,...,d_k)^T) = (\bar{a} + A\cdot\bar{b}) + (A\cdot B)\cdot(d_1,...,d_k)^T. \quad \bullet$$

Based on Theorem 7 one gets the following recursive algorithm for solving (P3):

LinSolve:

input: $f_{11},...,f_{1s},$ ··· $,f_{t1},...,f_{ts},$ $f_1,...,f_t,$ polynomials,

output: a polynomial vector $\bar{a}=(\bar{a}_1,...,\bar{a}_s)^T$, and a polynomial matrix of dimension (s,m) for some m, such that the set of solutions of (P3) is $\{\bar{a} + A\cdot(c_1,...,c_m)^T \mid c_1,...,c_m \text{ polynomials}\}$ or unsolvable.

if $t=0$

then return ((0,...,0), I_s)

else { $(\bar{a}',A')$:= LinSolve($f_{11},...,f_{1s},$ ··· $,f_{t-1\,1},...,f_{t-1\,s},$ $f_1,...,f_{t-1}$) (if the result is unsolvable then return unsolvable);

m := number of columns of A';

$(g_1,...,g_m)$:= $(f_{t1},...,f_{ts})\cdot A'$;

g := $f_t - (f_{t1},...,f_{ts})\cdot\bar{a}'$;

$(\bar{b},B)$:= LinSolve1($g_1,...,g_m,$ g) (if the result is unsolvable then return unsolvable);

return ($\bar{a}'+A'\cdot\bar{b}$, $A'\cdot B$) } $\quad \bullet$

Example 5: We want to find the solutions of the system

$$(6) \quad \begin{pmatrix} f_{11} & f_{12} & f_{13} \\ f_{11} & f_{12} & f_{13} \end{pmatrix} \cdot \begin{pmatrix} z_1 \\ z_2 \\ z_3 \end{pmatrix} = \begin{pmatrix} f_1 \\ f_2 \end{pmatrix}$$

where $f_{11}=x_1 x_3-x_2{}^2$, $f_{12}=x_1 x_4-x_2 x_3$, $f_{13}=x_2 x_4-x_3{}^2$,

$f_{21}=x_2{}^2-x_1 x_2$, $f_{22}=x_1 x_4-x_2 x_3$, $f_{23}=x_3 x_4-x_1 x_4$,

$f_1 = 2x_1 x_4 + x_1 x_3{}^3 - x_2{}^2 x_3{}^2 + 2x_2 x_3$,

$f_2 = 2x_1 x_4 - 2x_2{}^4 x_3{}^2 - 2x_1 x_2{}^3 x_3{}^2 - 2x_1{}^2 x_2{}^2 x_3{}^2 + x_1 x_2{}^2 x_3{}^2 + x_2{}^2 x_3{}^2 - x_1 x_2 x_3{}^2 - 2x_2{}^5 x_3 +$
$5x_1 x_2{}^4 x_3 - 2x_1{}^2 x_2{}^3 x_3 - 2x_2 x_3$.

The power products are ordered lexicographically, with $x_4 \gg x_3 \gg x_2 \gg x_1$.

(f_{11}, f_{12}, f_{13}) are already a Gröbner basis. Reducing the S-polynomials of this basis to 0, we get a basis for the solutions of the homogenous equation associated with the first equation in (6) as the columns of the matrix A'.

$$A' = \begin{pmatrix} x_4 & x_2 x_4 - x_3^2 & -x_3 \\ -x_3 & 0 & x_2 \\ x_2 & x_2^2 - x_1 x_3 & -x_1 \end{pmatrix}$$

The reduction of f_1 to normal form modulo this basis yields 0, and we get the particular solution $\bar{a}' = (x_3^2, 2, 0)^T$ of the first equation in (6).

Substitution of the general solution of the first equation into the second equation of (6) leads to

$$(7) \quad g_1 z_1 + g_2 z_2 + g_3 z_3 = g,$$

where

$g_1 = x_2 x_3 x_4 - x_1 x_3 x_4 + x_2^2 x_4 - 2 x_1 x_2 x_4 + x_2 x_3^2,$

$g_2 = -x_1 x_3^2 x_4 + x_2^2 x_3 x_4 + x_1^2 x_3 x_4 + x_2^3 x_4 - 2 x_1 x_2^2 x_4 - x_2^2 x_3^2 + x_1 x_2 x_3^2,$

$g_3 = -x_1 x_3 x_4 + x_1 x_2 x_4 + x_1^2 x_4 - 2 x_2^2 x_3 + x_1 x_2 x_3,$

and $g = x_1 x_2^2 x_3^3 - 2 x_2^4 x_3^2 + 2 x_1 x_2^3 x_3^2 - 2 x_1^2 x_2^2 x_3^2 - 2 x_2^5 x_3 + 5 x_1 x_2^4 x_3 - 2 x_1^2 x_2^3 x_3$.

A Gröbner basis for $ideal(g_1, g_2, g_3)$ is (g_1, g_3, g_4, g_5), where

$g_4 = x_1 x_2^2 x_4 - x_1^2 x_2 x_4 - \frac{1}{2} x_1^3 x_4 + \frac{1}{2} x_1 x_2 x_3^2 - x_2^3 x_3 + \frac{3}{2} x_1 x_2^2 x_3 - \frac{1}{2} x_1^2 x_2 x_3,$

$g_5 = x_1 x_2 x_3^3 - 2 x_2^3 x_3^2 + 2 x_1 x_2^2 x_3^2 - 2 x_1^2 x_2 x_3^2 - 2 x_2^4 x_3 + 5 x_1 x_2^3 x_3 - 2 x_1^2 x_2^2 x_3.$

Compare Example 3. The transformation matrices between the basis (g_1, g_2, g_3) and the Gröbner basis (g_1, g_3, g_4, g_5) are

$$X^T = \begin{pmatrix} 1 & 0 & 0 \\ 0 & 0 & 1 \\ \frac{1}{2} x_1 & 0 & -\frac{1}{2} x_1 + \frac{1}{2} x_2 \\ x_1 x_3 - x_1 x_2 - x_1^2 & 0 & x_2 x_3 - x_1 x_3 + x_2^2 - 2 x_1 x_2 \end{pmatrix} \qquad Y^T = \begin{pmatrix} 1 & 0 & 0 & 0 \\ x_2 & x_3 & 0 & 0 \\ 0 & 1 & 0 & 0 \end{pmatrix} .$$

The syzygies of (g_1, g_3, g_4, g_5) are the rows $R_1, ..., R_6$ of the matrix R, where

$R_1 = (-x_1, x_1 - x_2, 2, 0),$

$$R_2 = (x_1 x_2 , \tfrac{1}{2} x_1^2 , -x_3 - x_2 + x_1 , \tfrac{1}{2}),$$

$$R_3 = (x_1 x_3^2 - 2x_2^2 x_3 - x_1 x_2 x_3 + 2x_1 x_2^2 - 2x_1^2 x_2 - x_1^3 , -x_1 x_3^2 - x_1^2 x_3 , 4x_1 x_2 - 2x_2^2 , -x_4 - x_3 + x_2 - x_1),$$

$$R_4 = (x_1^2 , x_2^2 - \tfrac{3}{2} x_1^2 , x_3 - x_2 - 3x_1 , -\tfrac{1}{2}),$$

$$R_5 = (2x_2^2 x_3 - x_1 x_2 x_3 , x_2 x_3^2 , 0 , x_4),$$

$$R_6 = (x_1^2 x_3^2 - 2x_2^3 x_3 - 3x_1^2 x_2 x_3 + \tfrac{1}{2} x_1^3 x_3 + x_1 x_2^3 + 2x_1^2 x_2^2 - \tfrac{9}{2} x_1^3 x_2 - \tfrac{3}{2} x_1^4 , -\tfrac{3}{2} x_1^2 x_3^2 - 2x_1^3 x_3 - \tfrac{1}{2} x_1^4 , x_3^3 x_2 - x_1 x_2^2 + 7x_1^2 x_2 - x_1^3 , -x_2 x_4 - \tfrac{1}{2} x_3^2 - \tfrac{1}{2} x_2 x_3 - \tfrac{3}{2} x_1 x_3 + \tfrac{1}{2} x_2^2 + x_1 x_2 - 2x_1^2).$$

Collecting the linearly independent columns of

$$\left(\begin{array}{c} I_3 - Y^T X^T \\ \hline R X^T \end{array} \right)^T$$

as the columns B_1, B_2 of the matrix B, we get

$$B_1 = (-x_2 , 1 , -x_3)^T,$$

$$B_2 = (-x_1 x_3 x_4 + x_1 x_2 x_4 + x_1^2 x_4 - 2x_2^2 x_3 + x_1 x_2 x_3 , 0 , -x_2 x_3 x_4 + x_1 x_3 x_4 - x_2^2 x_4 + 2x_1 x_2 x_4 - x_2 x_3)^T.$$

Collecting the linearly independent columns of $A' \cdot B$ as the columns of the matrix A, the columns of A form a basis for the solutions of the homogeneous system associated with (6).

$$A = \left(\begin{array}{c} -x_1 x_3 x_4^2 + x_1 x_2 x_4^2 + x_1^2 x_4^2 + x_2 x_3^2 x_4 - x_1 x_3^2 x_4 - x_2^2 x_3 x_4 - x_1 x_2 x_3 x_4 + x_2 x_3^2 \\ x_1 x_3^2 x_4 - x_2^2 x_3 x_4 - x_1^2 x_3 x_4 - x_2^3 x_4 + 2x_1 x_2^2 x_4 + x_2^2 x_3^2 - x_1 x_2 x_3^2 \\ -x_1^2 x_3 x_4 + 2x_1 x_2^2 x_4 - x_1^2 x_2 x_4 + x_1 x_2 x_3^2 - 2x_2^3 x_3 + x_1 x_2^2 x_3 \end{array} \right)$$

Reducing g to 0 modulo the Gröbner basis (g_1, g_3, g_4, g_5), we get the particular solution $\bar{b}' = (0,0,0,x_2)^T$ of the equation

$$g_1 z_1 + g_3 z_3 + g_4 z_4 + g_5 z_5 = g.$$

So a particular solution of (7) is

$$\bar{b} = (\bar{b}'^T \cdot X^T)^T = \left(\begin{array}{c} x_1 x_2 x_3 - x_1 x_2^2 - x_1^2 x_2 \\ 0 \\ x_2^2 x_3 - x_1 x_2 x_3 + x_2^3 - 2x_1 x_2^2 \end{array} \right).$$

Thus, a particular solution of (6) is

$$\bar{a} = \bar{a}' + A' \cdot \bar{b} = \left(\begin{array}{c} x_1 x_2 x_3 x_4 - x_1 x_2^2 x_4 - x_1^2 x_2 x_4 - x_2^2 x_3^2 + x_1 x_2 x_3^2 + x_3^2 - x_2^3 x_3 + 2x_1 x_2^2 x_3 \\ -x_1 x_2 x_3^2 + x_2^3 x_3 + x_1^2 x_2 x_3 + x_2^4 - 2x_1 x_2^3 + 2 \\ x_1^2 x_2 x_3 - 2x_1 x_2^3 + x_1^2 x_2^2 \end{array} \right).$$

3.4. Membership in the radical of an ideal

The *radical* of an ideal I, $rad(I)$, consists of all polynomials f such that $f^m \in I$ for some $m \geq 1$. For a polynomial ideal I, the radical of I has a geometric meaning. If $f_1,...,f_s$ generate an ideal I, then $f \in rad(I)$ iff f vanishes on the algebraic curve described by I. The set of points in $\bar{K}^n$ on which all the polynomials of the ideal I vanish is called the *variety* of I.

Theorem 8: Let $f_1,...,f_s$, f be polynomials. $f \in rad(f_1,...,f_s)$ if and only if f vanishes on the variety of $ideal(f_1,...,f_s)$.

Proof: By Hilbert's Nullstellensatz [Lang 84], if f vanishes on the variety of $ideal(f_1,...,f_s)$ $= I$ then $f \in rad(I)$.

On the other hand assume that $f \in rad(I)$. So there is an $m \geq 1$ such that $f^m \in I$. So for every $\bar{x} \in \bar{K}^n$ in the variety of I we have $0 = f^m(\bar{x}) = (f(\bar{x}))^m$, and therefore $f(\bar{x})$ $= 0$. •

In order to test whether a polynomial f vanishes on the variety of an ideal I, we can adapt Rabinowitsch's method of proving Hilbert's Nullstellensatz.

Theorem 9: Let $f_1,...,f_s$, f be polynomials. f vanishes on the variety of $ideal(f_1,...,f_s) = I$ if and only if $1 \in ideal(f_1,...,f_s,f \cdot z\text{-}1)$, where z is a new variable.

Proof: f vanishes on the variety of I if and only if the system of equations
$$f_1=0, \ ... \ ,f_s=0, \ f \cdot z = 1$$
has no solution in $\bar{K}^{n+1}$. This is the case if and only if $ideal(f_1,...,f_s,f \cdot z\text{-}1)$ is the unit ideal [van der Waerden 67], i.e. I contains 1.   •

This geometric interpretation of the radical of an ideal I can be employed for automatizing the solution of a class of geometric problems. In the sequel we present a simple example. For further details the reader is referred to [Chou 84], [Wu 84], [Kutzler/Stifter 86], and [Kapur 86].

Example 6: We want to prove the geometric theorem that for every triangle ABC the lines which are orthogonal to the sides of the triangle and pass through the midpoints of the associated sides have a common point of intersection.

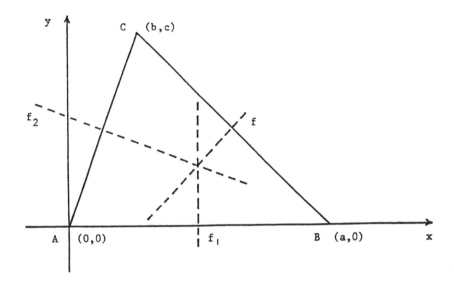

Before we can express this theorem algebraically, we have to place the triangle in a two dimensional coordinate system. Without loss of generality we can assume that A is placed at the origin and that the side AB is parallel to the x-axis. Now a point $(\bar{x},\bar{y})$ lies on the line that is perpendicular to AB and passes through the midpoint of AB if and only if the polyncmial

$$f_1(x,y) = x - \tfrac{1}{2}a$$

vanishes on $(\bar{x},\bar{y})$. A point $(\bar{x},\bar{y})$ lies on the line that is perpendicular to AC and passes through the midpoint of AC if and only if the polynomial

$$f_2(x,y) = b(x-\tfrac{1}{2}b) + c(y-\tfrac{1}{2}c)$$

vanishes on $(\bar{x},\bar{y})$. $(\bar{x},\bar{y})$ lies on the line that is perpendicular to BC and passes through the midpoint of BC if and only if the polynomial

$$f(x,y) = (a-b)(x-\tfrac{1}{2}(a+b)) + c(y-\tfrac{1}{2}c)$$

vanishes on $(\bar{x},\bar{y})$. So in order to prove the theorem, we have to prove that f vanishes on the variety of $ideal(f_1,f_2)$, or in other words that $f \in rad(f_1,f_2)$. Computation of a Gröbner basis for $ideal(f_1,f_2,f\cdot z-1)$ yields the basis (1). So f vanishes indeed on the variety of $ideal(f_1,f_2)$, and therefore the theorem holds.   •

References

[Buchberger 65] B. Buchberger: *Ein Algorithmus zum Auffinden der Basiselemente des Restklassenringes nach einem nulldimensionalen Polynomideal.* Ph.D. dissertation, Univ. Innsbruck (Austria) (1965)

[Buchberger 70] B. Buchberger: Ein algorithmisches Kriterium für die Lösbarkeit eines algebraischen Gleichungssystems, Aequationes Math. 4/3, 374-383 (1970)

[Buchberger 76a] B. Buchberger: A Theoretical Basis for the Reduction of Polynomials to Canonical Forms, *ACM SIGSAM Bull.* 10/3, 19-29 (1976)

[Buchberger 76b] B. Buchberger: Some Properties of Gröbner-Bases for Polynomial Ideals, *ACM SIGSAM Bull.* 10/4, 19-24 (1976)

[Buchberger 81] B. Buchberger: H-Bases and Gröbner-Bases for Polynomial Ideals, Techn. Rep. CAMP 81-2.0, Inst. f. Math., Univ. Linz (Austria) (1981)

[Buchberger 83] B. Buchberger: Gröbner Bases: A Method in Symbolic Mathematics, 5th Internat. Conf. on Simulation, Programming and Math. Methods for Solving Physical Problems, Joint Inst. for Nuclear Research, Dubna, USSR (1983)

[Buchberger 85] B. Buchberger: Gröbner Bases: An Algorithmic Method in Polynomial Ideal Theory, in *Recent Trends in Multidimensional Systems Theory*, N.K. Bose (ed.), D. Reidel Publ. Comp. (1985)

[Chou 84] S.C. Chou: Proving Elementary Geometry Theorems Using Wu's Algorithm, in *Contemporary Mathematics* 29, 243-286 (1984)

[Gröbner49] W. Gröbner: Moderne algebraische Geometrie, Springer-Verlag, Wien-Innsbruck, (1949)

[Hermann 26] G. Hermann: Die Frage der endlich vielen Schritte in der Theorie der Polynomideale, *Math. Ann.* 95, 736-788 (1926)

[Hilbert 1890] D. Hilbert: Über die Theorie der algebraischen Formen, *Math. Ann.* 36, 473-534 (1890)

[Hironaka 64] H. Hironaka: Resolution of Singularities of an Algebraic Variety over a Field of Characteristic Zero: I, II, *Ann. Math.* 79, 109-326 (1964)

[Kandri-Rody/Kapur 83] A. Kandri-Rody and D. Kapur: On Relationship between Buchberger's Gröbner Basis Algorithm and the Knuth-Bendix Completion Procedure, Rep. No. 83CRD286, General Electric Research and Development Center, Schenectady, New York (1983)

[Kapur 86] D. Kapur: Geometry Theorem Proving Using Hilbert's Nullstellensatz, Proc. 1986 ACM-SIGSAM Symp. on Symbolic and Algebraic Computation (SYMSAC'86)

[Kutzler/Stifter 86] B. Kutzler and S. Stifter: Automated Geometry Theorem Proving Using Buchberger's Algorithm, Proc. 1986 ACM-SIGSAM Symp. on Symbolic and Algebraic Computation (SYMSAC'86)

[Lang 84] S. Lang: *Algebra*, 2nd ed., Addison-Wesley (1984)

[Llopis 83] R. Llopis de Trias: Canonical Forms for Residue Classes of Polynomial Ideals and Term Rewriting Systems, Techn. Rep., Univ. Aut. de Madrid, Division de Matematicas (1983)

[Macauley 16] F.S. Macauley: Algebraic Theory of Modular Systems, *Cambridge Tracts in Mathematics and Mathematical Physics* **19**, Cambridge Univ. Press (1916)

[Möller/Mora 86] H.M. Möller and F. Mora: New Constructive Methods in Classical Ideal Theory, *J. of Algebra* **100**/1, 138-178 (1986)

[Trinks 78] W. Trinks: Über B. Buchbergers Verfahren, Systeme algebraischer Gleichungen zu lösen, *J. of Number Theory* **10**/4, 475-488 (1978)

[van der Waerden 67] B.L. van der Waerden: *Algebra II*, Springer-Verlag (1967)

[Winkler 84] F. Winkler: *The Church-Rosser Property in Computer Algebra and Special Theorem Proving: An Investigation of Critical-Pair/Completion Algorithms*, Ph.D. dissertation, Univ. Linz (Austria) (1984)

[Winkler/Buchberger 83] F. Winkler and B. Buchberger: A Criterion for Eliminating Unnecessary Reductions in the Knuth-Bendix Algorithm, Proc. Colloquium on Algebra, Combinatorics and Logic in Computer Science, Györ (Hungary) (1983)

[Winkler et al 85] F. Winkler, B. Buchberger, F. Lichtenberger, and H. Rolletschek: An Algorithm for Constructing Canonical Bases of Polynomial Ideals, *ACM Trans. on Math. Software* **11**/1, 66-78 (1985)

[Wu 84] W.T. Wu: Some Recent Advances in Mechanical Theorem Proving of Geometries, in *Contemporary Mathematics* **29**, 235-241 (1984)

Index

Accessory parameter, 101, 134,
    137-143, 145, 148-149
ALTRAN, 355, 361, 363-364
Analytic continuation, 129-131,
    133, 169-170
Automated reasoning, 83
Automated theorem proving, 83,
    85, 88, 98-102

Basic hypergeometric series, 42

Clebsch-Gordon decomposition,
    348-349
Complexity of computations, 109,
    112, 118, 121, 123, 130,
    132, 159, 166
  algebraic computational, 159,
    164, 167
  binary-splitting method, 115,
    117
  bit, 124, 128, 133
  bit-burst method, 129, 131,
    132
  boolean, 110, 112
  factorization, 121, 173-174
  logical, 109
  operational, 109, 112-113, 115,
    118-120
  Pade approximations, 135

Complexity of computations
  parallel polynomial root
    finding, 174, 175
  parallel power series
    computation, 136
  polynomial multiplication, 156
  polynomial root finding, 158,
    160, 172-174
  power series computation, 111,
    112, 120, 122, 136
  solution of linear differential
    equations, 110, 122-124, 128,
    129, 132
  sparse polynomial root finding,
    171, 172
Computation of PI, 263
Conjecture by
  Grothendieck-Katz, 151-153
  Hardy-Littlewood, 243
  Macdonald, 45, 55, 59, 77
  Macdonald-Morris, 77
  Siegel, 150
  Tate, 154
  Whittaker, 111, 143, 146-148
Continued fractions, 273, 312

Dyson-Gunson-Wilson identity,
    43, 56, 60-62

Expert Systems, 99

Factorization, 121, 154, 155, 261
    311, 315-317, 319, 370, 371
    by Pollard rho-method, 121, 319
    by Shanks' quadratic form
      decomposition, 121, 137
FORMAC, 351, 355, 359-361,
    363-364
Formal groups, 150, 153-154, 156
Fusion, inertial confinement,
    325

Grobner bases, 383, 384, 386-392,
    394, 402

Hard hexagon model, 18
Homotopy, regular, 321, 322, 324,
    334, 336-338, 345
Hypergeometric series, 39, 42

Lame equation, 137-139, 152, 153
Lattice bases reduction
    algorithm, 288-291, 294
Linear programming, 233, 240

MACSYMA, 61, 85, 87, 364
Machine learning, 101
Modular forms, 18, 156
Monodromy group computation, 133,
    138-140, 143, 149
Monte-Carlo method, 35

Negapolylogs, 282-283

Pade approximation, 111, 134, 135,
    151, 152, 164, 175
Path invariance, 262-264, 267-268,
    274
Polynomial
    root finding, 109, 111, 160-166,
      172-173
    for sparse polynomials, 167-173
    Gaussian, 20, 23, 27-28

Polynomial factorization, 285-286,
    302-305
    Berlekamp, 304
    bivariate, 295, 297
    Kaltofen algorithm, 287
    Lenstra algorithms, 286
    Lenstra-Lenstra-Lovasz, 288-291,
      294
    multivariate, 299, 304
    Zippel algorithm, 287
Primality testing, 154, 155
PROLOG, 84
Proof checking, 84

q-Trigonometry, 279, 281

REDUCE, 75-76, 363-364
Rewrite rule, 84
Risch algorithm, 35, 376, 379

Salamin-Brent algorithm, 124
SCRATCHPAD, 18, 20-22, 29-31, 78,
    110, 115, 138, 141, 143, 149,
    364
Selberg's integral, 38, 45, 55-57,
    65, 67, 68, 72, 75, 78
    q-extension, 45-46, 48, 51
Summation, Euler-Maclaurin, 271
Sums of cubes, 243-244

Telescopy, 269-271
Theta
    constant, 157
    function, 17, 23, 175

Uniformization, 110, 134, 137,
    139, 140, 142, 144, 146-148,
    152
UNIVAC 1, 236, 364